T0269272

THERAPEUTIC DRUG MONITORING DATA

THERAPEUTIC DRUG MONITORING DATA

A Concise Guide

Fourth Edition

AMITAVA DASGUPTA, Ph.D, DABCC
Professor of Pathology and Laboratory Medicine
University of Texas McGovern Medical School at Houston
Houston, TX, United States

MATTHEW D. KRASOWSKI, MD, Ph.D
Vice Chair of Clinical Pathology and Laboratory Services
University of Iowa Roy J. and Lucille A. Carver College of Medicine
Iowa City, IA, United States

Academic Press is an imprint of Elsevier
125 London Wall, London EC2Y 5AS, United Kingdom
525 B Street, Suite 1650, San Diego, CA 92101, United States
50 Hampshire Street, 5th Floor, Cambridge, MA 02139, United States
The Boulevard, Langford Lane, Kidlington, Oxford OX5 1GB, United Kingdom

Notices
Knowledge and best practice in this field are constantly changing. As new research and experience broaden our understanding, changes in research methods, professional practices, or medical treatment may become necessary.

Practitioners and researchers must always rely on their own experience and knowledge in evaluating and using any information, methods, compounds, or experiments described herein. In using such information or methods they should be mindful of their own safety and the safety of others, including parties for whom they have a professional responsibility.

To the fullest extent of the law, neither the Publisher nor the authors, contributors, or editors, assume any liability for any injury and/or damage to persons or property as a matter of products liability, negligence or otherwise, or from any use or operation of any methods, products, instructions, or ideas contained in the material herein.

Library of Congress Cataloging-in-Publication Data
A catalog record for this book is available from the Library of Congress

British Library Cataloguing-in-Publication Data
A catalogue record for this book is available from the British Library

ISBN 978-0-12-815849-4

For information on all Academic Press publications
visit our website at https://www.elsevier.com/books-and-journals

Publisher: Stacy Masucci
Acquisition Editor: Tari K. Broderick
Editorial Project Manager: Megan Ashdown
Production Project Manager: Joy Christel Neumarin Honest Thangiah
Cover Designer: Greg Harris

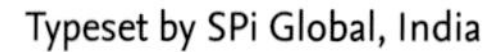

Typeset by SPi Global, India

Contents

Preface

The first edition of this pocket guide was published in 1980 by AACC Press when therapeutic drug monitoring (TDM) was still a fledgling area of laboratory medicine, and it was still unclear as to how laboratories and physicians could collaborate to optimize drug therapy. Clearly, all drugs did not need monitoring but it was shown that for many, monitoring allowed a more rapid achievement of efficacy and a reduction in adverse side effects or toxicity. Today, TDM is an established area of clinical laboratory testing, with the clinical utility of monitoring certain drugs with narrow therapeutic range now well established. The third edition of this book was also published by AACC Press in 2007. Because the co-editor of the third edition Dr. Catherine A. Hammett-Stabler has retired, the 4th edition of the book is authored by Amitava Dasgupta (co-editor of the 3rd edition) and Matthew Krasowski. We have extensively revised content of the 3rd edition and also added three new chapters (Chapters 5, 7 and 8) in the 4th edition. In addition, we have added monographs of seven relevant drugs (antiepileptic drugs tiagabine, topiramate and zonisamide; cardioactive drugs flecainide, mexiletine and tocainide; and the anticancer drug tamoxifen) where TDM is beneficial.

We acknowledge contributors of 3rd edition: Dr. Larry Broussard (now retired), Dr. Anthony Butch (now retired), Dr. Uttam Garg, Dr. Catherine Hammett-Stabler (now retired) and Dr. Christopher McCudden. We expanded the 4th edition by approximately 50% compared to the third edition. If readers enjoy reading this book and find the information helpful for their practice, our effort will be dully rewarded.

Respectfully submitted by
Amitava Dasgupta
Matthew Krasowski

CHAPTER 1

Pharmacokinetics and therapeutic drug monitoring

1.1 Introduction

There are over 6000 prescription and over-the counter pharmaceutical formulations available in the United States (U.S.) The total U.S. prescription sales in the 2015 calendar year were $419.4 billion, which was 11.7% higher than sales in 2014 [1].

When a drug enters the body, three systems come together to determine and describe its fate:

- Pharmacokinetics describes the absorption, distribution, metabolism, and excretion of the drug.
- Pharmacodynamics describes the action of the drug on the body.
- Pharmacogenomics forms the genetic basis for the differences observed between individuals in terms of drug metabolism and response.

This chapter will provide a review of basic pharmacokinetic principles and their use in therapeutic drug monitoring (TDM). Pharmacogenomics is capable of predicting possible pharmacokinetic behavior of a drug prior to administration. In contrast, TDM can provide pharmacokinetic information after administration of a drug based on concentration of the drug in blood or other body fluid. Therefore, TDM is a phenotype approach for personalized medicine, while pharmacogenomics is a genotype approach and they complement each other. Please see Chapter 7 for more in-depth discussion on this topic.

Fortunately few drugs require routine TDM. In general, approximately 26 drugs are routinely subjected to TDM due to narrow gap between therapeutic and toxic blood levels. Immunoassays are commercially available for most of these drugs which could be easily adapted to various automated clinical chemistry analyzers for simplicity of operation as well as good turnaround time. In addition, approximately 25–30 drugs are subjected to TDM less frequently and immunoassays are available only for a few such drugs. Therefore, chromatographic techniques are used for monitoring of these

Therapeutic Drug Monitoring Data
https://doi.org/10.1016/B978-0-12-815849-4.00001-3

drugs and such tests are only available in clinical laboratories of major medical centers and academic centers as well as reference laboratories.

TDM not only consists of measuring the concentration of a drug in a biological matrix but also involves the proper interpretation of the value using pharmacokinetic parameters, drawing appropriate conclusion regarding the drug concentration and dose adjustment. The International Association for Therapeutic Drug Monitoring and Clinical Toxicology adopted the following definition for drug monitoring; "Therapeutic drug monitoring is defined as the measurement made in the laboratory of a parameter that, with appropriate interpretation, will directly influence prescribing procedures. Commonly, the measurement is in a biological matrix of a prescribed xenobiotic, but it may also be of an endogenous compound prescribed as a replacement therapy in an individual who is physiologically or pathologically deficient in that compound" [2]. Traditionally TDM involves measuring drug concentration in a biological matrix (most commonly serum or plasma) and interpreting these concentrations in terms of relevant clinical parameters. Whole blood is the preferred matrix for TDM of immunosuppressants (cyclosporine, tacrolimus, sirolimus and everolimus) except for mycophenolic acid. A successful TDM program requires good communication between clinicians, laboratory professionals, and pharmacists. One report clearly documented that intervention of pharmacist significantly improved appropriateness of TDM use and significantly reduced unnecessary cost [3]. Commonly monitored and less commonly monitored drugs are listed in Table 1.1. Although not routinely monitored, TDM may be beneficial when applied to antiretrovirals. Even though TDM can be a valuable strategy in HIV management, its role remains controversial [4].

1.2 Compliance and TDM

Compliance (adherence) to drug therapy is an important issue for successful patient management. It is imperative that the correct drug must be prescribed and prepared. But if the patient is not taking the drug properly, i.e., they skip or add doses, then even the best analytical method would produce misleading drug concentration based on expected dosage because steady state may not be attained nor it could be maintained. Non-compliance with a prescribed drug is a commonly encountered problem especially for medications taken chronically. TDM can be a useful approach to identify non-compliant patients. Non-compliance and over-dosing/under dosing is common with digoxin therapy. In one study based on analysis of data from 10 different studies, the authors reported that in patients with heart failure

Table 1.1 Commonly and less commonly monitored therapeutic drugs

Class of drug	Frequently monitored	Less-frequently monitored
Anticonvulsants	Carbamazepine Phenobarbital Phenytoin Primidone Valproic acid	Clonazepam Lamotrigine Levetiracetam Topiramate
Cardioactive drugs	Digoxin Procainamide/*N*-acetyl procainamide Quinidine	Amiodarone Flecainide Mexiletine Tocainide Verapamil
Antiasthmatic	Theophylline Caffeine (mostly in neonates)	None
Antibiotics	Amikacin Gentamicin Tobramycin Vancomycin	Chloramphenicol Ethambutol Isoniazid Rifampin
Antidepressants	Amitriptyline Nortriptyline Doxepin Imipramine Desipramine Lithium	Amoxapine Fluoxetine/nor-fluoxetine Sertraline paroxetine
Antifungals	None	Fluconazole Itraconazole Voriconazole
Antineoplastic	Methotrexate	Busulfan 5-Fluorouracil
Antiretroviral agents	No antiretroviral drug is routinely monitored	TDM is currently used for protease inhibitors such as atazanavir, fosamprenavir, indinavir, nelfinavir, ritonavir and saquinavir
Immunosuppressants	Cyclosporine Tacrolimus Sirolimus Everolimus Mycophenolic acid	

and/or atrial fibrillation, the prevalence rate of non-compliance with digoxin was 38.7%; the corresponding prevalence rates of overdosing and underdosing were 33.04% and 33.8% respectively [5]. A high non-compliance rate of 63% with phenytoin and theophylline was reported in a study based on a study of 80 epileptic and asthmatic patients [6]. Calcagno commented that non-compliance with antiretroviral therapy may cause treatment failure in AIDS patients. TDM of protease inhibitors including ritonavir in plasma can uncover incomplete compliance with treatment. Therefore, TDM may represent a useful tool for identifying patients in need of adherence-promoting interventions [7].

1.3 Basic pharmacokinetics

A drug after oral administration undergoes various processes including liberation of the drug from formulation (tablet or capsule), absorption from the gastrointestinal (GI) track, distribution in blood and tissue, binding with appropriate receptor/tissue for the intended pharmacological action metabolism and finally elimination from the body.

1.3.1 Liberation

Before anything else can happen, drugs must be released from their solid form and made soluble (the exception are those administered intravenously). A number of drugs are found in forms that release the drug over a prolonged period of time, i.e., sustained release formulas. Such formulas allow the attainment of efficacy but at lower peak plasma concentrations and therefore with fewer adverse events. The rate at which such a drug is liberated depends upon the formulation, dosage, and ionization state, as well as the pH of the environment in which it enters.

Ionization and pH of the gastric fluid have major influence on drug liberation. Drugs are typically weak acids or bases possessing functional groups that become charged or ionized depending upon the pH in which they are placed. By knowing the pKa of a drug, one can predict the behavior of a drug in a given pH of the environment. The pKa is the pH at which the amounts of ionized and unionized forms are equal. An acidic drug will favor the ionized form when placed in an environment where the pH is higher than the pKa. Similarly, a basic drug will favor the ionized form as the pH decreases to a value less than the drug's pKa. This concept is important because ionized molecules cannot easily cross cell membranes.

1.3.2 Absorption

After liberation, absorption describes the movement of a drug from the site of administration into the circulation. How readily absorption occurs depends on the physicochemical characteristics of the drug (solubility, pKa, ionization) and on the local physiology (pH, perfusion, GI motility, etc.). Absorption is most efficient at the pH where the drug is predominately in the un-ionized form. If a drug enters a compartment and becomes ionized, it is likely to be trapped there. Factors which can alter the absorption of an orally administered drug include changes in the transit time through the GI track, ionization at a point along the GI track, metabolism by intestinal enzymes, and interactions with other compounds present to form insoluble compounds.

An additional term that should be discussed at this point is bioavailability. This term is used to describe the fraction of a dose that is absorbed and therefore available to elicit a pharmacological effect. Bioavailability depends upon the previously discussed characteristics of the drug, physiology (especially that of the GI track), and other factors such as first pass metabolism, fever, perfusion, etc. It may be an issue when a patient switches from one brand of medication to another brand or from a brand-name formula to a generic.

Another factor that affects bioavailability is the presence of food in the stomach. In order to avoid food-drug interaction, certain drugs should be taken on an empty stomach (drug instructions should specify that) but other drugs may be taken with food. Alcohol should be avoided while taking certain drugs. Nevertheless, approximately 71% of American adults consume alcoholic beverages, leading to potential alcohol-drug interactions. Numerous commonly prescribed medications interact negatively with alcoholic beverages including cardiovascular agents, diuretics, central nervous system agents, narcotics, psychotherapeutic agents such as antidepressants, and others. In one study based on 26,657 adults, the authors calculated that total prevalence of alcohol interactive medication use was 41.5% among current drinkers younger than 65 years of age. Among participants aged ≥65, total prevalence of such medication use was 78.6% and adjusted prevalence among current drinkers was 77.8% [8].

1.3.3 Distribution

Distribution describes the delivery of drug via the circulation to the rest of the body. How readily distribution takes place depends on the integrity of the circulatory system (how readily tissues are perfused), protein binding characteristics of the drug, and the ability of the drug to cross cellular membranes. Some drugs distribute extensively to organs and tissues while others remain confined within the vascular space.

1.3.4 Binding of drugs with serum proteins

Most drugs in the circulation bind at least some extent to serum proteins (there are also cases in which drugs are bound to erythrocytes and lipids) but the extent of protein binding may vary from being negligible (e.g., lithium) to over 99% (e.g., ketorolac). Acidic drugs are typically bound to albumin while basic drugs are predominantly bound by α_1-acid glycoprotein. Only the portion of the drug that is not bound to serum proteins (unbound or free drug) can cross cell membranes and subsequently interact at the intracellular level. Protein binding varies between drugs, but under healthy conditions should be relatively consistent between individuals for a given drug. Conditions that alter the concentration of the binding proteins naturally alter the amount of free drug and the impact of such a change will be greatest for a drug with high protein binding, typically 80% or more.

1.3.5 Metabolism

One of the most important and interesting aspects of pharmacology is that of metabolism. The simplest description of metabolism is that this is the process of chemically preparing a drug for excretion by converting it into a more polar, water-soluble form. As pharmacogenetics has shown, metabolism is a relatively complex issue that contributes to the differences observed between individuals prescribed the same drug and dose. While the vast majority of metabolism takes place in the liver through the cytochrome P450 (CYP) system, other tissues and enzymes also contribute to the process. Many drugs are metabolized through multiple pathways. Usually drugs are metabolized by one or two steps. In phase I drug metabolism, a variety of enzymes (most commonly various isoforms of CYP enzymes) act to introduce reactive and polar groups into their substrates in order to transform relatively non-polar drugs to polar metabolites that are more readily excreted in urine. In subsequent phase II reactions, some drug or certain drug metabolites are conjugated with charged species such as glutathione, sulfate, glycine, or glucuronic acid in order to make such species more water-soluble. This process is also catalyzed by various enzymes mainly transferases such as UDP (Uridine 5′-diphosphate) glucuronosyltransferases, sulfotransferases, *N*-acetyltransferases, glutathione *S*-transferases and methyltransferases (mainly thiopurine *S*-methyl transferase and catechol *O*-methyl transferase) [9]. A minority of drugs (e.g., digoxin) are not metabolized at all by liver enzymes.

The CYP isoenzymes that mediate the Phase I oxidative metabolism of many drugs include CYP1A2, CYP2C9, CYP2D6, CYP2E1 and CYP3A4.

These enzymes show marked variations in different populations including genetic polymorphisms (CYP2C9, CYP2C19 and CYP2D6) and a subset of the population may be deficient in enzyme activity (poor metabolizers). Therefore, if a drug is administered to a patient who is a poor metabolizer, drug toxicity may be observed even with a standard dose of the drug. The resulting metabolites may be inactive, active, partially active, or even toxic. Metabolites may also inhibit or induce the formation of other metabolic products of the same drug or of another drug. Metabolic processing may be necessary to transform an inactive, prodrug into an active drug.

Certain orally administered drugs undergo first pass metabolism. This occurs when an orally administered drug is absorbed through the intestinal mucosa and directly transported via the portal vein to the liver where it undergoes metabolism. In some cases, a significant portion of the drug is lost before it becomes available to the circulation. In such cases it might be possible to administer the drug in a higher oral dose or to administer the drug using a non-enteral route (intravenously, transdermally or intramuscularly). Some drugs are administered as inactive, prodrugs which are metabolized to the active drug. For example, tamoxifen, an antineoplastic drug is a prodrug which is converted into its primary active metabolite endoxifen (4-hydroxy-*N*-desmethyl-tamoxifen) by the action of CYP2D6.

Drugs may follow zero-order, first-order or second-order kinetics during metabolism. Most drugs follow first order kinetics, while a small number of drugs follow zero order kinetics.

- A drug follows *first-order kinetics* if the amount of metabolite produced steadily increases in proportion to the amount of parent drug.
- A drug follows *zero-order kinetics* when the amount of metabolite formed is relatively constant and independent of the amount of parent drug. In this system the enzyme is working at capacity producing a fixed amount of metabolites and the rate of metabolism approaches V_{max}, the maximum capacity of metabolism at saturating substrate concentration.
- Some drugs follow first-order kinetics at low concentrations but change to zero-order kinetics at higher (sometimes overdose) concentrations. These drugs are said to follow Michaelis-Menton or capacity limited kinetics, and notable examples include ethanol, phenytoin, and salicylates. The most dramatic dose-dependent effects are seen with drugs that are primarily metabolized by a single metabolic enzyme or where saturation of the main metabolism for the drug toxicity produced by secondary metabolism pathway. An example of the latter situation is seen with

acetaminophen. In acetaminophen overdoses, important Phase II reactions key to detoxification become saturated, the conjugating enzyme depleted, and as a result, toxic intermediates accumulate via a CYP-mediated pathway that normally plays only a minor role in acetaminophen metabolism.

Smoking can alter metabolism of several drugs because polycyclic aromatic hydrocarbons (PAHs) present in tobacco smoke are not only carcinogens but also are potent inducers of the hepatic CYP isoenzymes CYP1A1, CYP1A2, and, possibly, CYP2E1. The primary pharmacokinetic interactions with smoking occur with drugs that are CYP1A2 substrates, such as caffeine, clozapine, fluvoxamine, olanzapine, tacrine, and theophylline. Therefore, smokers taking a medication that interacts with smoking may require higher dosages than nonsmokers. Conversely, upon smoking cessation, smokers may require a reduction in the dosage of an interacting medication [10]. TDM may be very useful for dosage adjustment in smokers, for example, theophylline [11]. Certain drug-food and drug-herb interactions are clinically very significant and are not limited to drug-grapefruit juice and drug-St. John's wort interactions [12]. This topic is addressed in Chapter 6.

1.3.6 Excretion

Drugs are excreted through many routes. The primary route is renal (urine) but others include skin (sweat), ductal tissue (breast milk), lungs (breath), and intestinal (feces). Very few drugs reach this point without first being metabolized to some degree. The fraction of a drug excreted unchanged can be determined by measuring the amount of the parent drug present in the excretion medium (urine, sweat, etc.) and comparing it to the dose administered.

1.3.7 Pharmacokinetic calculations

The volume of distribution (V_d) is a hypothetical estimation that relates the amount of drug in the body to the concentration of drug in the blood or plasma. Although V_d cannot be measured physically it can be calculated from dosage and blood concentration of the drug. This parameter represents the amount of fluid that would be necessary to contain all of a dose after distribution, and at times this parameter is greater than the actual body volume. A drug found completely within the vascular compartment will have a small V_d, while one distributed in an extravascular tissue, such as adipose tissue, will have a very large V_d that may exceed the patient's plasma volume.

The V_d for a specific drug is useful when calculating the loading dose necessary to achieve a targeted drug concentration, in estimating the body burden of a drug, and when determining if dialysis will be an effective means of decontaminating a toxic amount of the drug. Dialysis is not usually effective in removing a drug with a large V_d. Published values for V_d, as those in the specific drug information sections in this book, are typically determined using healthy, young volunteers. Deviations from these values should be expected with ageing, when there are conditions resulting in alterations in fluid compartments, and due to differences in body fat composition. The formulas to calculate V_d:

$$V_d = \text{Dose/plasma concentration of drug}$$

The amount of a drug that interacts with the receptor or target site is usually a small fraction of the total drug administered. Muscle and fat tissues may serve as a reservoir for lipophilic drugs. For neurotherapeutics, penetration of blood brain barrier by the drug is essential.

The half-life of a drug is the time required for the serum concentration to be reduced by 50%. Half-life of a drug can be calculated from elimination rate constant (K) of a drug.

$$\text{Half-life} = 0.693/K$$

Elimination rate constant can be easily calculated from the serum concentrations of a drug at two different time points using the formula where Ct_1 is the concentration of drug at a time point t_1 and Ct_2 is the concentration of the same drug at a later time point t_2:

$$K = \frac{\ln Ct_1 - \ln Ct_2}{t_2 - t_1}$$

Once the half-life of the drug is known, the time required for clearance can be estimated. Approximately 97% of the drug is eliminated by 5 half-lives, while ~99% is eliminated by 7 half-lives. Most drugs are administered in multiple doses at intervals approximating the drug's half-life. When dosed in this manner, the drug begins to accumulate until the amount of drug administered and the amount metabolized and cleared are equal (5–7 half-lives), at which time steady state is said to be achieved. TDM is typically applied when a drug reaches its steady state.

1.3.8 Clearance

Clearance reflects excretion and is linear (first-order) for most drugs. It allows for the estimation of the fraction of the plasma volume being cleared of drug for each unit of time. It is expressed in units of mL/m or L/kg/h and may be calculated in several ways:

$$CL = \text{dose}/\text{dosing interval} \div C_{ss}$$

where C_{ss} is the concentration measured at steady state.

Other calculations include:

$$CL = 0.693/t_{1/2} \times V_d$$

1.4 Therapeutic ranges

Once a drug has reached steady state, pharmacokinetic principles can be applied in order to adjust and individualized the dosage regimen to maintain the patient within the therapeutic range. The therapeutic range spans the concentrations between the lowest concentration at which the drug is effective for most patients (the minimum effective concentration, MEC) and the lowest concentration above which most patients are likely to experience adverse or toxic effects (minimum toxic concentrations, MTC). If serum concentrations are measured before steady state is achieved, they may be misinterpreted as too low for efficacy. Following a dosage change, an additional five half-lives will be required for the patient to achieve the new steady state concentration. For practical purposes, especially for drugs with long half-lives, monitoring may begin as early as 3–4 half-lives, but it is important to recognize that steady state is reached after five half-lives. Therapeutic ranges of commonly monitored drugs are given with individual drugs discussed later in the book.

1.5 Guidelines for TDM

As mentioned earlier in the chapter, only a small fraction of drugs used clinically today requires TDM. The characteristics of a drug where TDM is beneficial include:

- Narrow therapeutic range where the dose of a drug which produces the desired therapeutic concentrations in one patient may cause toxicity in another patient.

- Good correlation between serum or whole blood drug concentration and clinical efficacy/toxicity but poor correlation between dosage and clinical outcome.
- Well established therapeutic range and drugs levels associated with adverse reaction/toxicity.
- Toxicity of a drug that may lead to hospitalization, irreversible organ damage and even death may be avoided by TDM.

For strongly protein bound drugs (protein binding >80%), a better correlation may be observed between unbound (free) drug concentration and clinical outcome rather than between traditionally monitored total drug concentration, because it is only the unbound drug that is responsible for pharmacological action of a drug. For example, adjusting phenytoin dosing in patients based on their serum phenytoin concentrations rather than seizure frequencies not only decrease the morbidity but also prevent phenytoin toxicity in these patients. Peterson et al. reported that in their study involving 114 patients, free phenytoin concentrations correlated better with clinical picture than total phenytoin concentrations [13]. Please see Chapter 3 for more detail.

Usually trough drug concentration is determined in TDM because trough concentration (15–30 min before the next dose) shows least variation and most reliability for TDM. Most published therapeutic ranges are based on trough concentrations. However, for very toxic drugs such as vancomycin and aminoglycosides, both trough and peak concentrations may be monitored.

Timing of proper specimen collection is important for proper interpretation of drug concentration. In one report, the authors interpreted serum digoxin levels in patients who took digoxin at the same time daily (40 patients, 300 tests) and observed that 52% of tests were performed on inappropriate samples drawn within 6 h of the last dose. No patient who took digoxin after 1700 (5 p.m.) had inappropriate tests. The authors concluded that phlebotomy for serum digoxin determinations before distribution of digoxin is complete is a common problem in outpatients, leading to clinically uninterpretable test results. However, nighttime dosing of digoxin eliminated such problem [14].

Usually a drug may be monitored several times within a week or first month after initial prescription in order to achieve desired therapeutic condition as determined by the clinician. However, when a steady state is achieved, TDM may be conducted monthly or less often provided dosage and factors that may alter pharmacokinetics have not be changed. An illustration of this point was demonstrated in a study of TDM for azathioprine, methotrexate and cyclosporine, oral immunomodulatory drugs used in pediatric dermatology. In one study the authors reported that less frequent monitoring (3 months interval versus 1 month intervals) did not result in any significant adverse events over a 15-year period based on study of 242 children receiving azathioprine, methotrexate and cyclosporine [15].

1.6 Special population and TDM

Drug metabolism is significantly affected by the age of a patient. In the fetus, CYP3A7 is the major hepatic enzyme of CYP family of enzymes, although CYP3A5 may also present in significant level in half of the children. In adults, CYP3A4 is the major functional CYP enzyme responsible for metabolism of many drugs. After birth, hepatic CYP2D6, CYP2C8/9 and CYP2C18/19 are activated. CYP1A2 becomes active during the fourth to fifth months after birth [16]. In general, drug metabolizing capacity by the liver enzymes is reduced in newborns particularly in premature babies but increases rapidly during the first few weeks and months of life to reach values which are generally higher than adult metabolizing rates and then declines with old age. Renal function at the time of birth is reduced by more than 50% of adult value but then increases rapidly in the first two to 3 years of life. Renal function, however, starts declining with advanced age. Oral clearance of lamotrigine, topiramate, levetiracetam, oxcarbazepine, gabapentin, tiagabine, zonisamide, vigabatrin and felbamate is significantly higher (20–120%) in children compared to adults depending on the drug and the age distribution of the patients. On the other hand, clearance of these drugs is reduced (10–50%) in the elderly population compared to middle-aged adults [17].

Although the basic rules of TDM are applicable to neonates receiving aminoglycosides, vancomycin, phenobarbital or digoxin, additional factors must also be considered. First, due to both pharmacokinetic variability and non-pharmacokinetic factors, the correlation between dosage and drug concentration may be poor in neonates, but that limitation can be overcome with the use of more complex, validated dosing regimens. Second, the time to reach steady state is prolonged, especially when no loading dose is used. Consequently, the timing of TDM sampling is important in this population. Third, the target concentration may be uncertain (vancomycin) or depend on specific factors (phenobarbital during whole body cooling). Finally, because of differences in matrix composition (serum proteins, bilirubin), assay-related inaccuracies may be different in neonates. Pauwels and

Allegaert concluded that complex validated dosing regimens, with subsequent TDM sampling and Bayesian forecasting, are the next step in tailoring pharmacotherapy to individual neonates [18]. The science of TDM in children remains relatively underdeveloped especially due to lack of sufficient clinical reports dealing with TDM of various drugs in children [19]. Zakrzewski-Jakubiak et al. reported that elderly patients are at higher risk of adverse drug reaction because most elderly patients receive multiple drugs that increase risk of pharmacokinetic drug interactions [20].

Men and women may show both pharmacokinetic and pharmacodynamic differences in response to certain drugs. In addition, pregnancy may also significantly alter disposition of a limited number of drugs. Gender difference affects bioavailability, distribution, metabolism and elimination of drugs due to variations between men and women in body weight, blood volume, gastric emptying time, drug protein binding, activities of drug metabolizing enzymes, drug transporter function and excretion activity [21]. In general women are also more susceptible to adverse drug reactions compared to men. Women are at increased risk of QT prolongation with many antiarrhythmic drugs which may even lead to critical condition such as torsade de pointes compared to men even at same levels of serum drug concentrations [22]. Other classes of drugs that may cause more adverse drug reaction in women compared to men include anesthetics and antiretrovirals. Hormonal effects may be one of the underlying causes in majority of adverse drug reactions observed in women [23].

1.6.1 Drug disposition in uremia

Renal disease causes impairment in the clearance of many drugs by the kidney. Correlations have been established between creatinine clearance and clearance of digoxin, lithium, procainamide, aminoglycoside and several other drugs, but creatinine clearance does not always predict renal excretion of all drugs. Dose adjustments based on renal function is recommended for many medications in elderly patients as well as patients with impaired renal function even for medications that exhibit large therapeutic windows [24]. Renal disease also causes impairment of drug protein binding, because uremic toxins compete with drugs for binding to albumin. Such interaction leads to increases in concentration of pharmacologically active free drug concentration especially for classical anticonvulsants such as phenytoin, carbamazepine and valproic acid. Therefore, monitoring free phenytoin, free valproic acid and to some extent free carbamazepine is recommended in uremic patients in order to avoid drug toxicity [25].

1.6.2 Drug disposition in hepatic disease

Liver dysfunctions not only reduces clearance of a drug metabolized through hepatic enzymes or eliminated by biliary excretion but also affect serum protein binding due to reduced synthesis of albumin and other drug binding proteins. Even mild to moderate hepatic disease may cause unpredictable effects on drug metabolism. Portal-systemic shunting present in patients with advanced liver cirrhosis may cause significant reduction in first pass metabolism of high extraction drugs thus increasing bioavailability as well as risk of drug overdose and toxicity [26]. In addition, activities of several isoenzymes of CYP enzymes (CYP1A1, CYP2C19 and CYP3A4/5) are reduced due to liver dysfunction while activities of other isoenzymes such as CYP2D6, CYP2C9 and CYP2E1 may not be affected significantly. Therefore, drugs that are metabolized by CYP1A1, CYP3A4/5 and CYP2C19 may show increased blood levels in patients with hepatic dysfunction requiring dosage adjustment in order to avoid toxicity [27]. Although the Phase I reaction involving CYP enzymes may be impaired in liver disease, the Phase II reaction (glucuronidation) seems to be affected to a lesser extent. Mild to moderate hepatitis infection may also alter clearance of drugs. Trotter et al. reported that total mean tacrolimus dose in year one after transplant was lower by 39% in patients with hepatitis C infection compared to patients without infection. [28]. Zimmermann et al. reported that oral dose clearance of sirolimus was significantly decreased in subjects with mild to moderate hepatic impairment compared to controls. These authors stressed the need for careful monitoring of trough whole blood sirolimus concentrations in renal transplant recipients exhibiting mild to moderate hepatic impairment [29].

Hypoalbuminemia is often observed in patients with hepatic dysfunction which impairs protein binding of many drugs. Because free (unbound) drugs are responsible for pharmacological action, careful monitoring of free concentrations of strongly albumin bound antiepileptic drugs such as phenytoin, carbamazepine and valproic acid is recommended in patients with hepatic dysfunction in order to avoid drug toxicity [25].

1.6.3 Drug disposition in cardiovascular diseases

Cardiac failure is often associated with disturbances in cardiac output, influencing the extent and pattern of tissue perfusion, sodium and water metabolism and GI motility that eventually may affect absorption and disposition of many drugs. Hepatic elimination of drugs via oxidative

Phase I metabolism is impaired in patients with congestive heart failure due to decreased blood supply in the liver [30]. Theophylline metabolism is reduced in patients with severe cardiac failure, and dose reduction is strongly recommended. Digoxin clearance is also decreased. Quinidine plasma level may also be high in these patients due to lower volume of distribution [31]. Therefore, TDM is crucial in avoiding drug toxicity in these patients.

1.6.4 Drug disposition in critically ill patients

Physiological changes in critically ill patients can significantly affect the pharmacokinetics of many drugs. These changes include absorption, distribution, metabolism and excretion of drugs in critically ill patients. Understanding these changes in pharmacokinetic parameters is essential for optimizing drug therapy in critically ill patients.

1.6.5 Drug disposition in thyroid diseases

Patients with thyroid disease may have an altered drug disposition because thyroxine activates CYP enzyme system. Therefore, lower levels of drugs may result from high thyroxine levels due to induction of hepatic oxidative metabolism pathway causing lower than expected drug level. In contrast, hypothyroidism is associated with inhibition of hepatic oxidative metabolism of many drugs causing elevated serum levels of drugs. Therefore, TDM is useful in adjusting proper dosage of drugs in patients with thyroid disorders. Haas et al. reported a case where a patient developed hypothyroidism 6 months after single lung transplantation and was admitted to the hospital for anuric renal failure. The patient showed a toxic blood level of tacrolimus which was resolved with the initiation of thyroxine replacement therapy and dose reduction of tacrolimus [32].

1.7 Conclusions

TDM is very useful not only for individualized drug dosing but also to monitor compliance and avoiding adverse drug effects. Many factors other than age, sex and various pathological conditions such as uremia, renal disease, cardiovascular disease and critical illness also affect drug disposition. Proper understanding of altered pharmacokinetics under various diseases is essential for proper interpretation of drug concentrations during TDM.

References

[1] Schumock GT, Li EC, Suda KJ, Wiest MD, et al. National trends in prescription drug expenditures and projections for 2016. Am J Health Syst Pharm 2016;73:1058–75.
[2] Watson I, Potter J, Yatscoff R, Fraser A, et al. Ther Drug Monit 1997;19:125. [Editorial].
[3] Ratanajamit C, Kaewpibal P, Setthawacharavanich S, Faroongsarng D. Effect of pharmacist participation in the health care team on therapeutic drug monitoring utilization for antiepileptic drugs. J Med Assoc Thail 2009;92:1500–7.
[4] Punyawudho B, Singkham N, Thammajaruk N, Dalodom T, et al. Therapeutic drug monitoring of antiretroviral drugs in HIV-infected patients. Expert Rev Clin Pharmacol 2016;9:1583–95.
[5] Kongkaew C, Sakunrag I, Jianmongkol P. Non-compliance with digoxin in patients with heart failure and/or atrial fibrillation: a systematic review and meta-analysis of observational studies. Arch Cardiovasc Dis 2012;105:507–16.
[6] Dowse R, Futter WT. Outpatient compliance with theophylline and phenytoin therapy. S Afr Med J 1991;80:550–3.
[7] Calcagno A, Pagani N, Ariaudo A, Arduino G, et al. Therapeutic drug monitoring of boosted PIs in HIV-positive patients: undetectable plasma concentrations and risk of virological failure. J Antimicrob Chemother 2017;72:1741–4.
[8] Breslow RA, Dong C, White A. Prevalence of alcohol-interactive prescription medication use among current drinkers: United States, 1999 to 2010. Alcohol Clin Exp Res 2015;39:371–9.
[9] Jancova P, Anzenbacher P, Anzenbacherova E. Phase II drug metabolizing enzymes. Biomed Pap Med Fac Univ Palacky Olomouc Czech Repub 2010;154:103–16.
[10] Kroon LA. Drug interactions with smoking. Am J Health Syst Pharm 2007;64:1917–21.
[11] Goseva Z, Gjorcev A, Kaeva BJ, Janeva EJ, et al. Analysis of plasma concentrations of theophylline in smoking and nonsmoking patients with asthma. Open Access Maced J Med Sci 2015;3:672–5.
[12] Mouly S, Lloret-Linares C, Sellier PO, Sene D, et al. Is the clinical relevance of drug-food and drug-herb interactions limited to grapefruit juice and Saint-John's Wort? Pharmacol Res 2017;118:82–92.
[13] Peterson GM, Khoo BH, von Witt RJ. Clinical response in epilepsy in relation to total and free serum levels of phenytoin. Ther Drug Monit 1991;13:415–9.
[14] Bernard DW, Bowman RL, Grimm FA, Wolf BA, et al. Nighttime dosing assures postdistribution sampling for therapeutic drug monitoring of digoxin. Clin Chem 1996;42(1):45–9.
[15] Yee J, Orchard D. Monitoring recommendations for oral azathioprine, methotrexate and cyclosporin in a pediatric dermatology clinic and literature review. Aust J Dermatol 2018;59:31–40.
[16] Oesterheld JR. A review of developmental aspects of cytochrome P 450. J Child Adolesc Psychopharmacol 1998;8:161–74.
[17] Perucca E. Pharmacokinetics variability of new antiepileptic drugs at different age. Ther Drug Monit 2005;27:714–7.
[18] Pauwels S, Allegaert K. Therapeutic drug monitoring in neonates. Arch Dis Child 2016;101:377–81.
[19] Macleod S. Therapeutic drug monitoring in pediatrics: how do children differ? Ther Drug Monit 2010;32:253–6.
[20] Zakrzewski-Jakubiak H, Doan J, Lamoureux P, Singh D, et al. Detection and prevention of drug-drug interactions in the hospitalized elderly: utility of new cytochrome P450 based software. Am J Geriatr Pharmacother 2011;9:461–70.
[21] Gandhi M, Aweeka F, Greenblatt RM, Blaschke TE. Sex difference in pharmacokinetics and pharmacodynamics. Annu Rev Pharmacol Toxicol 2004;44:499–523.

[22] Makkar RR, Fromm BS, Steinman RT, Meissner MD, et al. Female gender as a risk factor for torsade de pointes associated with cardiovascular drugs. JAMA 1993;270:2590–7.
[23] Nicolson TJ, Mellor HR, Roberts RR. Gender difference in drug toxicity. Trends Pharmacol Sci 2010;31:108–14.
[24] Terrell KM, Heard K, Miller DK. Prescribing to older ED patients. Am J Emerg Med 2006;24:468–78.
[25] Dasgupta A. Usefulness of monitoring free (unbound) concentrations of therapeutic drugs in patient management. (Review) Clin Chem Acta 2007;377:1–13.
[26] Verbeeck RK. Pharmacokinetics and dosage adjustment in patients with hepatic dysfunction. Eur J Clin Pharmacol 2008;64:1147–61.
[27] Villeneuve JP, Pichette V. Cytochrome P450 and liver disease. Curr Drug Metab 2004;5:273–82.
[28] Trotter JF, Osborne JC, Heller N, Christians U. Effect of hepatitis C infection on tacrolimus does and blood levels in liver transplant recipients. Aliment Pharmacol Ther 2005;22:37–44.
[29] Zimmermann JJ, Lasseter KC, Lim HK, Harper D, et al. Pharmacokinetics of sirolimus (rapamycin) in subjects with mild to moderate hepatic impairment. J Clin Pharmacol 2005;45:1368–72.
[30] Ng CY, Ghabrial H, Morgan DJ, Ching MS, et al. Impaired elimination of propranolol due to right heart failure: drug clearance in the isolated liver and its relationship to intrinsic metabolic capacity. Drug Metab Dispos 2000;28:1217–21.
[31] Benowitz NL, Meister W. Pharmacokinetics in patients with cardiac failure. Clin Pharmacokinet 1976;1:389–405.
[32] Haas M, Kletzmayer J, Staudinger T, Bohmig G, et al. Hypothyroidism as a cause of tacrolimus intoxication and acute renal failure: a case report. Wien Klin Wochenschr 2000;112:939–41.

CHAPTER 2

Before and after: The pre-analytical and post-analytical phases of TDM

2.1 Introduction

The renewed interest in therapeutic drug monitoring (TDM) is reflected by the flurry of new assay development. Some of these are immunoassay-based, while others are based upon chromatographic methods including liquid chromatography combined with mass spectrometry (LC/MS) or tandem mass spectrometry (LC/MS/MS). Although major attention has been focused on assay development and avoiding interferences, it is important to note that most errors in clinical laboratories occur during pre- and post-analytical phases. Plebani commented that most errors in clinical laboratories are due to pre-analytical factors (46–68.2% of total errors), while a high error rate (18.5–47% of total errors) has also been found in the post-analytical phase. Errors due to analytical problems have been significantly reduced over time by improvements in instrumentation and assay design, but there is evidence that, particularly for immunoassays, interference may have a serious impact on patients, as seen with TDM of digoxin and immunosuppressants [1]. Digoxin is routinely monitored using immunoassays and such assays are subjected to significant interferences from various cross-reactants including herbal supplements. The gold standard for monitoring immunosuppressants (e.g., cyclosporine, sirolimus, tacrolimus) is LC/MS or LC/MS/MS, because metabolites have significant cross-reactivity with immunoassays causing falsely elevated drug levels. However, many clinical laboratories use immunoassays for TDM of immunosuppressants due to factors such as test complexity, cost, and turnaround time. Issues of interferences are discussed with monographs of individual drugs.

Therapeutic Drug Monitoring Data
https://doi.org/10.1016/B978-0-12-815849-4.00002-5

2.2 Proper timing of specimen collection

As mentioned in Chapter 1, the preferred specimen for most TDM applications is trough specimens. Trough specimens are collected at the time point when concentrations are expected to be in the lowest part of the therapeutic range and are used when efficacy is the main reason for monitoring. Strictly speaking, trough samples should be collected immediately before the administration of a dose. The timing of phlebotomy presents a practical challenge, especially when different personnel are involved in drug dosing and phlebotomy. Many phlebotomy or nursing services find coordination of this exact timing difficult, but every attempt should be made to collect the sample as close to the dose as possible. Many laboratories allow a window of 30–60 min for collection, but even 30 min may introduce inaccuracies of 10–15% for drugs with short half-lives such as aminoglycosides and caffeine. Collecting trough samples 5–6 h before dosing is not acceptable even for drugs with relatively long half-lives.

When a drug is effective over a very narrow concentration range or toxicity is a concern, peak concentrations may be monitored. This type of collection is challenging to time and requires consideration of absorption and distribution phases. The time of collection is influenced by drug formulation, dosing route, and other factors.

Aminoglycosides, digoxin, and lithium presents special considerations for TDM. For the aminoglycosides, peak samples are used to judge efficacy when using conventional dosing regimens, and trough samples are used to judge the potential for toxicity for either conventional or once daily dosing (ODD) regimens. In addition when using an ODD protocol, samples used to adjust the next dose are collected at defined points within the dosing cycle. Some laboratories call these "random" samples since they are neither peak nor trough and the time of collection may vary, but this terminology is not technically correct since the collection time must be controlled and known—i.e., not truly random. The therapeutic ranges used in the conventional dosing protocols must never be applied to samples collected from a patient receiving these drugs on ODD protocols. For these dosing protocols, pharmacy services use nomograms to compare the drug concentration and the time since initiation of the dose to assess efficacy and modify subsequent dosing intervals.

Lithium and digoxin present challenges due to their long absorption and distribution phases. For either drug, a sample collected before distribution is complete will often be elevated, leading to potential misinterpretation of a toxic drug concentration in the absence of clinical toxicity [2, 3]. Samples for lithium should be collected no earlier than 10–12 h after dosing. Samples for digoxin measurement should be collected no earlier than 6–8 h after dosing, preferably after 12 h.

It is also important to perform TDM when a drug has reached its steady state. Allergra conducted TDM in children receiving voriconazole by collecting blood samples immediately before drug intake (trough) and under steady-state conditions (5 days after both intravenous and oral administration). Plasma samples were obtained by centrifugation, and voriconazole concentrations were determined using LC/MS. Sex significantly influenced drug levels: males had higher median drug concentrations than females. Close voriconazole TDM should help individualize antifungal therapy for children [4].

2.3 Issues with blood collection tubes

Plastic gel-barrier tubes (plasma or serum separators) are frequently used for clinical chemistry analyses, but these devices may cause falsely lower concentrations for certain drugs due to absorption by the gel. For early generation gel-barrier tubes, drug adsorption was an insidious and unpredictable problem, with significant lot-to-lot variation. Drug adsorption also varied with the amount of sample in contact with the gel, the duration of contact, drug concentration, and storage temperature. As a result, many TDM laboratories moved to gel-free tubes. The introduction of a new generation of plastic gel-barrier tubes has laboratories once again considering these as an option. The report from Bush and colleagues [5] was encouraging in that they found <10% loss for only two of the drugs tested, carbamazepine and phenytoin, and only after in contact for 7 days.

Recently, Schrapp et al. explored the impact of Becton Dickinson (BD) PST II (plasma separator tube) and Barricor separator tubes on the stability of 167 therapeutic compounds and common drugs of abuse in plasma samples using LC/MS/MS. In 2016, BD released a new heparin collection tube with a novel technology of separation, a mechanical separator called BD Vacutainer Barricor. Instead of a gel, it uses a high-density plastic connected on elastomer top that stretches during centrifugation, finally creating a seal. Blood cells are allowed to flow around this separator reducing the time of centrifugation and providing high-quality plasma. 40 drugs were significantly affected by the use of the gel barrier PST II tubes, including antidepressants (11/26), neuroleptics (9/13), cardiovascular drugs (5/26), anxiolytics and hypnotics (4/25) and some drugs of abuse (5/26). In contrast, only six compounds exhibited significant reduction by the mechanical

Barricor tubes. The study showed the Barricor non-gel tubes cause less drug interference and are recommended for the drugs studied [6].

From time to time, other chemical additives in blood tubes (e.g., clot activators, plasticizers) have been found to interfere with TDM analyses. [7–11] All other tubes or devices, e.g., microfuge tubes and filtration devices, used in the method should also be considered as potential sources of adsorption or interference. [12] These type of issues can affect chromatography, mass spectrometry, electrochemical, and immunoassay-based methods [11, 12].

2.4 Processing and storage of specimens

It is often assumed that once collected and processed, samples for TDM analysis are stable; however, serum, plasma, and whole blood are complex matrices and far from static after processing. Samples should be tested as soon as possible after processing or should be stored refrigerated or frozen. This is especially true for aminoglycoside testing. Often patients receiving these drugs are also treated with additional antibiotics, including beta-lactam antibiotics. It is well known in pharmacy that beta-lactams will inactivate aminoglycosides, rendering the aminoglycoside ineffective against the microorganism for which it is intended. Consequently, aminoglycosides and beta-lactam antibiotics should not be co-administered but instead administered sequentially. However, for the clinical laboratory, both drugs may be present in the serum sample. If both are indeed present and the sample is left at room temperature, the aminoglycoside concentration will decline significantly at a rate dependent upon the aminoglycoside and the beta-lactam antibiotic, as well as the time and temperature of exposure. Tindula et al. studied the degree of in vitro inactivation of gentamicin, tobramycin, and amikacin by various penicillins and cephalosporins and observed that the beta-lactams can be divided into three groups: (A) cefazolin and cefamandole, which caused little inactivation; (B) nafcillin, cephapirin, and cefoxitin, which caused moderate inactivation; and (C) penicillin, ampicillin, carbenicillin, and ticarcillin, which cause marked inactivation. In general, tobramycin was the most reactive of the three aminoglycosides studied while amikacin was the most stable drug. Frozen samples were much less affected than those left at room temperature [13]. Since laboratory may not have information whether patient is also receiving beta-lactam antibiotic, it is advisable to analyze specimens submitted for aminoglycosides as soon as possible or, if necessary, keep specimens frozen.

Most therapeutic drugs are stable for several hours at room temperature and freeze–thaw cycles have no effect on drug stability, for example, no obvious degradation was observed under various storage conditions including room temperature for 12h, three freeze–thaw cycles and −80°C for 1 month for valproic acid [14].

2.5 Troubleshooting

It is very important to properly identify the specimen, because misidentification may have serious impact on patient safety such as erroneous dosing changes. Specimen must be labeled at the bedside or immediately after collection in the phlebotomy station. Collecting several blood specimens from different patients and then labelling all specimens later is not a wise practice because it may significantly increase the risk of mislabeling of specimens. Wrong blood collection tube may falsely elevate or decrease drug level causing confusion. Lithium should be collected in a heparin sodium tube but not lithium heparin tube. An erroneous toxic plasma concentration of lithium in an infant due to use of lithium containing container has been reported [15].

Sometimes the clinical service may contact the clinical laboratory complaining about an unexpectedly higher concentration of a therapeutic drug in a patient who previously had drug concentrations in the therapeutic range. One common pre-analytical error is that the patient took the medication shortly before blood draw, with the sample thus not representing a trough collection. In contrast, lower than expected drug level may indicate non-compliance therapy or various drug–drug or drug-herbal product interactions.

During method evaluation and validation, it is important not only to assess effects of hemolysis, lipemia, and icterus on test results, but also to evaluate the tubes and devices used in the sample collection, storage, or preparation. As stated before, drug concentrations must be interpreted with knowledge of the time at which the dose was administered. The ideal delivery of TDM results includes the time of collection, and all pertinent dosing information. It has always been recommended that TDM laboratories request this information, but this information may not be provided by the clinical team. However, dose administration times should be accessible through the medical record. Fortunately, we are finally seeing hospital information systems capable of linking pharmacy, nursing, and laboratory information. Integration of these systems has the potential of significantly

Table 2.1 Information requested with specimens for therapeutic drug monitoring

Patient related information	Essential information	Other useful information
• Name of the patient • Hospital identification number • Age • Gender (if female pregnant?)	• Time of last dosage • Identification of peak vs trough specimen (for aminoglycosides and vancomycin) • Special request (such as free phenytoin, carbamazepine or valproic acid) • Type of specimen (if other than blood)	• Dosage regime • If aminoglycosides is ordered patient taking any beta-lactam antibiotic? • Is the patient critically ill or suffer from hepatic, cardiovascular or renal disease? • Albumin level, creatinine clearance for special population

reducing errors and adverse events as commonly encountered problems can be flagged to alert the medical team. This type of system can alert clinicians to potential drug interactions and laboratory signs of toxicity such as rising plasma creatinine. Optimal patient related information which should be submitted with TDM specimens are summarized in Table 2.1.

References

[1] Plebani M. Errors in clinical laboratories or errors in laboratory medicine? Clin Chem Lab Med 2006;447:750–9.

[2] Hammett-Stabler CA, Johns T. Laboratory guidelines for monitoring of antimicrobial drugs. National Academy of Clinical Biochemistry. Clin Chem 1998;44:1129–40.

[3] Bernard DW, Bowman RL, Grimm FA, Wolf BA, Simson MB, Shaw LM. Nighttime dosing assures postdistribution sampling for therapeutic drug monitoring of digoxin. Clin Chem 1996;42:45–9.

[4] Allegra S, Fatiguso G, De Francia S, Favata F, et al. Therapeutic drug monitoring of voriconazole for treatment and prophylaxis of invasive fungal infection in children. Br J Clin Pharmacol 2018;84:197–203.

[5] Bush V, Blennerhasset J, Wells A, Dasgupta A. Stability of therapeutic drugs in serum collected in vacutainer serum separator tubes containing a new gel (SST II). Ther Drug Monit 2001;23:259–62. [Erratum in: Ther Drug Monit 2001;23:738].

[6] Schrapp A, Mory C, Duflot T, Pereira T, et al. The right blood collection tube for therapeutic drug monitoring and toxicology screening procedures: standard tubes, gel or mechanical separator? Clin Chim Acta 2019;488:196–201.

[7] Murthy VV. Unusual interference from primary collection tube in a high-performance liquid chromatography assay of amiodarone. J Clin Lab Anal 1997;11:232–4.

[8] Drake SK, Bowen RAR, Bemaley AT, Hortin GL. Potential interferences from blood collection tubes in mass spectrometric analyses of serum polypeptides. Clin Chem 2004;50:2398–401.

[9] Wu SL, Wang YJ, Hu J, Leung D. The detection of the organic extractables in a biotech product by liquid chromatography on-line with electrospray mass spectrometry. PDA J Pharm Sci Technol 1997;51:229–37.

[10] Sampson M, Ruddel M, Albright S, Elin RJ. Positive interference in lithium determinations from clot activator in collection container. Clin Chem 1997;43:675–9.
[11] Bowen RAR, Chan Y, Ruddel ME, Hortin GL, Csako G, Demosky SJ, Remaley AT. Immunoassay interference by a commonly used blood collection tube additive, the organosilicone surfactant silwet L-720. Clin Chem 2005;51:1874–82.
[12] Yen HC, Hsu YT. Impurities from polypropylene microcentrifuge tubes as a potential source of interference in simultaneous analysis of electrochemical detection. Clin Chem Lab Med 2004;42:390–5.
[13] Tindula RJ, Ambrose PJ, Harralson AF. Aminoglycoside inactivation by penicillins and cephalosporins and its impact on drug-level monitoring. Drug Intell Clin Pharm 1983;17:906–8.
[14] Zhao M, Zhang T, Li G, Qiu F, et al. Simultaneous determination of valproic acid and its major metabolites by UHPLC-MS/MS in Chinese patients: application to therapeutic drug monitoring. J Chromatogr Sci 2017;55:436–44.
[15] Arslan Z, Athiraman NK, Clark SJ. Lithium toxicity in a neonate owing to false elevation of blood lithium levels caused by contamination in a lithium heparin container: case report and review of the literature. Paediatr Int Child Health 2016;36:240–2.

CHAPTER 3

Clinical utility of monitoring free drug levels

3.1 Introduction

Therapeutic drug monitoring (TDM) involves the determination of drug concentrations in serum, plasma, whole blood, or less commonly another body fluid such as saliva. These concentrations are interpreted in relation to the dose, the expected therapeutic range, and the clinical presentation. For drug concentrations to be useful, there must be a clear correlation between the measured concentration and clinical efficacy, that is the desired pharmacological effect. A key assumption of TDM is that the concentration measured in the sample reflects that available to elicit the pharmacological response at the receptor. In the periphery, drugs circulate either as bound to proteins or free (unbound). Protein binding can be low (<20%), moderate (20–80%) or high (>80%), and there are some drugs such as lithium or levetiracetam with very low (<10%) protein binding. Monitoring unbound (free) drug concentration is clinically very useful for certain drugs which are strongly bound to serum proteins in patients with certain pathological conditions.

3.2 Binding of drugs with serum proteins

The major drug binding protein in serum is albumin, followed by α_1-acid glycoprotein and lipoproteins. Drugs exist in peripheral circulation as free (unbound) and bound to protein forms following the principle of reversible equilibrium and law of mass action. Only free drug is capable of crossing the plasma membrane and binding with the receptor for pharmacological action [1]. Moreover, for central nervous system drugs, mostly free fraction could cross blood brain barrier [2].

In general, there is equilibrium between free drug and protein-bound drug:

$$[D]+[P]=[DP]$$
$$K=[DP]/[D][P]$$

Therapeutic Drug Monitoring Data
https://doi.org/10.1016/B978-0-12-815849-4.00003-7

where [D] is the drug concentration, [P] is the binding protein concentration, [DP] is the drug/protein complex and K is the association constant (liters/mole). The greater the affinity of the protein for the drug, the higher is the K value. The free fraction of a drug represents the relationship between bound and free drug concentration and is often referred as "α."

$$\alpha = \text{Free drug concentration}/\text{total drug concentration}\left(\text{bound} + \text{free}\right)$$

Free fraction (α) of a drug typically does not vary with total drug concentration because protein-binding sites usually exceed the number of drug molecules present. For example, normal reference range of molar concentration of albumin is 527–784 μmol/L (3.5–5.2 g/dL), while reference range of phenytoin, which is strongly bound to albumin, is 40–79 μmol/L (10–20 μg/mL). Therefore, normal molar concentration of the binding protein albumin significantly exceeds the expected phenytoin concentration in serum. However, for certain drugs that achieve high molar concentrations in serum, the number of protein binding sites in albumin may approach or be less than the number of drug molecules at the upper end of therapeutic range, for example valproic acid. The normal albumin concentration in serum is 527–784 μmol/L while the therapeutic range of valproic acid is 50–100 μg/mL or 347–693 μmol/L. Because 1 mol of albumin can bind 1 mol of valproic acid, at the upper end of therapeutic concentration of valproic albumin binding sites may be saturated or nearly saturated causing disproportionate increase in free valproic acid concentration which may not be reflected when only total valproic acid concentration is monitored.

Hepatic drug clearance depends on hepatic blood flow rate, protein binding of the drug as well as ability of hepatic enzymes to effectively metabolize the drug (intrinsic clearance). The ratio of hepatic clearance of the drug and the hepatic blood flow is termed as "extraction ratio" of the drug. Extraction ratio is considered as low if the value is <0.3 or high if the value is >0.7. The metabolism of low clearance drugs (phenytoin, valproic acid, quinidine, etc.) depends only on the unbound concentration of the drug passing through the liver because these drugs usually have higher affinity for their binding protein than for the liver enzymes responsible for their metabolism. Therefore, if protein binding is decreased, metabolism is increased, and a decrease in total drug concentration at steady state is observed. In contrast, metabolism of high clearance drugs (lidocaine, propranolol, etc.) does not depend on the free drug concentration because the affinity of the hepatic enzymes for the drugs is greater, resulting in the liver clearing both free and protein bound drug. Systematic clearance depends only on hepatic blood flow and changes in protein binding will not affect clearance.

3.3 Drugs requiring free drug monitoring

It is usually considered that if the protein binding of a drug is <80%, it is not a candidate for free drug monitoring. For example, if the total concentration of phenytoin (90% bound to serum albumin) is 10 μg/mL the free drug concentration should be 1 μg/mL. Now if the protein binding is reduced to 80%, which is often observed in patients with uremia, the free drug concentration will increase to 2.0 μg/mL, which is a clinically significant 100% increase in free concentration! In contrast, for another drug which is only 40% protein bind, free drug level should be 6 μg/mL if the total drug concentration is also 10 μg/mL. A 10% decrease in protein binding will increase the free concentration from 6.0 μg/mL to 6.4 μg/mL, a change which is likely to be clinically insignificant. An exception is free digoxin monitoring (digoxin is only 25% protein bound), which is very useful in patients overdosed with digoxin and being treated with Digibind or DigiFab, the FAB fragment of anti-digoxin antibody. Free digoxin monitoring is discussed in more detail later in this chapter. Protein binding of some commonly monitored therapeutic drugs is given in Table 3.1.

In today's practice, free drug monitoring is most common with classical anticonvulsants; phenytoin, carbamazepine and valproic acid because these drugs are strongly bound to serum albumin. Soldin has reported that in his personal experience, free phenytoin is the mostly requested free drug levels by clinicians [3]. The first comprehensive report demonstrating the clinical utility of free drug monitoring was published in 1973. In a population of 30 epileptic patients, the authors found a better correlation between toxicity and free phenytoin concentrations compared to toxicity and total phenytoin concentrations [4]. Kits are available commercially for monitoring free phenytoin and free valproic acid concentrations. Moreover, the College of American Pathologists (CAP, the accrediting agency for clinical laboratories) also have free anticonvulsant levels in their external survey specimens. Therapeutic range of total and free reference ranges of certain drugs are given in Table 3.2.

In addition to monitoring free concentrations of these anticonvulsants, clinical utility of monitoring free mycophenolic acid has been demonstrated for certain patient population (patients with uremia, liver disease, hypoalbuminemia and hyperbilirubinemia). Moreover, there are also some

Table 3.1 Protein binding of commonly monitored therapeutic drugs

Class of drug	Individual drug	Average protein binding under normal pathophysiology
Anticonvulsants	Carbamazepine	80%
	Ethosuximide	0%
	Phenobarbital	40%
	Phenytoin	90%
	Primidone	15%
	Valproic acid	90–95%
Antibiotics	Amikacin	Negligible
	Gentamicin	Negligible
	Kanamycin	Negligible
	Tobramycin	Negligible
	Vancomycin	40%
Cardioactive drugs	Digoxin	25%
	Lidocaine	60–80%
	Procainamide	10–15%
	Quinidine	80%
Antiasthmatic	Theophylline	40%
	Caffeine	35%
Antidepressants	Amitriptyline	90%
	Nortriptyline	90%
	Doxepin	80%
	Imipramine	90%
	Desipramine	80%
	Lithium	Negligible
Antineoplastic	Methotrexate	46%
Immunosuppressants	Cyclosporine	98%
	Tacrolimus	97%
	Sirolimus	92%
	Everolimus	74%
	Mycophenolic acid	97–98%

indications of monitoring free concentrations of certain protease inhibitors. Free drug monitoring of mycophenolic acid and protease inhibitors is discussed in more detail below.

3.4 When to monitor free drug concentration?

As long as there is no pathological condition that may alter concentration of the drug binding protein or impaired protein binding of the drug, monitoring free drug concentration for even strongly protein bound drugs is

Table 3.2 Therapeutic range of total and free phenytoin, valproic acid carbamazepine and mycophenolic acid

Drug	Therapeutic range		Free drug routinely monitored?
	Total drug	Free drug	
Phenytoin	10–20 μg/mL	1.0–2.0 μg/mL	Yes (most commonly ordered free drug)
Valproic acid	50–100 μg/mL	4.3–10.8 μg/mL	Yes
Carbamazepine	4–12 μg/mL	1.4–3.1 μg/mL	Less frequently monitored
Mycophenolic acid	1.0–3.5 μg/mL	Not established	No

not necessary because free drug level can be predicted from the total drug concentration. However, when pathological conditions are known to exist, there should be concern that the free fraction may be changed even though the total drug concentration remains unaltered. Free drug monitoring is useful in these situations.

3.5 Monitoring free anticonvulsant concentration

In general, free anticonvulsant monitoring is applicable for phenytoin, valproic acid and carbamazepine. Significant interindividual variations can be observed in the free fraction of phenytoin, carbamazepine and valproic acid, especially in the presence of uremia and liver disease. Drug–drug interactions can also lead to elevated free drug concentration. When protein binding is changed, the total concentration no longer reflects the pharmacologically active free drug in the plasma. Measuring free drug concentration for anti-epileptic drugs eliminates a potential source of interpretative errors in TDM using traditional total drug concentrations. In general free concentrations of classical anticonvulsants (phenytoin, carbamazepine and valproic acid) should be monitored under following circumstances:

- Uremic patients
- Patients with chronic liver disease
- Patients with hypoalbuminemia (critically ill patients, burn patients, elderly, pregnancy, human immunodeficiency virus (HIV) infection, etc.)
- Suspected drug–drug interactions where one strongly protein bound drug such as valproic acid can displace another strongly protein bound anticonvulsant.

Table 3.3 Special patient population where free anticonvulsant monitoring free may be useful

Pathophysiological condition	Comments
Uremia	• Elevated free phenytoin and free valproic acid have been reported in uremia due to hypoalbuminemia, impaired protein binding and displacement of drugs from protein binding by uremic toxins • Less variation of free carbamazepine concentrations in uremia has also been reported
Chronic liver disease	• Hyperbilirubinemia is usually observed in chronic liver disease which may also cause elevated free anticonvulsant levels
Any patient with hypoalbuminemia	• Critically ill patients usually have hypoalbuminemia that may cause elevated free anticonvulsant levels. However, in severely ill head trauma patient free phenytoin may be elevated despite absence of hypoalbuminemia, renal and hepatic failure • Elevated free phenytoin level in burn patients due to hypoalbuminemia has also been reported
AIDS patients	• Elevated free phenytoin may be related to hypoalbuminemia as well as displacement of phenytoin by other strongly protein bound drugs
Condition that increases binding of drugs with alpha-1-acid glycoprotein	• Although strongly alpha-1-acid glycoprotein bound drugs are less commonly monitored (lidocaine, quinidine etc.), protein binding of these drugs are increased in pathological conditions such as uremia, myocardial infarction, rheumatoid arthritis, infection, liver disease etc. because alpha-1-acid glycoprotein is an acute phase reactant

Patient populations where monitoring free anticonvulsants may be useful are listed in Table 3.3.

3.5.1 Clinical utility of monitoring free phenytoin concentrations

In general, free phenytoin concentration correlates better with therapeutic efficacy as well as toxicity compared to total phenytoin. Kilpatrick et al. reported that unbound phenytoin concentration reflected the clinical status of a patient equally or better than the total phenytoin concentration [5]. Booker reported that free phenytoin concentration correlated better with toxicity, and the authors observed no toxicity at free phenytoin concentration of 1.5 μg/mL or less [6]. Dutkiewicz et al. demonstrated that in hypercholesterolemia and in mixed hyperlipidemia, the serum level of free phenytoin was elevated. The effect was probably related to displacement of phenytoin by free fatty acids [7].

The binding of phenytoin to serum albumin can be altered significantly in uremia. The lower protein binding capacity of phenytoin in uremia can be related to hypoalbuminemia, structural modification of albumin and accumulation of uremic compound in blood such as hippuric acid, indoxyl sulfate, and 3-carboxy-4-methyl-5-propyl-2-furanpropionate (CMPF) that displaces phenytoin from protein binding sites. Otagiri commented that reduced protein binding of drugs in uremia can be explained by a combined mechanism that involves a combination of direct displacement by free fatty acids and a cascade effect initiated by free fatty acids and uremic toxins [8, 9]. Therefore, monitoring free phenytoin concentration in uremic patients is essential. Elevated free phenytoin concentration also occurs in patients with hepatic disease mainly due to hypoalbuminemia [10]. In hepatic failure, the hepatic clearance of unbound phenytoin may also be reduced as a result of hepatic tissue destruction and a reduction in hepatic enzyme activities responsible for metabolism of phenytoin. When this occurs, a reduction of phenytoin dose is necessary to maintain unbound phenytoin concentration below toxic level. Prabhakar and Bhatia reported that free phenytoin levels are elevated in patients with hepatic encephalopathy [11].

Monitoring free phenytoin concentration is clinically important in critically ill patients and also in patients with hypoalbuminemia in order to avoid drug toxicity. Lindow et al. described severe phenytoin toxicity associated with hypoalbuminemia in critically ill patients that was confirmed by direct measurement of free phenytoin [12]. Zielmann et al. reported that in 76% of 38 trauma patients, the free phenytoin fraction was increased to as high as 24% compared to 10% free levels observed in otherwise healthy subjects. The authors recommended monitoring of free phenytoin in such patients [13]. Wolf et al. observed that mean free to total phenytoin ratio was 0.13 (range: 0.06–0.42) in critically ill children and commented that total phenytoin is unreliable in directing phenytoin therapy in these children [14].

In pregnancy, plasma volume usually increases and hypoalbuminemia may also be observed. The pharmacokinetics of many anticonvulsants undergo important changes in pregnancy due to modification in body weight, altered plasma composition, hemodynamic alteration, hormonal influence and contribution of fetoplacental unit to drug distribution and disposition.

Therefore, monitoring total phenytoin during pregnancy may provide misleading information. Monitoring free phenytoin level is recommended in pregnant woman receiving phenytoin [15].

Seizures are a common manifestation of central nervous system disease in patients who have HIV infection. Despite receiving higher phenytoin doses, Acquired Immune Deficiency Syndrome (AIDS) patients have been found to have lower total phenytoin concentrations compared to non-HIV infected patients receiving phenytoin. Since AIDS patients are frequently found to be hypoalbuminemic, free phenytoin concentrations may be increased. Toler et al. described a case in which profound hypoalbuminemia caused toxic free phenytoin concentration in an HIV-positive 25-year-old woman [16]. Therefore, AIDS patients may also benefit from monitoring using free phenytoin concentrations.

A strongly albumin bound drug can displace phenytoin from protein binding site causing phenytoin toxicity due to elevated free phenytoin level. Displacement of phenytoin from protein binding site by valproic acid resulting in phenytoin toxicity has been reported [17]. In another study with cancer patients receiving phenytoin, comedication with valproic acid or carbamazepine increased free fraction of phenytoin 52.5% and 38.5% respectively. A reduction of serum albumin by 1.0 g/dL resulted in 15% increase in free phenytoin fraction. Authors recommended monitoring free phenytoin concentrations in patients comedicated with valproic acid or carbamazepine [18]. Nonsteroidal antiinflammatory drugs such as salicylate, ibuprofen, tolmetin, naproxen, mefenamic acid and fenoprofen as well as penicillins including oxacillin and dicloxacillin also displace phenytoin from protein binding sites. In vivo, the total phenytoin concentration in serum decreased during penicillin administration while the free phenytoin concentration increased. In vitro and in vivo displacement of phenytoin by antibiotics ceftriaxone, nafcillin and sulfamethoxazole also have been reported [19].

3.5.2 Clinical utility of monitoring free valproic acid concentration

Valproic acid (therapeutic range of 50–100 μg/mL) is extensively bound to serum proteins, mainly albumin. Fluctuations in protein binding occur within the therapeutic range due to saturable binding phenomenon leading to wide variation of free fraction from 10% to 50% [20]. In addition, unbound valproic acid concentration may also vary during dosing interval in patients already stabilized on valproic acid [21]. Several studies reported problems associated with predicting a therapeutic response of valproic acid from total serum concentrations [22, 23].

As expected, protein binding of valproic acid is significantly reduced in uremia. In one study based on four uremic patients, the average free fraction of valproic acid was 31% which was significantly higher than observed in non-uremic patients. During hemodialysis, the free fraction increased significantly (mean free level: 64%) [24]. Hepatic disease can also alter pharmacokinetic parameters of valproic acid. Klotz et al. reported that alcoholic cirrhosis and viral hepatitis decreased valproic acid protein binding from 88.7% to 70.3% and 78.1%, respectively, with a significant increase in volume of distribution. Elimination half-life was also prolonged [25].

Monitoring free valproic acid concentration is essential in hypoalbuminemic patients. Gidal et al. reported a case where markedly elevated plasma free valproic acid in a hypoalbuminemic patient contributed to neurotoxicity. The total valproic acid concentration was 103 μg/mL, but the free valproic acid concentration was 26.8 μg/mL. This unexpected elevation was due to a low albumin level (3.3 g/dL) of the patient [26]. Haroldson et al. reported a case demonstrating the importance of monitoring free valproic acid in a heart transplant recipient with hypoalbuminemia. When the valproic acid dose was adjusted based on free valproic acid concentration rather than total valproic acid concentration, the patient improved and was eventually discharged from the hospital [27]. A very high free fraction of valproic acid (>60%) in a critically ill patient has also been reported mostly due to extremely low serum albumin concentration of 1.2 g/dL [28].

3.5.3 Clinical utility of monitoring free carbamazepine concentrations

Although carbamazepine is approximately 80% bound to serum albumin, the primary and active metabolite 10,11-epoxide is only 50% bound to serum proteins. There seems to be less variability in the protein binding of carbamazepine compared to phenytoin and valproic acid in various pathophysiological conditions [29]. As a result, free carbamazepine concentration is monitored less frequently compared to free phenytoin and valproic acid. Froscher et al. showed that in patients with carbamazepine monotherapy, there was no closer relationship between free concentration and pharmacological effects compared to total concentration and pharmacological effects [30]. Nevertheless, an increase in unbound concentration of carbamazepine has been reported in patients with hepatic disease [31].

3.6 Drugs bound to alpha-1-acid glycoprotein and free drug monitoring

Although albumin is the major drug binding protein present in the serum, α_1-acid glycoprotein, an acute phase reactant protein, binds basic drugs in serum. Lidocaine is bound to α_1-acid glycoprotein in blood. Routledge et al. reported wide inter-individual variation in free lidocaine concentration. In contrast to elevated free phenytoin in uremic patients, free lidocaine concentration is decreased in uremia (20.8% in uremic patients vs 30.8% in control) as well as in renal transplant recipients due to increased concentrations of α_1-acid glycoprotein (134.9 mg/dL in patients vs 66.3 mg/dL in controls) because this protein is an acute phase reactant [32]. Displacement of lidocaine from protein binding by disopyramide may result in elevated free lidocaine concentration because disopyramide has a stronger binding affinity for α_1-acid glycoprotein [33]. Quinidine is also bound to α_1-acid glycoprotein and genetic polymorphism of the gene encoding this protein may influence binding of quinidine to α_1-acid glycoprotein [34]. In clinical practice, monitoring of free lidocaine or quinidine concentrations is rarely done.

3.7 Monitoring free concentrations of immunosuppressants

Immunosuppressants such as cyclosporine, tacrolimus, sirolimus, everolimus and mycophenolic acid are strongly bound to serum proteins. While cyclosporine, tacrolimus, sirolimus and everolimus are measured in whole blood (due to high distribution of these drugs in various cellular components of blood but mostly erythrocytes), mycophenolic acid is mostly distributed in serum (>99%). Therefore, mycophenolic acid is the only immunosuppressant which is monitored in serum or plasma. Although there are several publications indicating clinical utility of monitoring free cyclosporine and tacrolimus, due to technical difficulty (equilibrium dialysis using stainless steel equipment is needed), free cyclosporine and tacrolimus are not monitored in clinical laboratories. In contrast, free mycophenolic acid can be monitored in protein free ultrafiltrate.

Traditionally total mycophenolic acid level is monitored with an assumption that approximately 97–99% would be protein bound. In general, in transplant recipients with normal albumin concentration as well as normal renal and liver function, the free fraction of mycophenolic acid is 2–3% and free mycophenolic acid concentration can be predicted from total mycophenolic acid concentration. However, free mycophenolic acid concentration is significantly increased in uremic patients as well as patients with any pathological condition that may cause hypoalbuminemia. For these patients, monitoring free mycophenolic acid should be more beneficial.

In uremic patients, free fraction of mycophenolic acid may be elevated due to hypoalbuminemia as well as accumulation of mycophenolic acid glucuronide metabolite (protein binding approximately 82%) in plasma which is capable of displacing mycophenolic acid from protein binding sites. In patients with chronic renal failure, both total and free mycophenolic acid should be measured in order to avoid mycophenolic acid toxicity induced by increased free mycophenolic acid concentration. Kaplan et al. studied 8 renal transplant recipients (one patient with both kidney and pancreas transplant) with chronic renal insufficiency and 15 renal transplant patients with preserved renal function and observed that average free mycophenolic acid fractions were more than double in renally compromised patients compared to patients with normal renal function (5.8 + 2.7% vs 2.5 + 0.4). Such differences were both clinically and statistically significant [35]. Atcheson et al. analyzed data from 42 renal transplant recipients who received mycophenolic acid and observed that patients with albumin concentration below 3.1 g/dL showed elevated percentage of free mycophenolic acid (>3%). These authors commented that clinicians should consider monitoring free mycophenolic acid concentrations in hypoalbuminemic patients with plasma albumin levels below 3.1 g/dL [36]. However, liquid chromatography combined with mass spectrometry (LC/MS) or tandem mass spectrometry (LC/MS/MS) is needed to monitor free mycophenolic acid in protein free ultrafiltrate, because commercially available immunoassays do not have enough sensitivity to measure free mycophenolic acid concentrations. Nevertheless, Rebollo et al. modified the EMIT mycophenolic acid immunoassay for application on the Viva-E analyzer for determination of free mycophenolic acid. The limit of quantitation for free mycophenolic acid was 5 ng/mL, which was comparable to limit of quantitation achieved by an LC/MS/MS method [37].

3.8 Monitoring free concentrations of protease inhibitors

Many antiretroviral drugs are used today in treating patients with AIDS using HAART (highly active antiretroviral therapy). Although TDM of several antiretroviral agents has been suggested, it is not standard of care in this patient population. Clinically used protease inhibitors except indinavir are strongly protein bound (>90%), mainly to α_1 acid glycoprotein. The pharmacological effect of antiretroviral drugs is dependent on unbound concentration of drugs capable of entering cells that harbor HIV [38]. Fayet et al. described a modified ultrafiltration method for monitoring of free concentrations of several antiretroviral agents and commented that monitoring free concentrations of lopinavir, saquinavir and efavirenz may increase its clinical usefulness due to high variability in the free fraction [39]. Calcagno commented that the measurement of protease inhibitors and ritonavir plasma levels can uncover incomplete compliance with treatment; TDM may represent a useful tool for identifying patients in need of adherence-promoting interventions [40]. However TDM of antiretroviral drugs are at this point performed only at reference laboratories and relatively few medical centers.

3.9 Special situation: Monitoring free digoxin

Digoxin is a cardioactive drug which is only 25% bound to serum proteins (mainly albumin). Therefore, monitoring free digoxin is not indicated for patients with uremia, liver disease or any other pathophysiological conditions that may cause hypoalbuminemia. However, free digoxin monitoring is very useful in monitoring progress of therapy in patients overdosed with digoxin and being treated with anti-digoxin FAB fragments (Digibind and more recently introduced DigiFab). McMillin et al. studied effect of Digibind and DigiFab on 13 different digoxin immunoassays. Increasing concentrations of Digibind or DigiFab reduced the levels of total digoxin in all immunoassays except fluorescence polarization immunoassay (FPIA) on the TDx analyzer, Abbott Laboratories, Abbott Park, IL (FPIA digoxin assay is no longer commercially available). However, no interference was observed when free digoxin concentrations in the protein free ultrafiltrates were measured; the authors concluded that ultrafiltration remains the best strategy for accurate determination of free digoxin concentrations in the presence of FAB products [41]. In addition, it is expected that free digoxin level should be reduced significantly after initiation of Digibind or DigiFab therapy. Therefore, free digoxin level also correlates with success of therapy. Chan commented that in their experience, one to two vials of anti-digoxin Fab initially bound all free digoxin confirming Fab efficacy in treating digoxin overdose [42].

3.10 Saliva for monitoring of free drug concentration

Saliva is an ultrafiltrate of blood, and salivary drug levels are typically reflective of free drug levels for certain drugs that are not ionizable within the salivary pH range. Za'abi et al. concluded that monitoring of salivary phenytoin and carbamazepine proved to be a realistic alternative to plasma free level monitoring because excellent correlations were found between salivary levels and serum unbound levels of both phenytoin and carbamazepine [43]. Saliva may be suitable for TDM of other drugs. Please see Chapter 8 for more detail.

3.11 Methods for monitoring free drug concentration

Although equilibrium dialysis is the gold standard for separation of bound drug from free drug, ultrafiltration using Centrifree Micropartition System is the most common technique for preparation of protein free ultrafiltrate for monitoring free drug concentration in clinical laboratories. Usually 0.8–1.0 mL of serum is centrifuged for 15–20 min to prepare the ultrafiltrates. Then free drug concentration can be measured in the protein free ultrafiltrates using appropriate immunoassay or preferably by using a chromatographic method such as LC/MS/MS. The time of centrifuging to prepare ultrafiltrate is crucial for measuring free drug concentrations. Liu et al. demonstrated that there is a significant difference between measured free valproic acid concentration in ultrafiltrates prepared by centrifuging specimens for 5 min vs 10 min or 20 min. The measured free concentrations were low if the specimen was centrifuged for 5 min. Therefore, authors recommended centrifugation of specimens for at least 15 min [44]. McMillin et al. reported that ultrafiltrate volumes were directly proportional to the centrifugation time (15–30 min) and were inversely proportional to albumin concentrations of serum. Although ultrafiltrate volume was significantly increased with increasing centrifugation time, free phenytoin values did not change significantly indicating that equilibrium was maintained between the ultrafiltrate and serum retained in the ultrafiltration device. The authors recommended 15–20 min centrifugation time for preparing protein free ultrafiltrate [45].

Free drug concentration can be measured in the ultrafiltrate using appropriate immunoassay.

For other common drugs, more sophisticated methods such as high performance LC/MS/MS may be required to measure the free drug level in the protein free ultrafiltrate. In fact, even for monitoring free phenytoin (where immunoassays are commercially available), measuring free phenytoin using LC/MS or LC/MS/MS offers better precision and accuracy compared to immunoassay [46].

3.12 Conclusions

Therapeutic drug monitoring of strongly protein bound antiepileptic drugs such as phenytoin, valproic acid and carbamazepine is useful for patients with uremia, liver disease as well as any pathophysiological condition that may cause hypoalbuminemia. Monitoring free concentration of immunosuppressant drug such as mycophenolic acid also has clinical value. Monitoring free concentration of certain protease inhibitors may be useful but further studies are needed for establishing guidelines.

References

[1] Chan S, Gerson B. Free drug monitoring. Clin Lab Med 1987;7:279–87.

[2] Maurer TS, Debartolo DB, Tess DA, Scott DO. Relationship between exposure and nonspecific binding of thirty three central nervous system drugs in mice. Drug Metab Dispos 2005;33:175–81.

[3] Soldin SJ. Free drug measurements when and why? An overview. Arch Pathol Lab Med 1999;123:822–3.

[4] Booker HE, Darcey B. Serum concentrations of free diphenylhydantoin and their relationship to clinical intoxication. Epilepsia 1973;(2)177–84.

[5] Kilpatrick CJ, Wanwimolruk S, Wing LMH. Plasma concentrations of unbound phenytoin in the management of epilepsy. Br J Clin Pharmacol 1984;17:539–46.

[6] Booker HE, Darcey B. Serum concentrations of free diphenylhydatoin and their relationship to clinical intoxication. Epilepsia 1973;(2)177–84.

[7] Dutkiewicz G, Wojcicki J, Garwronska-Szklarz B. The influence of hyperlipidemia on pharmacokinetics of free phenytoin. Neurol Neurochir Pol 1995;29:203–11.

[8] McNamara PI, Lalka D, Gibaldi M. Endogenous accumulation products and serum protein binding in uremia. J Lab Clin Med 1981;98:730–40.

[9] Otagiri M. A molecular functional study on the interactions of drugs with plasma proteins [Review]. Drug Metab Pharmacokinet 2005;20:309–23.

[10] Reidenberg MM, Affirme M. Influence of disease on binding of drugs to plasma proteins. Ann NY Acad Sci 1973;226:115–26.

[11] Prabhakar S, Bhatia R. Management of agitation and convulsions in hepatic encephalopathy. Indian J Gastroenterol 2003;22(Suppl. 2):S54–8.

[12] Lindow J, Wijdicks EF. Phenytoin toxicity associated with hypoalbuminemia in critically ill patients. Chest 1994;105:602–4.

[13] Zielmann S, Mielck F, Kahl R, et al. A rational basis for the measurement of free phenytoin concentrations in critically ill trauma patients. Ther Drug Monit 1994;16:139–44.

[14] Wolf GK, McClain CD, Zurakowski D, Dodson B, et al. Total phenytoin concentrations do not accurately predict free phenytoin concentrations in critically ill children. Pediatr Crit Care Med 2006;7:434–9.

[15] Yerby MS, Friel PN, McCormick K. Antiepileptic drug disposition during pregnancy. Neurology 1992;42(Suppl. 5):12–6.
[16] Toler SM, Wilkerson MA, Porter WH, Smith AJ, et al. Severe phenytoin intoxication as a result of altered protein binding in AIDS. DICP 1990;24:698–700.
[17] Carvalho IV, Carnevale RC, Visacri MB, Mazzola PG, et al. Drug interaction between phenytoin and valproic acid in a child with refractory epilepsy: a case report. J Pharm Pract 2014;27(2):214–6.
[18] Joerger M, Huitema AD, Boogerd W, van der Sande JJ, et al. Interactions of serum albumin, valproic acid and carbamazepine with the pharmacokinetics of phenytoin in cancer patients. Basic Clin Pharmacol Toxicol 2006;99:133–40.
[19] Dasgupta A. Usefulness of monitoring free (unbound) concentrations of therapeutic drugs in patient management. Clin Chim Acta 2007;377:1–13.
[20] Bowdle TA, Patel IH, Levy RH, Wilensky AJ. Valproic acid dosage and plasma protein binding and clearance. Clin Pharmacol Ther 1980;28:486–92.
[21] Marty JJ, Kilpatrick CJ, Moulds RFW. Intra-dose variation in plasma protein binding of sodium valproate in epileptic patients. Br J Clin Pharmacol 1982;14:399–404.
[22] Gugler R, Von Unruh GE. Clinical pharmacokinetics of valproic acid. Clin Pharmacokinet 1980;5:67–83.
[23] Chadwick DW. Concentration-effect relationship of valproic acid. Clin Pharmacokinet 1985;10:155–63.
[24] Bruni J, Wang LH, Marbury TC, Lee CS, et al. Protein binding of valproic acid in uremic patients. Neurology 1980;30:557–9.
[25] Klotz U, Rapp T, Muller WA. Disposition of VPA in patients with liver disease. Eur J Clin Pharmacol 1978;13:55–60.
[26] Gidal BE, Collins DM, Beinlich BR. Apparent valproic acid neurotoxicity in a hypoalbuminemic patient. Ann Pharmacother 1993;27:32–5.
[27] Haroldson JA, Kramer LE, Wolff DL, Lake KD. Elevated free fractions of valproic acid in a heart transplant patient with hypoalbuminemia. Ann Pharmacother 2000;34:183–7.
[28] de Maat MM, van Leeuwen HJ, Edelbroek PM. High unbound fraction of valproic acid in a hypoalbuminemic critically ill patient on renal replacement therapy. Ann Pharmacother 2011;45:e18.
[29] Bertilsson L, Tomson T. Clinical pharmacokinetics and pharmacological effects of carbamazepine and carbamazepine 10,11-epoxide. Clin Pharmacokinet 1986;11:177–98.
[30] Froscher W, Burr W, Penin H, Vohl J, et al. Free level monitoring of carbamazepine and valproic acid: clinical significance. Clin Neuropharmacol 1985;8:362–71.
[31] Hooper W, Dubetz D, Bochner F, et al. Plasma protein binding of carbamazepine. Clin Pharmacol Ther 1975;17:433–40.
[32] Routledge PA, Barchowsky A, Bjornsson TD, Kitchell BB, Shand DG. Lidocaine plasma protein binding. Clin Pharmacol Ther 1980;27:347–51.
[33] Bonde J, Jenen NM, Burgaard P, Angelo HR, et al. Displacement of lidocaine from human plasma proteins by disopyramide. Pharmacol Toxicol 1987;60:151–5.
[34] Li JH, Xu JQ, Cao XM, Ni L, et al. Influence of the ORM1 phenotypes on serum unbound concentration and protein binding of quinidine. Clin Chim Acta 2002;317:85–92.
[35] Kaplan B, Meier-Kriesche HU, Friedman G, Mulgaonkar S, et al. The effect of renal insufficiency on mycophenolic acid protein binding. J Clin Pharmacol 1999;39:715–20.
[36] Atcheson BA, Taylor PJ, Kirkpatrick CM, Duffull SB, et al. Free mycophenolic acid should be monitored in renal transplant recipients with hypoalbuminemia. Ther Drug Monit 2004;26:284–6.
[37] Rebollo N, Calvo MV, Martin-Suarez A, Dominguez-Gil A. Modification of the EMIT immunoassay for the measurement of unbound mycophenolic acid in plasma. Clin Biochem 2011;44:260–3.

[38] Boffito M, Black DJ, Blaschke TF, Rowland M, et al. Protein binding in antiretroviral therapies. AIDS Res Hum Retrovir 2003;19(9):825–35.
[39] Fayet A, Beguin A, de Tejada BM, Colombo S, et al. Determination of unbound antiretroviral drug concentrations by a modified ultrafiltration method reveals high variability in free fraction. Ther Drug Monit 2008;30:511–22.
[40] Calcagno A, Pagani N, Ariaudo A, Arduino G, et al. Therapeutic drug monitoring of boosted PIs in HIV-positive patients: undetectable plasma concentrations and risk of virological failure. J Antimicrob Chemother 2017;72:1741–4.
[41] McMillin GA, Owen WE, Lambert TL, De BK, et al. Comparable effects of DIGIBIND and DigiFab in thirteen digoxin immunoassays. Clin Chem 2002;48:1580–4.
[42] Chan BS, Isbister GK, O'Leary M, Chiew A, et al. Efficacy and effectiveness of anti-digoxin antibodies in chronic digoxin poisonings from the DORA study (ATOM-1). Clin Toxicol (Phila) 2016;54:488–94.
[43] Za'abi M, Deleu D, Batchelor C. Salivary free concentrations of anti-epileptic drugs: an evaluation in a routine clinical setting. Acta Neurol Belg 2003;103:19–23.
[44] Liu H, Montoya JL, Forman LJ, et al. Determination of free valproic acid: evaluation of centrifree system and comparison between high performance liquid chromatography and enzyme immunoassay. Ther Drug Monit 1992;14:513–21.
[45] McMillan GA, Juenke J, Dasgupta A. Effect of ultrafiltrate volume on the determination of free phenytoin concentration. Ther Drug Monit 2005;27:630–3.
[46] Peat J, Frazee C, Garg U. Quantification of free phenytoin by liquid chromatography tandem mass spectrometry (LC/MS/MS). Methods Mol Biol 2016;1383:241–6.

CHAPTER 4

Effects of bilirubin, lipemia, hemolysis, paraproteins and heterophilic antibodies on immunoassays for therapeutic drug monitoring

4.1 Introduction

While there has been a recent trend back to chromatography-based methods in order to achieve better specificity and accuracy, many TDM (therapeutic drug monitoring) assays are still performed using immunoassay-based methods on automated clinical chemistry analyzers. Since most drugs are small molecules, the immunoassays employed are typically based on a competition format in which the analyte in the sample competes with labeled analyte for a limited number of antibody binding sites. The signal generated by the bound label or free label is converted into the analyte concentration in the assay. Bilirubin, lipids, hemolysis, proteins, and other antibodies have been well documented to interfere with immunoassays and should be evaluated when considering the use of a specific assay. The interference can be related to the spectral characteristics of the interferent or to a chemical or physical interaction of the interferent with components of the assay.

4.2 Bilirubin interferences in immunoassays used for TDM

Derived from hemoglobin of aged or damaged red blood cells, bilirubin exists in several forms in circulation, many of which bind to albumin. The spectral characteristics vary between the different forms. For example, glucuronide conjugated bilirubin absorbs between 400 and 500 nm with two small maximum wavelengths at ~430 nm and 480 nm, while unconjugated bilirubin has a much stronger maxima at ~480 nm (Fig. 4.1). Normally, total bilirubin concentrations are low (<1.2 mg/dL in serum), but in disorders

Therapeutic Drug Monitoring Data
https://doi.org/10.1016/B978-0-12-815849-4.00004-9

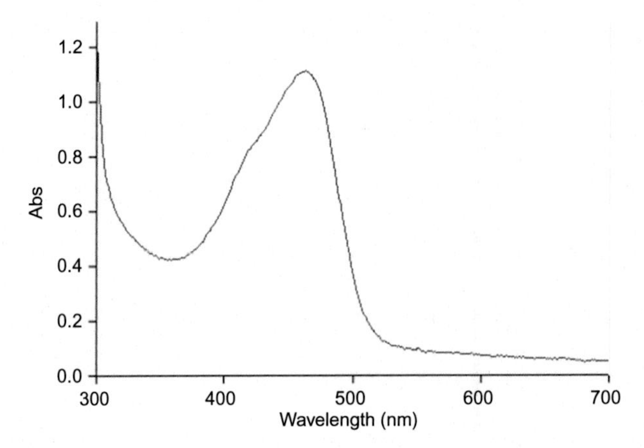

Fig. 4.1 UV–visible spectrum of bilirubin (0.9 mg/dL) obtained by 1:10 dilution of SYNCHRON System bilirubin calibrator (Beckman) using 1 M PBS buffer (authors own data).

affecting the hepatobiliary tract, concentrations may increase dramatically and exceed 20 mg/dL. The analytical interference from bilirubin observed for some immunoassays may be due to the spectral characteristics of the various forms or to chemical reactivity. Regardless, interference can result in either a positive or a negative bias depending upon the specific method involved.

A positive bias from bilirubin has been reported for an acetaminophen assay in which the drug was first hydrolyzed using aryl acylamidase to *p*-aminophenol, which was then condensed with o-cresol in the presence of periodate to form a blue indophenol chromophore [1]. The case reported involved a severely jaundiced 17-year-old male patient whose total bilirubin was determined to be 19.8 mg/dL. As part of the initial toxicological evaluation, the laboratory reported a low, but measurable serum acetaminophen concentration of 3.4 mg/dL using the described method. The patient, however, denied use of the drug for at least 1 week. The cause of the false-positive result was determined when acetaminophen-free specimens supplemented with bilirubin yielded similar results. This type of interference from bilirubin is not unique to this particular method as similar results have been reported for another acetaminophen assay which involved the reaction of acetaminophen with ferric-2,4,6-tripyridyl-s-triazine [2].

Recently, significance positive interference of bilirubin in the digoxin immunoassay on the Unicel DxI 800 analyzer (Beckman Coulter) has been reported at bilirubin concentration of 15 mg/dL and higher, although digoxin immunoassays on the Vitros 4600, and the Roche Cobas 8000 analyzers were not affected. The authors speculated that use of polyclonal rabbit antibodies to digoxin in the Beckman Coulter digoxin assay may be responsible for bilirubin interference, because bilirubin may be prohibiting the binding of the label to the anti-digoxin antibody, causing a decrease in signal (and, consequently, an increased reported digoxin concentration). Another possible explanation may be that bilirubin is directly binding to digoxin, thus preventing the binding of the label to the anti-digoxin antibody and falsely elevating detected digoxin in patient plasma. In contrast, more specific monoclonal antibodies are used in the digoxin assays for application on the Vitros 4600 and Cobas 8000 analyzers. The authors commented that positive interference by bilirubin in the Beckman Coulter digoxin assay may cause apparent therapeutic digoxin concentration in a patient who really has a subtherapeutic level or falsely elevated digoxin level in a patient actually showing a therapeutic level [3]. Either scenario could lead to incorrect clinical decisions.

Negative interferences by bilirubin have also been reported. Chong et al. reported that the apparent acetaminophen concentration was reduced by an average of 30% at a bilirubin concentration of 420 μmol/L (24.6 mg/dL) in the Beckman Coulter AU5822 analyzer [4]. Wood et al. demonstrated negative interference of bilirubin in the FPIA (fluorescence polarization immunoassay, Abbott) vancomycin assay but not in the microparticle enzyme immunoassay (MEIA) of vancomycin also manufactured by Abbott Laboratories. However, FPIA vancomycin assay is no longer commercially available [5].

4.3 Interference due to hemolysis

Hemolysis may occur when erythrocytes are damaged during venipuncture or sample processing or as part of an in vivo process. Most individuals can detect hemolysis by visual inspection when the hemoglobin concentration exceeds 20 mg/dL. Interference from hemolysis may be related not only to the release of hemoglobin, but to any of the other cellular constituents of the cell, with the magnitude of interference dependent upon the degree of hemolysate or to the concentration of the analyte [6, 7]. As with bilirubin, interference may be spectrophotometric or chemical, and cause either positive or negative bias. The typical absorption spectrum of hemoglobin shows an increase beginning ~340 nm, then further

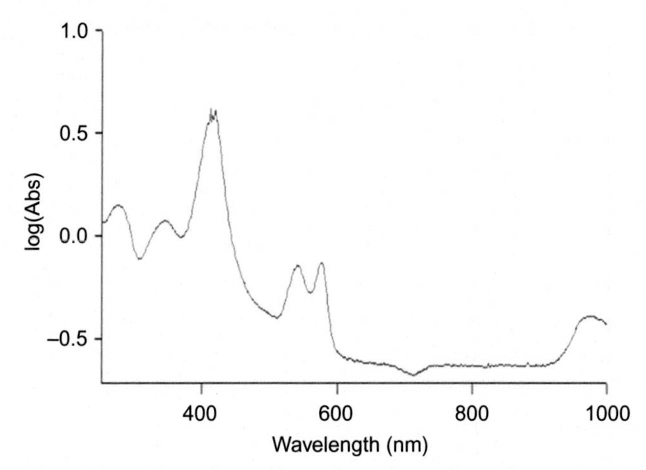

Fig. 4.2 UV–visible spectrum of purified human hemoglobin obtained after 1:200 dilution of Multi-4, Level 2 Co-oximeter control (Instrumentation Laboratory, Lexington, KY) using 1 M PBS buffer. This control contains hemoglobin, oxyhemoglobin and carboxy hemoglobin consistent with values observed in the normal population (author's own data).

increase in 400–430 nm and reaching maxima at 540–580 nm (Fig. 4.2). Complicating investigations, different results may be obtained depending upon the method used to prepare samples for testing. For example, evaluations conducted by simply adding known amounts of purified hemoglobin may demonstrate no interference, but when samples are prepared using hemolysates, the results may be quite different. Furthermore, different results may be obtained depending upon the method used to prepare the hemolysates, i.e., osmotic shock method vs mechanical vs ultrasonic disruption of the erythrocytes, and whether washed erythrocytes are used instead of whole blood [7].

Seldom considered, the hemoglobin-based oxygen carriers (used as blood substitutes) are a real source of potential interference as these products continue to be used on an investigational or emergent basis. Though publications regarding the impact of these products are limited, all have been shown to at least initially interfere with many chemistry analyses including immunoassays [8]. The effect of these on TDM and toxicology assays is largely unknown.

In a related issue, most immunosuppressant drugs are measured in whole blood. For such samples, hematocrit may affect the assay results. False positive tacrolimus results were reported in (MEIA) assay for patients with low hematocrit values especially at tacrolimus concentration <9 ng/mL [9]. The enzyme multiplied immunoassay technique (EMIT assay) for tacrolimus was not affected.

Unfortunately, the interference related to hemolysis is so complex that there is no simple solution other than to recollect the sample, if possible.

4.4 Lipid interference

All the lipids in plasma exist as complexes with proteins. Lipoproteins range from 10 to 1000 nm in size. The particle size and density are dependent upon the percentage of lipid with larger, less dense lipoproteins containing more lipids. Unlike bilirubin and hemoglobin, lipids normally do not participate in chemical reactions and mostly cause interference in assays by increasing the turbidity of the sample. Lipemic interference is most pronounced with spectrophotometric assays, less important with fluorometric methods and rarely a problem with chemiluminescent methods. Assays in which the antigen–antibody reaction is measured using turbidimetry are the ones most affected by lipid interference [10] (Table 4.1).

Although lipemic specimens rarely interfere with immunoassays for therapeutic drugs, lipemia may affect many routinely performed tests in clinical laboratory. Interfering triglyceride particles, chylomicron and VLDL, have lower density than the serum and can be removed by ultra-centrifugation (lipids remain in the supernatant, and the serum in bottom layers).

In addition, LipoClear is commercially available (StatSpin, Norwood, MS) for removing lipid interference; it is not a universal method for removal of lipid interference in all tests. Saracevic et al. demonstrated that LipoClear is not suitable for lipemia removal from samples designated for testing concentrations of glucose, sodium, potassium, chlorides, phosphates,

Table 4.1 Specimen appearances in the presence of various interfering substances

Specimen appearance	Interfering substance
Greenish, icteric	Bilirubin, high levels (20 mg/dL or more) may be found in patients with jaundice
Red, hemolyzed	Hemoglobin at concentration of 20 mg/dL or more may interfere with photometric, fluorometric of chemiluminescent based immunoassays
Turbid	Lipids, VLDL (very low density lipoprotein), chylomicrons. Lipids interfere by light scattering
Normal	A small clot or micro clot may be present

magnesium, CK-MB, ALP, GGT, total protein, albumin, CRP and troponin T. High speed centrifugation should be used for lipid removal [11]. High-speed centrifugation (10,000×g for 15 min) can also be used instead of ultracentrifugation to remove lipemia in serum/plasma samples [12].

4.5 Interferences from protein and paraproteins

Interference from proteins is possible both in hypo- and hyperproteinemia. Ezan et al. observed significant differences in drug recovery thought to be due to differences in plasma protein concentrations when developing an immunoassay for an experimental seven amino acid peptide drug. [13] Plasma specimens, which have been refrigerated for prolonged periods or which have undergone freeze–thaw cycles demonstrate another facet of protein interference. Fibrins precipitate under such conditions and may block auto-analyzer sample probes, generating incorrect results. Most modern auto-analyzers include clot-detection, but samples should still be centrifuged to remove such precipitates before assay.

The potential impact of paraproteins on immunoassays is demonstrated in a case reported by Hullin in which a 77-year-old man had ingested 100 acetaminophen tablets ~18 h earlier. At admission his serum acetaminophen concentration was reported to be 53 mg/L (Cambridge Life Science). Since the sample blank gave a higher than expected absorbance, the sample was reanalyzed using high performance liquid chromatography (HPLC) and found to have an acetaminophen concentration of 86 mg/L. The source of the bias was determined to be related to the patient's IgM monoclonal gammopathy [14].

A 69-year-old man was admitted to the hospital because of uncontrolled seizure. He had alcoholic liver disease (GGT = 296 U/L), but his renal function was normal. He also exhibited a malnourished condition with severe hypoalbuminemia (1.9 g/dL). A loading dose of fosphenytoin, (corresponding to 825 mg of phenytoin), was administered IV followed by maintenance dose of 412.5 mg/d for 6 days but the trough plasma levels measured by the particle-enhanced turbidimetric inhibition immunoassay (PETINIA: Siemens) were low (<1.0 μg/mL) and inconsistent with fosphenytoin therapy. In contrast, plasma levels measured using HPLC, showed values between 5 and 10 μg/mL consistent with fosphenytoin therapy. The patient had high IgM level (2–3 times over normal) but IgG level was within reference range. The authors concluded that the elevated IgM was responsible for negative interference [15]. Negative interference of IgM in vancomycin measurement using PETINIA assay (Siemens Diagnostics) has also been reported. However, EMIT vancomycin assay was not affected [16].

4.6 Interferences from heterophilic antibodies

Immunoassays may also suffer from interference of endogenous human (heterophilic) antibodies to the analyte. Such antibody interference can be categorized in four groups:

a. auto-antibodies
b. heterophilic antibodies
c. anti-animal antibodies
d. therapeutic antibodies

In general, competition immunoassays (using single antibody) are less affected by such interference than the sandwich immunoassays (which use separate capture and label antibodies). Because competitive immunoassays are commonly used in TDM, interference from heterophilic antibodies is relatively uncommon.

In one report, an endogenous anti-avidin antibody was found to interfere in a theophylline assay based on the avidin–biotin reaction. Through its reaction with the avidin reagent, the endogenous autoantibody interfered in complex formation, lowered the detection signal, and lead to a falsely increased result. The observed theophylline concentration using the avidin-based assay was 27.2 μg/mL compared to the theophylline value of 8.4 μg/mL observed using a non-avidin assay [17]. In another case, a high serum digoxin of 4.2 ng/mL was reported for a patient 24 h after his last digoxin dose. Due to the lack of clinical symptoms associated with digitalis toxicity, additional testing was performed using another method and the correct result found to be within the therapeutic range. Ultrafiltration of the specimen also eliminated the interference. The authors concluded that this false positive digoxin level was due to binding of a heterophilic antibody to the murine monoclonal antibody used in the digoxin assay [18].

Harmida-Cadahia et al. reported positive interference of heterophilic antibody in serum digoxin measurement using the Dimension DGNA digoxin immunoassay (rabbit capture antibody) in a Dimension XPand Plus analyzer (Siemens Healthcare Diagnostics Inc., Newark, DE, USA). However, no interference was observed using the Architect iDigoxin immunoassay (mouse capture antibody) in an Architect i1000 analyzer (Abbott Laboratories, Abbott Park, IL, USA). However, interference of heterophilic antibody in serum digoxin measurement can be eliminated by pre-treatment of specimen with HBT (Heterophilic Blocking Tubes; Scantibodies Laboratory, Inc., Santee, CA, USA). For example, in one specimen digoxin concentrations were 1.7 ng/mL (Dimension analyzer) and 0.9 ng/mL (Architect analyzer) but after treatment with HBT, serum digoxin was reduced to 0.9 ng/mL using Dimension analyzer but remained unchanged when measured again using Architect analyzer indicating that digoxin assay on the Architect analyzer is free from such interferences [19].

Interference of heterophilic antibodies in the antibody-conjugated magnetic immunoassays (ACMIA; Siemens) for tacrolimus and cyclosporine has been reported. However, immunoassays manufactured by other diagnostic companies for monitoring immunosuppressants are not affected by the presence of heterophilic antibodies. A 59 year old man underwent a kidney transplant and was managed with tacrolimus and corticosteroids. For the first 3 weeks after transplant the patient's tacrolimus whole blood concentrations were consistent with dosage and were below 12 ng/mL. Twenty five days after transplant, his tacrolimus level measured by the ACMIA tacrolimus assay was found to be highly elevated to 21.5 ng/mL. Tacrolimus was discontinued but tacrolimus level was still elevated. Suspecting interference, tacrolimus was analyzed using MEIA assay (Abbott Laboratories), and the observed value was below 2 ng/mL, indicating interference in tacrolimus measurement using the ACMIA assay most likely due to the presence of heterophilic antibody in his serum [20]. Falsely elevated blood cyclosporine levels due to presence of heterophilic antibody has also been reported to ACMIA cyclosporine assay run on Dimension RXL analyzer (Siemens). De Jonge et al. reported a falsely elevated cyclosporine level of 492 ng/mL in a 77 year old patient. However, using liquid chromatography combined with mass spectrometry, the cyclosporine level was undetectable. In addition, Architect cyclosporine assay also yielded a value lower than the detection limit. Treating specimen with polyethylene glycol and remeasuring cyclosporine in the supernatant by the same ACMIA assay showed no detected level of cyclosporine, confirming that the interfering substance was a protein most likely an endogenous antibody such as heterophilic antibody [21].

4.7 Biotin interferences in TDM

Biotin (vitamin H or B7) is a member of vitamin B complex which is essential for life. However, biotin requirement is only 30 microgram/day while daily biotin intake from Western diet is 35–70 μg/day. Supplements containing 2.5–10mg biotin/day have become popular as dietary supplements for healthy hair, nail and skin. People with inherited biotin deficiency may need 20–40mg/day. Even higher doses of biotin (100–300mg/day) may be effective in treating symptoms of multiple sclerosis. Biotin is not toxic. However, higher dose biotin can impact immunoassays that use biotinylated antibodies and streptavidin in assay design.

Normal biotin serum level is 0.4–1.2ng/mL. People taking multivitamins and biotin supplement up to 2.5mg/day shows no interference because serum levels are <10ng/mL. Biotin interference is observed in some assay when people take least 5mg/day. However, biotin at 500ng/mL (biotin dosage: 100–300mg/day prescribed as experimental therapy to patients with multiple sclerosis) affects all assays. Biotin falsely increases true analyte values in competitive immunoassays, but falsely lowers true analyte values in sandwich (immunometric or non-competitive) immunoassays [22]. One TDM assay affected by biotin is the LOCI digoxin assay (Siemens) if biotin concentration exceeds 250ng/mL [23].

4.8 Conclusions

Increased levels of bilirubin, hemoglobin, and lipids interfere in assays through their spectrophotometric, fluorometric, or chemiluminescence properties, or through side reactions. Heterophilic antibodies may also interfere with immunoassays for therapeutic drug monitoring but this type of interference is relatively uncommon for TDM assays. When suspected, the interferents may be removed from the specimen by specific agents, ultrafiltration, or centrifugation and reanalyzed. Alternatively the specimen may be analyzed by a different method known to be free from such interference.

References

[1] Bertholf RL, Johannsen LM, Bazooband A, Mansouri V. False-positive acetaminophen results in a hyperbilirubinemic patient. Clin Chem 2003;49:695–8.
[2] Kellmeyer K, Yates C, Parker S, Hilligoss D. Bilirubin interference with kit determination of acetaminophen. Clin Chem 1982;28:554–5.
[3] Simonson PD, Kim KH, Winston-McPherson G, Parakh RS, et al. Characterization of bilirubin interference in three commonly used digoxin assays. Clin Biochem 2019;63:102–5.
[4] Chong Y, Mak CM, Lam HL, Lau MH, et al. Bi-variate approach to negative interference of bilirubin towards an acetaminophen assay. Clin Biochem 2015;48:186–8.
[5] Wood FL, Earl JW, Nath C, Coakley JC. Falsely low vancomycin results using the Abbott TDx. Ann Clin Biochem 2000;37:411–3.

[6] Fonseca-Wolheim FD. Hemoglobin interference in the bichromatic spectrophotometry of NAD(P)H at 340/380 nm. Eur J Clin Chem Clin Biochem 1993;31:595–601.
[7] Snyder JA, Rogers MW, King MS, Phillips JC, et al. The impact of hemolysis on ortho-clinical diagnostic's ECi and Roche's Elecsys immunoassay systems. Clin Chim Acta 2004;348:181–7.
[8] Ma Z, Monk TG, Goodnough LT, McClellan A, et al. Effect of hemoglobin- and perflubron-based oxygen carriers on common clinical laboratory tests. Clin Chem 1997;43:1732–7.
[9] Armedariz Y, Garcia S, Lopez R, et al. Hematocrit influences immunoassay performance for the measurement of tacrolimus in whole blood. Ther Drug Monit 2005;27:766–9.
[10] Weber TH, Kaoyho KI, Tanner P. Endogenous interference in immunoassays in clinical chemistry. Scand J Clin Lab Investig Suppl 1990;201:77–82.
[11] Saracevic A, Nikolac N, Simundic AM. The evaluation and comparison of consecutive high speed centrifugation and LipoClear® reagent for lipemia removal. Clin Biochem 2014;47:309–14.
[12] Castro-Castro MJ, Candás-Estébanez B, Esteban-Salán M, Calmarza P, et al. Removing lipemia in serum/plasma samples: a multicenter study. Ann Lab Med 2018;38:518–23.
[13] Ezan E, Emmanuel A, Valente D, Grognet J-M. Effect of variability of plasma interferentes on the accuracy of drug immunoassays. Ther Drug Monit 1997;19:212–8.
[14] Hullin DA. An IgM paraprotein causing a falsely low result in an enzymatic assay for acetaminophen. Clin Chem 1999;45:155–6.
[15] Hirata K, Saruwatari J, Enoki Y, Iwata K, et al. Possible false-negative results on therapeutic drug monitoring of phenytoin using a particle enhanced turbidimetric inhibition immunoassay in a patient with a high level of IgM. Ther Drug Monit 2014;36:553–5.
[16] Cooper AA, Cowart K, Clayton A, Paul J. Undetectable vancomycin concentrations utilizing a particle enhanced turbidimetric inhibition immunoassay in a patient with an elevated IgM level. Clin Lab 2017;63:1527–32.
[17] Banfi G, Pontillo M, Sidoli A, et al. Interference from antiavidin antibodies in thyroid testing in a woman with multi endocrine neoplasia syndrome type 2B. J Clin Ligand Assay 1995;18:248–51.
[18] Liendo C, Ghali JK, Graves SW. A new interference in some digoxin assays: anti-murine heterophilic antibodies. Clin Pharmacol Ther 1996;60:593–8.
[19] Hermida-Cadahía EF, Calvo MM, Tutor JC. Interference of circulating endogenous antibodies on the dimension® DGNA digoxin immunoassay: elimination with a heterophilic blocking reagent. Clin Biochem 2010;43(18):1475–7.
[20] D'Alessandro M, Mariani P, Mennini G, Severi D, et al. Falsely elevated tacrolimus concentrations measured using the ACMIA method due to circulating endogenous antibodies in a kidney transplant recipient. Clin Chim Acta 2011;412:245–8.
[21] De Jonge H, Geerts I, Declercq P, de Loor H, et al. Apparent elevation of cyclosporine whole blood concentration in a renal allograft recipient. Ther Drug Monit 2010;32:529–31.
[22] Piketty ML, Souberbielle JC. Biotin: an emerging analytical interference. Ann Biol Clin (Paris) 2017;75(4):366–8.
[23] Dasgupta A, Belousova T, Bourgeois L, Wahed A. Taking advantage of assay harmonization, biotin interference in the LOCI digoxin assay could be eliminated by using the ADVIA centaur digoxin assay. Ann Clin Lab Sci 2018;48:614–7.

CHAPTER 5

Application of chromatographic techniques for therapeutic drug monitoring

5.1 Introduction

Therapeutic drug monitoring (TDM) is performed using either commercial methods or laboratory-developed tests (LDTs) [1]. The most widely used commercial methods are competitive homogeneous immunoassays performed on clinical chemistry analyzers commonly used in hospital-based or reference clinical laboratories. Commercially available immunoassays are available for a number of drugs and drug classes including anti-epileptic drugs (e.g., carbamazepine, lamotrigine, levetiracetam, phenobarbital, phenytoin, valproic acid), aminoglycoside antibiotics (e.g., amikacin, gentamicin, tobramycin), digoxin, methotrexate, transplant immunosuppressants (cyclosporine, everolimus, sirolimus, tacrolimus), and vancomycin. Immunoassays have the advantage of convenience, low technical complexity, and fast turnaround time.

Limitations of immunoassays for TDM include interferences (discussed in Chapter 4), cross-reactivity with structurally similar molecules, and, for some drugs, inability to reach a desired lower limit of quantitation (LOQ) [1]. Commercially available immunoassays for TDM are generally approved for venipuncture-collected specimens (plasma, serum, or whole blood) and not for alternative specimens such as saliva (oral fluid), dried blood spots, or cerebrospinal fluid (discussed in Chapter 8). There are also many drugs and drug metabolites for which immunoassays are currently unavailable.

The most common alternate analytic methodologies for TDM are chromatographic methods, either used alone or coupled to mass spectrometry (MS; note that methods that combine chromatography with MS can be referred to as 'hyphenated' techniques or methods) [1–4]. The predominant chromatographic methods for TDM use either high-performance liquid chromatography (HPLC, or often simply LC) or gas chromatography (GC) coupled to a variety of detectors [1, 5]. Sensitive detectors

Therapeutic Drug Monitoring Data
https://doi.org/10.1016/B978-0-12-815849-4.00005-0

(including MS) can often attain a lower LOQ compared to immunoassays. Chromatographic methods offer great flexibility in the analysis of drugs and drug metabolites. Mass spectrometry-based methods, especially liquid chromatography coupled to tandem mass spectrometry (LC/MS/MS), can achieve extremely high degrees of specificity compared to immunoassay or even chromatographic detection using detectors such as ultraviolet or fluorescence [2–4, 6]. Chromatographic TDM methods are generally LDTs, although there are some commercially marketed methods that have received approval from regulatory agencies such as the United States Food and Drug Administration (FDA). This chapter will discuss the application, benefits, and limitations of chromatographic methods, with description of various methodologies.

5.2 Limitations of immunoassays

For some drugs, TDM is possible using either immunoassays or chromatographic-based methods [1]. It is therefore worthwhile to consider some limitations of immunoassays that might impact their use in TDM. Immunoassays incorporate monoclonal or polyclonal antibodies that can capture drugs and/or drug metabolites. While some immunoassays used for TDM are very specific, cross-reactivity with compounds structurally related to the target compound(s) of the immunoassay can be problematic [7]. Structurally related compounds can include other medications, metabolites of the parent drug, herbal products, or endogenous compounds.

The transplant immunosuppressants (cyclosporine, everolimus, sirolimus, and tacrolimus) illustrate the challenge of cross-reactivity well [8, 9]. For example, immunoassays currently marketed for TDM of everolimus or sirolimus cannot distinguish between these two drugs that are structurally very similar to one another [7]. This complicates TDM for patients in whom both drugs are prescribed or who are transitioning from one drug to the other. In addition, marketed immunoassays for cyclosporine and tacrolimus have historically had significant cross-reactivity for both the parent drug and at least some of the metabolites, generally leading to higher apparent drug concentrations for immunoassays compared to chromatographic assays targeting only the parent drug. While newer generation immunoassays have reduced cross-reactivity compared to earlier assay generations, cross-reactivity towards metabolites is still an issue. Another limitation that can impact immunoassays are interferences such as hemolysis, icterus, and lipemia (discussed in detail in Chapter 4). Depending on assay design, these interferences may either decrease or increase apparent drug concentrations. In general, these interferences tend to have minimal or no impact on chromatographic methods, although, as discussed below, chromatographic methods may be impacted by other types of interference.

5.3 Overview of chromatographic methods

When used in TDM, chromatographic assays physically separate drugs or drug metabolites of interest from components of the specimen matrix including potentially interfering substances or structurally similar drugs or metabolites [1, 5]. The sensitivity and specificity of a chromatographic method is influenced by sample preparation, chromatographic conditions (e.g., choice of analytical column and mobile phase for HPLC), and the detector. Sample preparation methods include protein precipitation, chemical derivatization, extraction, and/or filtration. Perhaps the highest impact advance in the last decade for quantitative TDM has been the development of LC/MS/MS, a technique that will be described in more detail below.

5.4 High-performance liquid chromatography methods

The main components of HPLC methods that achieve compound separation are the analytical column (packed with the solid phase) and the mobile phase [5]. There are many options for columns and mobile phase schemes to achieve desired separation. Traditional HPLC methods often use detectors that measure single or multiple wavelengths of light; examples include ultraviolet/visible or diode array detectors. These type of detectors are typically less expensive and easier to maintain than MS detectors but may be more limited in terms of lower LOQ and vulnerability to interference by other compounds. Use of HPLC without MS generally requires longer run times to achieve resolution of compounds of interest, a factor that can be rate-limiting for high-volume testing. One approach to shorten run times without compromising compound separation has been the use of analytical columns with small (<2 μm) solid phase particle sizes compared to the larger particle sizes (3–5 μm) in conventional columns [2]. These type of columns require LC instruments designed for very high back pressures. Such analyzers are sometimes referred to as ultra-high pressure liquid chromatography (UHPLC or simply UPLC) and are more expensive than conventional HPLC instruments. UPLC instruments can be used on their own or coupled to mass spectrometers and are typically more expensive than conventional HPLC instruments.

5.5 Gas chromatography methods

The main components of GC analysis are the analytical column (containing the solid phase chemically bonded to the internal walls of the column), inert carrier gas, and the detector [5]. Unlike HPLC, GC requires the analytes of interest to be volatile. This can be achieved by chemical derivatization and/or heating the sample, if needed. A lack of volatility or the inability to make the analytes volatile precludes GC analysis. Common detectors for GC analysis include flame ionization, nitrogen–phosphorus, and electron capture. Flame ionization detectors are inexpensive but are broadly specific and thus require adequate chromatographic separation between the analytes of interest and other compounds with similar retention times.

GC/MS methods have a long history for clinical toxicology given their ability to provide high probability qualitative identification of a wide array of compounds [10]. Typical GC/MS analysis for toxicology uses 'hard ionization' techniques such as electron impact ionization that fragment molecules in the mass spectrometer after their elution from the GC. The detailed spectra of mass/charge ratios often contain multiple fragments from parent compounds and metabolites that can be identified with a high degree of certainty by comparison with internally developed or commercially available libraries. GC/MS is not commonly used for quantitative TDM.

5.6 Liquid chromatography coupled to mass spectrometry

LC/MS methods have become increasingly popular due to high specificity and the potential for short run-times and multi-analyte analysis [2–4]. LC/MS methods achieve specificity by combining chromatographic retention time with detection and quantification of ions of defined mass/charge ratios associated with the drug or drug metabolites of interest. LC/MS/MS methods further increase specificity by transfer of parent ions within a defined range of mass/charge ratios to a collision chamber that fragments compounds to daughter ion(s) that can be detected and quantified. A common configuration for tandem mass spectrometry used in TDM is the triple quadrupole. This section will focus primarily on LC/MS/MS, but there are also applications that use LC/MS with a single quadrupole.

Advantages of LC/MS/MS include analytical sensitivity, specificity, and often extended range of linearity [2–4]. LC/MS/MS systems can achieve relatively high throughput by needing only short HPLC separations prior to introduction of the eluates into the mass spectrometers. The transitions induced and monitored within triple quadrupole systems are fast (nearly instantaneous) and much shorter than the HPLC runtime. This allows for specific analysis of multiple HPLC peaks of interest that may not be adequately resolved by HPLC analysis alone. Given that HPLC run times are in essence the time-limiting component of LC/MS/MS analysis, more sophisticated configurations for high test volume laboratories link multiple HPLC instruments with a synchronized sample injection/delivery system to a single tandem mass spectrometer, increasing overall throughput several-fold [11].

Another advantage of LC/MS/MS methods is the ability to analyze multiple drugs and/or metabolites simultaneously on small sample volumes (e.g., 10–50 μL whole blood for analysis of cyclosporine, sirolimus, and tacrolimus) [11]. Multi-analyte analysis is possible by GC or HPLC alone but often requires extended run times to achieve adequate separation. LC/MS/MS allows laboratories to increase throughput by having a single method handle what would otherwise be separate methods for various analytes. Small sample sizes are also particularly helpful in TDM for pediatric patients or for alternative TDM specimen types such as dried blood spots or saliva (discussed in Chapter 8) [12]. LC/MS/MS methods can also achieve the low LOQ needed for analysis of free/unbound concentrations of some drugs.

5.7 Challenges with liquid chromatography tandem mass spectrometry systems

There are limitations to LC/MS/MS systems that impact use by clinical laboratories [4]. The methods are predominantly LDTs that require extensive method optimization and validation, in addition to being subject to higher regulatory standards. The expense of instrumentation can be a barrier to adoption. For example, capital costs of LC/MS/MS triple quadrupole analyzers in the United States can be approximately $250,000 USD or higher. Some clinical laboratories opt for less expensive LC/MS single quadrupole systems if LC/MS/MS instrumentation is not essential. Clinical laboratories that have high test volumes (e.g., regional or reference laboratories, larger medical centers) can achieve favorable return on investment by use of LC/MS/MS systems compared to immunoassay due to lower per sample costs of LC/MS/MS once instruments and methods are in place. In contrast, institutions with lower test volumes may not be able to achieve cost savings with LC/MS/MS sufficient to recover cost of capital investment in instrumentation.

LC/MS/MS methods also require skilled and robustly trained technical and supervisory staff [4, 11]. This can be a significant practical barrier when clinical laboratories are already facing staffing challenges due to factors such as aging demographics of the workforce combined with insufficient new people entering the clinical laboratory field. Clinical laboratory administrators faced with limited numbers of highly skilled staff may opt to focus on other areas requiring specialized skill (e.g., bone marrow, flow cytometry, molecular pathology) rather than developing LC/MS/MS for TDM.

Sample preparation is a critical and significant challenge for LC/MS/MS TDM methods [2–5, 11]. Sample treatment prior to introduction into the HPLC is an important step to remove lipids, proteins, and salts that can clog the HPLC system or disrupt the MS analysis [2]. Insufficient pretreatment can degrade the sensitivity and specificity of the method. Common sample preparation methods include protein filtration, protein precipitation, liquid–liquid extraction, and solid-phase extraction. For some matrices such as whole blood or saliva, freeze–thaw cycles can improve extraction efficiency and analytic consistency [4, 12]. Sample pretreatment often requires multiple pipetting and centrifugation steps, each with opportunity for errors. These steps can be automated, although this requires investment in expensive liquid handling and robotic pipetting instrumentation [2, 3, 13]. Clinical laboratories with sample pretreatment automation may still need to do some steps such as centrifugation manually.

There are a number of effects that can specifically impact LC/MS/MS analysis [14]. Several common ones include matrix effect (ion enhancement or ion suppression), isobaric interferences, and in-source fragmentation. These effects can be subtle and require detailed optimization and validation of methods prior to clinical use. The recently published C62-A guideline from the Clinical and Laboratory Standards Institute (CLSI) provides helpful guidance for clinical laboratories performing LC/MS/MS analysis [15]. Matrix effects impacting LC/MS/MS result from compounds that co-elute with the targeted analytes and affect ionization [14]. A variety of constituents from the sample or reagents (e.g., anticoagulants, mobile phase additives, phospholipids, proteins, reagent contaminants, salts) can cause ion suppression or enhancement. Methods to mitigate matrix effect include use of high purity reagents and materials, sample dilution, efficient chromatographic separation, post-extraction cleanup, and careful choice of internal standard. Matrix effect can become especially complicated for analyses targeting many analytes and using multiple internal standards.

Isobaric interference (equal mass isotopes in the sample) impacting TDM by LC/MS/MS typically results from drug metabolites or structurally similar medications co-administered to the patient [16]. One method to detect isobaric interference is the simultaneous monitoring of the target drug and one or more metabolites to identify unexpected signals [17]. Isobaric interference can also occur with endogenous compounds. More challenging analyses such as these may require longer HPLC run times to avoid co-elution of isobaric compounds.

Another form of interference can occur when compounds undergo reactions or decomposition within the MS/MS system [14, 16]. This scenario can produce a variety of conjugates (e.g., glucoside, glucuronide, sulfate, glutathione) and oxide derivatives that may produce interfering ions to the targeted analytes. Optimization of the chromatographic separation and/or mass spectrometry conditions can limit this interference.

5.8 Therapeutic drug monitoring applications

Chromatographic methods greatly expand the number of TDM applications. To date, while immunoassays are available for some drugs and drug classes discussed above, there are many drugs that lack commercially available immunoassays [4]. Examples include antibacterial agents other than aminoglycosides or vancomycin (e.g., cephalosporins, linezolid, quinolones), antifungals (e.g., caspofungin, flucytosine, posaconazole, voriconazole), antivirals (e.g., HIV medications, ganciclovir, ribavirin), antidepressants (e.g., selective serotonin reuptake inhibitors, tricyclic antidepressants, venlafaxine), antipsychotics (e.g., aripiprazole, clozapine, risperidone, ziprasidone), and newer generation anti-epileptic medications (e.g., clobazam, lacosamide).

An area of future development is for anti-cancer medications, a diverse group of drugs for which TDM has been historically relatively limited with a few exceptions such as methotrexate. Opportunities for TDM exist for both traditional cytotoxic agents and also for therapies targeted at specific molecular variants [18, 19]. In some cases, TDM can play a basic yet important role in assessing patient adherence to therapy, a factor of particular importance with continual emergence of high cost targeted therapies. An example of this is in TDM for imatinib, the prototype agent targeting the specific tyrosine kinase alteration found in chronic myelogenous leukemia and some other neoplasms [20]. For a patient not responding as expected to imatinib, distinguishing between treatment failure (e.g., due to resistance mutations) and therapy non-adherence can influence the decision to switch to higher cost second-generation tyrosine kinase inhibitors.

There are also drugs whose TDM is performed by either immunoassay or chromatographic methods. Transplant immunosuppressants perhaps represent the best current example of this. LC/MS/MS methods for this group of drugs has clear advantages in analytical specificity (especially distinguishing parent drug from metabolites or closely related drugs) but may not be practical for some institutions in terms of cost and turnaround time, as discussed above [8]. Even for drugs and drug metabolites for which immunoassays are eventually developed, the first analytical methodologies for measuring drug concentrations in body fluids are almost always some form of chromatographic method by the company or institute performing animal or human pharmacokinetic studies during the drug development process. Lastly, there are also drugs for which TDM by immunoassay is so well-established that it is unlikely that chromatographic methods will significantly displace immunoassays given current technology [7]. Examples include carbamazepine, digoxin, gentamicin, methotrexate, phenobarbital, phenytoin, valproic acid, and vancomycin. Table 5.1 summarizes the common methodologies for various drugs and drug classes, including those for which either immunoassay or chromatographic methods are used.

5.9 Future directions

There are a number of areas of future improvement and development for chromatographic methods, especially those coupled to mass spectrometry. This includes technologies such as high-resolution mass spectrometry systems that can simultaneously analyze the parent drugs and array of metabolites [21]. This allows for so-called "pharmacometabolomics" that can help identify patterns related to patient adherence to medication, drug–drug interactions, and even biomarkers for drug adverse effects. In terms of adoption of technologies such as LC/MS/MS by a wider array of clinical laboratories, issues such as cost and technical complexity need to be addressed. Research has also developed methods such as Laser Diode Thermal Desorption-Atmospheric Pressure Chemical Ionization – Tandem Mass Spectrometry and Paper Spray ambient temperature ionization that can potentially eliminate the chromatography step [22, 23].

Table 5.1 Methods used for therapeutic drug monitoring (TDM) of selected therapeutic drug categories

Therapeutic category of drug	Immunoassays predominantly used for TDM	Immunoassays or chromatographic methods for TDM	Chromatographic methods predominantly or solely used for TDM
Anti-bacterial agents	Amikacin, gentamicin, tobramycin, vancomycin		Cephalosporins, linezolid
Anti-cancer agents	Methotrexate		Imatinib
Anti-depressants or mood stabilizers	Lithium		Selective serotonin reuptake inhibitors, tricyclic antidepressants, venlafaxine
Anti-epileptic drugs	Carbamazepine, phenobarbital, phenytoin, valproic acid	Lamotrigine, levetiracetam, topiramate	Clobazam, lacosamide, pregabalin, tiagabine
Anti-fungal agents			Caspofungin, flucytosine, posaconazole, voriconazole
Anti-psychotics			Aripiprazole, clozapine, risperidone, ziprasidone
Anti-viral agents			Ganciclovir, HIV medications, ribavirin
Cardiovascular drugs	Digoxin, disopyramide, quinidine		Amiodarone
Immunosuppressants		Cyclosporine, everolimus, sirolimus, tacrolimus	Mycophenolic acid

Lastly, a practical challenge is that many clinical laboratories lack the infrastructure and personnel skill to undertake optimization and validation of complex LDTs. Development and marketing of systems with pre-developed methodologies and straightforward workflows would be a positive step towards making the technology more widely accessible. There is also opportunity to improve automation and workflow of LC/MS/MS and other similar systems to reduce manual steps that can lead to errors. Overall, the coming decade will undoubtedly see substantial strides in the analytic methodologies for TDM.

References

[1] Snozek CL, McMillin GA, Moyer TP. Therapeutic drugs and their management. In: Burtis CA, Ashwood ER, Bruns DL, editors. Tietz textbook of clinical chemistry and molecular diagnostics. 5th ed. Elsevier Saunders; 2012. p. 1057–108.
[2] Adaway JE, Keevil BG. Therapeutic drug monitoring and LC-MS/MS. J Chromatogr B Anal Technol Biomed Life Sci 2012;883–884:33–49.
[3] Adaway JE, Keevil BG, Owen LJ. Liquid chromatography tandem mass spectrometry in the clinical laboratory. Ann Clin Biochem 2015;52:18–38.
[4] Shipkova M, Svinarov D. LC-MS/MS as a tool for TDM services: where are we? Clin Biochem 2016;49:1009–23.
[5] Hortin G, Goldberger BA. Chromatography and extraction. In: Burtis CA, Ashwood ER, Bruns DL, editors. Tietz textbook of clinical chemistry and molecular diagnostics. 5th ed. Elsevier Saunders; 2012. p. 307–28.
[6] Rockwood AL, Annesley TM, Sherman NE. Mass spectrometry. In: Burtis CA, Ashwood ER, Bruns DL, editors. Tietz textbook of clinical chemistry and molecular diagnostics. 5th ed. Elsevier Saunders; 2012. p. 329–54.
[7] Krasowski MD, Siam MG, Iyer M, Ekins S. Molecular similarity methods for predicting cross-reactivity with therapeutic drug monitoring immunoassays. Ther Drug Monit 2009;31:337–44.
[8] McShane AJ, Bunch DR, Wang S. Therapeutic drug monitoring of immunosuppressants by liquid chromatography-mass spectrometry. Clin Chim Acta 2016;454:1–5.
[9] Taylor PJ, Tai CH, Franklin ME, Pillans PI. The current role of liquid chromatography-tandem mass spectrometry in therapeutic drug monitoring of immunosuppressant and antiretroviral drugs. Clin Biochem 2011;44:14–20.
[10] Maurer HH. Mass spectrometry for research and application in therapeutic drug monitoring or clinical and forensic toxicology. Ther Drug Monit 2018;40:389–93.
[11] Jannetto PJ, Fitzgerald RL. Effective use of mass spectrometry in the clinical laboratory. Clin Chem 2016;62:92–8.
[12] Ghareeb M, Akhlaghi F. Alternative matrices for therapeutic drug monitoring of immunosuppressive agents using LC-MS/MS. Bioanalysis 2015;7:1037–58.
[13] Vogeser M, Kirchhoff F. Progress in automation of LC-MS in laboratory medicine. Clin Biochem 2011;44:4–13.
[14] Vogeser M, Seger C. Pitfalls associated with the use of liquid chromatography-tandem mass spectrometry in the clinical laboratory. Clin Chem 2010;56:1234–44.
[15] CLSI. Liquid chromatography-mass spectrometry methods; approved guideline. C62-A, Wayne, PA: Clinical and Laboratory Standards Institute; 2014.

[16] Sauvage FL, Gaulier JM, Lachatre G, Marquet P. Pitfalls and prevention strategies for liquid chromatography-tandem mass spectrometry in the selected reaction-monitoring mode for drug analysis. Clin Chem 2008;54:1519–27.
[17] Furlong M, Bessire A, Song W, Huntington C, Groeber E. Use of high-resolution mass spectrometry to investigate a metabolite interference during liquid chromatography/tandem mass spectrometric quantification of a small molecule in toxicokinetic study samples. Rapid Commun Mass Spectrom 2010;24:1902–10.
[18] Paci A, Veal G, Bardin C, Leveque D, Widmer N, Beijnen J, Astier A, Chatelut E. Review of therapeutic drug monitoring of anticancer drugs part 1—cytotoxics. Eur J Cancer 2014;50:2010–9.
[19] Widmer N, Bardin C, Chatelut E, Paci A, Beijnen J, Leveque D, Veal G, Astier A. Review of therapeutic drug monitoring of anticancer drugs part two—targeted therapies. Eur J Cancer 2014;50:2020–36.
[20] D'Aronco S, D'Angelo E, Crotti S, Traldi P, Agostini M. New mass spectrometric approaches for the quantitative evaluation of anticancer drug levels in treated patients. Ther Drug Monit 2019;41:1–10.
[21] Jiwan JL, Wallemacq P, Herent MF. HPLC-high resolution mass spectrometry in clinical laboratory? Clin Biochem 2011;44:136–47.
[22] Shi RZ, El Gierari el TM, Faix JD, Manicke NE. Rapid measurement of cyclosporine and sirolimus in whole blood by paper spray-tandem mass spectrometry. Clin Chem 2016;62:295–7.
[23] Shi RZ, El Gierari el TM, Manicke NE, Faix JD. Rapid measurement of tacrolimus in whole blood by paper spray-tandem mass spectrometry (PS-MS/MS). Clin Chim Acta 2015;441:99–104.

CHAPTER 6

Drug–herb and drug–food interactions and effects on therapeutic drug monitoring

6.1 Introduction

Complementary and alternative medicines (CAM) represent an interesting, but often difficult, challenge for the therapeutic drug monitoring (TDM) laboratory. The range of CAM practices and products is quite broad, covering not only herbal and dietary supplements but also energy therapies, massage therapies, Reiki, mind–body interventions, and manipulative techniques. According to the World Health Organization (WHO), approximately 80% of the world population relies on herbal medicines. In addition, many patients take herbal medicines concurrently with conventional drugs. In the US, concurrent use of herbals and conventional drugs occurs in 20–30% of patients. Moreover, 60–85% of Native Africans use herbals in combination with other drugs. In Taiwan, 94% of patients and in Korea 83% of patients simultaneously use prescribed drugs and herbal supplements [1].

In the US, herbal remedies are subject to much more lenient regulations than prescription medications. Herbal remedies and dietary supplements are currently regulated under the Dietary Supplement Health and Education Act of 1994. This act allows manufacturers of dietary supplements to claim health benefits but not that the product cures, treats, or prevents specific diseases. Dietary supplements can be manufactured and sold without demonstrating safety and efficacy to US Food and Drug Administration (FDA) as required for drugs. Although the general population considers herbal medicines safe, scientific studies have shown significant toxicity and even fatality associated with certain herbal medicines. Periodic warnings regarding safety and effectiveness have turned a few individuals away from these practices, but surveys still find that up to half of patients use CAM in addition or in place of conventional medical therapy, and that botanicals and biologicals represent about a quarter of these [2].

Therapeutic Drug Monitoring Data
https://doi.org/10.1016/B978-0-12-815849-4.00006-2

From the TDM or toxicology perspective, the use of therapies employing herbal products presents challenges. These products are readily available to the general population without prescription. They are obtained commercially, raised in private gardens, or foraged in the wild. The manufacturing of products is not regulated in the same manner as pharmaceuticals, and numerous studies have found substantial variability in the "active" compounds of preparations. For this reason, some people have turned to raising or foraging their own herbs, but the chemical composition of plants varies with growing conditions and geographic setting. To date, most products have not proven effective in treating the disorders for which they are touted; never the less, many have been shown to elicit some pharmacological actions and to interact with therapeutic drugs.

6.2 Drug–herb interactions

Clinically significant drug–herb interactions may be encountered in outpatient or emergency room settings depending on the severity of such interactions. Drug–herb interactions may cause fatality. Kupiec and Raj reported the case of a 55-year-old male who suffered a fatal breakthrough seizure, with no evidence of non-compliance with his anticonvulsant medications (phenytoin and valproic acid) due to drug–herb interactions involving *Ginkgo biloba* which is known to induce cytochrome P450 (CYP) enzymes especially CYP2C19 [3].

Most drug–herb interactions are pharmacokinetic in nature, with the herbal supplement affecting the metabolism of drugs. Drug–herb interactions may be also pharmacodynamic in nature whereby one herbal product may augment or inhibit the pharmacological response of a drug. Sood et al. surveyed 1818 patients and identified 107 drug–herb interactions that had clinical significance. The five most common herbal supplements (St. John's wort, ginkgo, garlic, valerian and kava) accounted for 68% of such interactions and four different classes of prescription drugs (antidepressants, antidiabetic, sedatives and anticoagulation medications) accounted for 94% of all clinically significant interactions [4]. TDM is useful in identifying certain clinically significant drug–herb interactions, because a patient may not disclose the use of herbal supplement to the clinician. Shi and Klotz commented that drug–herb interaction certainly increases the risk of therapy with drugs with narrow therapeutic range such as warfarin, cyclosporine and digoxin [5]. Fortunately, therapy with warfarin is monitored routinely using prothrombin time/international normalized ratio (PT/INR), while digoxin and cyclosporine are subjected to routine TDM.

6.2.1 Interactions of St. John's wort with drugs

St. John's wort, a perennial herb with bright yellow flowers, is used as an herbal antidepressant. It is one of the most commonly used herbal supplements in the US Hyperforin and hypericin are two of the pharmacologically active chemicals present in its stems and leaves. Hyperforin has been shown to induce CYP enzymes through activation of a nuclear steroid/pregnane and xenobiotic receptor [6]. In general it is considered that St. John's wort induces CYP3A4, CYP2E1 and CYP2C19, with no effect on CYP1A2, CYP2D6 and CYP2C9. St. John's wort also induces the *P*-glycoprotein drug transporter (also known as multidrug resistance protein 1, MDR1), which can reduce efficacy of drugs where hepatic metabolism is not the major pathway of clearance. There is evidence that hypericin may be the active component that modulates *P*-glycoprotein. Therefore, it is likely that St. John's wort will interact with drugs that are metabolized via cytochrome liver enzymes and/or are substrates of *P*-glycoprotein transporters. However, significant herb–drug interactions with St John's wort only occur with products that contain more than 3 mg or more of hyperforin per daily dose. Low-hyperforin containing extracts (<1 mg hyperforin per day) have demonstrated efficacy and safety in the management of people with depression, prompting the call to restrict hyperforin content in the daily dosage to avoid the potential for serious herb–drug interactions [7].

Published reports indicate that St. John's wort significantly reduces steady state plasma concentrations of many drugs including cyclosporine, tacrolimus, amitriptyline, digoxin, fexofenadine, indinavir, methadone, midazolam, nevirapine, saquinavir, simvastatin, theophylline, warfarin, oral contraceptives, and others. Important pharmacokinetic drug–St. John's wort interactions are listed in Table 6.1. However, perhaps the most clinically important pharmacokinetic drug interactions with St. John's wort involve immunosuppressants, warfarin and antiretrovirals. Transplant recipients taking cyclosporine or tacrolimus may face acute organ rejection due to self-medication with St. John's wort, because St. John's wort induces metabolism of both drugs, thus reducing whole blood concentrations of these drugs by more than 50%. Ernst commented that St. John's wort can endanger the success of organ transplantation due to its interaction with cyclosporine, leading to several cases of organ rejection [8].

Warfarin therapy should be carefully controlled by measuring the clotting capacity of blood (using PT/INR). Patients taking warfarin should avoid St. John's wort because of potential failure of warfarin therapy due to

Table 6.1 Clinically significant pharmacokinetic drug interactions with St. John's wort

Class of drug	Specific drug(s)	Clinical effect of interaction	Therapeutic drug monitoring routinely performed?
Immunosuppressant	Cyclosporine Tacrolimus	Possibility of organ rejection due to lower blood level (CYP3A4 induction)	Yes
HIV protease inhibitors	Atazanavir Indinavir Lopinavir	Possible treatment failure due to increased clearance (CYP3A4 induction)	No
HIV non-nucleotide reverse transcriptase inhibitor	Nevirapine	Possible treatment failure due to increased clearance (CYP3A4 induction)	No
Anticoagulant	Warfarin	Possible treatment failure due to increased clearance (CYP2C9 induction)	Warfarin is routinely monitored through PT/INR
Cardioactive	Digoxin	Reduced efficacy due to lower serum level (*P*-glycoprotein modulation)	Yes
Antianginal	Ivabradine	Possible treatment failure due to increased clearance (CYP2C9 induction)	No
Anticonvulsants	Carbamazepine Phenobarbital Phenytoin	Reduced efficacy due to induction of CYP3A4	Yes
Calcium channel blocker	Nifedipine, Verapamil	Reduced level due to induction of CYP3A4.	Verapamil is infrequently monitored
Antihistamine	Fexofenadine	Decreased level due to induction of P-glycoprotein	No
Benzodiazepines	Alprazolam Midazolam Quazepam	Reduced blood level due to induction CYP3A4	May be monitored for compliance

Anticancer	Imatinib Irinotecan	Reduced efficacy due to lower serum level (CYP3A4 induction)	No
Tricyclic antidepressant	Amitriptyline Nortriptyline	Reduced efficacy due to lower serum level (CYP3A4 induction)	Yes
Statins	Atorvastatin Simvastatin	Reduced efficacy due to lower serum level (CYP3A4 induction)	No
Oral contraceptives	Ethinyl Estradiol Norethindrone	Reduced efficacy and possible contraception failure due to lower serum level (CYP1A2 and CYP3A4 induction)	No
Antiasthmatic	Theophylline	Reduced efficacy due to lower serum level (CYP1A2 induction)	Yes
Proton Pump Inhibitor	Omeprazole	Reduced efficacy due to lower serum level (CYP2C19 induction)	No
Hypoglycemic agent	Gliclazide	Reduced efficacy due to lower serum level (CYP2C9 induction)	Mo
Opioid	Methadone	Reduced efficacy due to lower serum level (CYP3A4 induction)	May be monitored in urine for compliance
Opioid	Oxycodone	Reduced efficacy due to lower serum level (CYP1A2 induction)	May be monitored in urine for compliance

increased clearance of warfarin. Jiang et al. studied interaction of warfarin and St. John's wort using 12 healthy subjects and concluded that St. John's wort significantly induces clearance of both *S*- and *R*-warfarin, resulting in a significant reduction in the pharmacological effect of racemic warfarin [9].

Patients with human immunodeficiency virus/acquired immunodeficiency syndrome (HIV/AIDS) receiving highly active antiretroviral therapy (HAART) should avoid St. John's wort or other herbal supplements due to possibility of treatment failure from drug–herb interactions. Clinically significant interactions of antiretroviral agents with St. John's wort have been documented. Therefore, patients with HIV/AIDS taking amprenavir, atazanavir, zidovudine, efavirenz, indinavir, lopinavir, nelfinavir, nevirapine, ritonavir, and saquinavir should avoid concomitant use of St. John's wort [10]. St. John's wort was shown to reduce the area under the curve (AUC) of the HIV-1 protease inhibitor indinavir by a mean of 57% [11]. Reduced concentration of nevirapine due to administration of St. John's wort has also been reported [12].

Interaction between St. John's wort and digoxin is also of clinical significance. Johne et al. reported that 10 days usage of St. John's wort resulted in a 33% decrease of peak and 26% decrease in trough serum digoxin concentrations. The mean peak digoxin concentration was 1.9 ng/mL in the placebo group and 1.4 ng/mL in the group taking St. John's wort [13]. Clearance of imatinib mesylate, an anticancer drug use primarily for the treatment of chronic myelogenous leukemia, is also increased due to administration of St. John's wort, potentially resulting in reduced clinical efficacy of the drug [14]. St. John's wort also showed significant interaction with another anticancer drug irinotecan. [15]. Many benzodiazepines are metabolized by CYP3A4 and, as expected, St. John's wort reduces plasma concentrations of alprazolam, midazolam and quazepam. Theophylline is metabolized by CYP1A2, CYP2E1, and CYP3A4. Reduced plasma concentration of theophylline due to intake of St. John's wort has also been reported [16].

St. John's wort has significant interaction with oral contraceptives, possibly leading to contraception failure [17]. St. John's wort also induces both CYP3A4 catalyzed sulfoxidation and 2C19 dependent hydroxylation of omeprazole [18]. Tannergren et al. reported that repeated administration of St. John's wort significantly decreases bioavailability of *R*- and *S*-verapamil. This effect is caused by induction of first pass metabolism by CYP3A4, most likely in the small intestine [19]. Xu et al. reported that treatment with St. John's wort significantly increases the apparent clearance of gliclazide, a drug which is used in treating patients with type II diabetes mellitus [20].

Sugimoto et al. reported interactions of St. John's wort with the cholesterol lowering drugs simvastatin and pravastatin. [21]. Several reviews dealing with drug–herb interactions discuss St. John's wort–drug interactions in detail [22, 23].

Although pharmacokinetic interactions of drugs with St. John's wort are more common, there are also clinically significant pharmacodynamic interactions of certain antidepressants with St. John's wort. Hyperforin, an active component of St. John's wort, is primarily responsible for its pharmacodynamic effect by a mechanism shared by many prescription antidepressant drugs. Therefore, if a patient takes antidepressant medications such as fluoxetine, sertraline, paroxetine and venlafaxine along with St. John's wort, serotonin syndrome may occur [24]. Taking St. John's wort along with buspirone may also cause serotonin syndrome [25]. Important pharmacodynamic drug interactions are listed in Table 6.2.

6.2.2 Interactions of warfarin with herbal supplements

Warfarin therapy must be critically monitored by measuring PT/INR, because warfarin is known to interact with many drugs, herbal supplements and food. Warfarin acts by antagonizing vitamin K epoxide reductase (VKOR), an enzyme involved in the reduction of vitamin K. The clinical efficacy of warfarin varies with intake of vitamin K and genetic polymorphisms that modulate expression of CYP2C9, the isoform responsible for clearance of *S*-warfarin. Various herbal supplements also have significant effects on metabolism of warfarin. Herbal supplements that may potentiate the effect of warfarin and thus increase the risk of bleeding include angelica root, arnica flower, anise, borage seed oil, bromelain, chamomile, fenugreek, feverfew, garlic, ginger, horse chestnut, licorice root, lovage root, meadowsweet, passionflower herb, poplar and willow bark. The anticoagulant effect

Table 6.2 Pharmacodynamic interaction with St. John's wort that may cause serotonin syndrome

Drug class	Examples
SSRI (selective serotonin reuptake inhibitor)	Fluoxetine Paroxetine Sertraline
SNRI (serotonin and norepinephrine reuptake inhibitor)	Venlafaxine
Other antidepressant	Bupropion Buspirone
Anti-migraine agent	Eletriptan

of warfarin also increases if combined with herbs with antiplatelet properties such as danshen, and *G. biloba*. Conversely, vitamin K containing supplements such as green tea extract may antagonize the anticoagulant effect of warfarin [26]. Leite et al. reviewed published articles on warfarin–herb interaction and commented that 84% of published articles describe the potentiation of warfarin while 16% relate to the inhibition of warfarin. Thus, the most likely risk that can occur due to inadequate or unsupervised use of herbs by anticoagulated patients on warfarin is bleeding, which increases the morbidity, mortality, and resource use [27].

6.2.3 Interactions of ginkgo with drugs

G. biloba is prepared from dried leaves of the ginkgo tree and sold as a dietary supplement to improve blood flow in brain (and thus sharpen mental focus) and peripheral circulation. Yang et al. studied the bioavailability of cyclosporine in the presence of ginkgo and onion in rats. Oral administration of ginkgo and onion significantly decreased the maximum serum concentration by 62% and 60% and also reduced AUC by 51% and 68% respectively. In contrast, no effect was seen on pharmacokinetics of intravenous cyclosporine by ginkgo and onion [28].

Several recent cases have demonstrated reduced serum concentrations of phenytoin and valproic acid following initiation of ginkgo. The post-mortem investigation of one of the cases revealed sub-therapeutic concentrations of both phenytoin (2.5 μg/mL) and valproic acid (<26 μg/mL) at the time of death. Interestingly, the patient's phenytoin serum concentrations had been within therapeutic range in the last 6 months (range: 9.6–21.2 μg/mL) and the last phenytoin value prior to his death was 13.9 μg/mL. Investigations revealed the patient had begun taking a variety of herbal supplements including some with gingko as the main component [3]. Granger reported cases of two patients who were stable with valproic acid but developed seizures within 2 weeks of using ginkgo products. After discontinuation the products, both patients were again seizure free without any need for increases in dose of valproic acid [29].

6.2.4 Miscellaneous drug–herb interactions

Ginger and garlic supplements are popular but both supplements increase bleeding risk in patients taking warfarin. Ginger increases antiplatelet activity of nifedipine, a calcium channel blocker. Feverfew reduces immunosuppression of cyclosporine and tacrolimus. Interestingly, milk thistle, which has protective effects on the liver, is often used in patients with liver disease or hepatitis C infection to protect the liver from toxicity of acetaminophen, cisplatin and cyclosporine [22].

Kava, an herbal sedative with antidepressant properties, is the most commonly cited herb related to liver toxicity due to the presence of kavalactones in the alcoholic extract of kava. In January 2003, kava was banned in the European Union and Canada; and the FDA issued another warning. By 2009, more than 100 cases of hepatotoxicity have been linked to kava exposure. Many have followed co-ingestion with alcohol which appears to potentiate the hepatoxicity [30]. Although liver damage is the most widely documented toxicity of kava, several cases involving severe central nervous system depression have been reported when the herbal is combined with other sedatives and hypnotics. One such case report described a 54 year old male who became comatose after 3 days of kava ingestion while also taking alprazolam, cimetidine and terazosin. It was thought that the adverse reaction was related to pharmacodynamic interaction between kava and alprazolam [31]. Kava lactones also inhibit CYP1A2, CYP2C9, CYP2D6 and CYP3A4 but have no effect on CYP2A6, and CYP2E1. Because of such inhibition, toxicities of drugs that are metabolized through these enzymes are increased if taken with kava [24]. Due to the potential for toxicity, kava-containing product should be avoided. Miscellaneous important drug–herb interactions are summarized in Table 6.3.

6.3 Food–drug interactions

It has also long been recognized that intake of food and fluid can alter the extent of drug absorption, either directly or indirectly. This change may be related to alteration of physiological factors in the gut such as gastric pH, gastric emptying time, intestinal motility, hepatic portal blood flow, and bile flow rate. In addition, smoking, as well as the intake of charcoal broiled food or cruciferous vegetables is known to induce the metabolism of many drugs. Fegan et al. reported increased clearance of propranolol and theophylline in the presence of high protein/low carbohydrate diet compared to a low protein/high carbohydrate diet using six volunteers. When the diet was switched from a low protein/high carbohydrate to a high protein/low carbohydrate, the clearance of propranolol was increased by an average of 74% and clearance of theophylline was increased by an average of 32% [32].

Table 6.3 Miscellaneous clinically important drug–herb interactions

Herbal supplement	Interacting drug	Clinical effect
Garlic, ginger	Warfarin	Potentiation of effect of warfarin (increased PT/INR), with increased risk of bleeding
Garlic	Ritonavir Saquinavir	Reduced plasma level due to induction of CYP3A4
Garlic	Chlorpropamide	Hypoglycemia because components of garlic have hypoglycemic properties
Ginger	Nifedipine	Ginger increases antiplatelet activity of nifedipine, a calcium channel blocker
Kava	Alprazolam	Pharmacodynamic interaction may cause serotonin syndrome
Feverfew	Iron tablets	Reduced absorption
Feverfew	Cyclosporine Tacrolimus	Reduced immunosuppression
Echinacea	Ketoconazole Methotrexate	Increased liver toxicity
Echinacea	Lansoprazole Simvastatin	Increased effect
Flaxseed oil, saw palmetto	Aspirin (325 mg)	Increased risk of bleeding
Bromelain	Naproxen	Increased risk of bleeding
Valerian	Sedative-hypnotics (barbiturates, benzodiazepines, ethanol, zolpidem)	Increased sedation
Milk thistle	Acetaminophen Cyclosporine Cisplatin	Reduced liver toxicity associated with these drugs

6.3.1 Grapefruit juice and drug interactions

It was reported in 1991 that a single glass of grapefruit juice caused a 2–3 fold increase in the plasma concentration of felodipine, a calcium channel blocker, after oral intake of a 5 mg tablet. A similar amount of orange juice showed no effect [33]. Subsequent investigations demonstrated that pharmacokinetics of approximately 40 other drugs are also affected by intake of grapefruit juice [34]. Furanocoumarins, especially bergamottin found in the grapefruit juice, was eventually determined to be responsible for these observations. The compounds inhibit CYP3A4 in the small intestine but not CYP3A4 in the liver. Grapefruit juice also inhibits *P*-glycoprotein, thereby decreases the absorption of certain drug by carrying the drug back to the intestinal lumen from enterocytes. Therefore, these combined mechanisms are responsible for increased bioavailability of drugs due to intake of grapefruit juice. Grapefruit juice does not alter metabolism or excretion of drugs after it enters into vascular circulation. In most cases, ingestion of a single glass (approximately 250 mL) of grapefruit juice is sufficient to increase the bioavailability of the affected drugs. The bioavailability of lovastatin has reported to double even when the drug was taken 12 h after intake of grapefruit juice; the effect becomes insignificant after 24 h. Other drugs for which similar interactions are reported include cyclosporine, tacrolimus, carbamazepine, saquinavir, verapamil, felodipine, nitrendipine, pranidipine, and nimodipine. Grapefruit juice increases bioavailability of several benzodiazepines including diazepam, triazolam and midazolam but has no effect on alprazolam even after repeated intake. Grapefruit juice also interacts with cholesterol lowering drugs. Simvastatin, a substrate for CYP3A4, is extensively metabolized during first pass. Grapefruit juice (200 mL once a day for 3 days) increased the AUC (0–24 h) of simvastatin by 3.6 fold and simvastatin acid by 3.3 fold. The peak concentrations were also increased significantly and only one glass of grapefruit juice taken daily is capable of producing such effects [35].

Furonocoumarins containing drinks such as Sundrop Citrus soda (registered trademark used under license of Dr Pepper/Seven Up Inc., 2002) may also produce grapefruit juice-like effect with cyclosporine. A 32-year old lung transplant recipient showed a trough cyclosporine concentration of 358 ng/mL 2 weeks after being discharged from the hospital. On the next four visits spanning 24 days, elevated cyclosporine concentrations (676 ng/mL and 761 ng/mL) were observed in two occasions. The higher cyclosporine concentrations correlated with consumption of Sundrop Citrus soda [36]. Important interactions between grape fruit juice and drugs are listed in Table 6.4.

6.3.2 Other juice–drug interactions

Although interactions between grapefruit juice and drugs are most common, interaction between other fruit juices and drugs causing potential adverse reactions have also been reported. The potential adverse interactions included decreased drug bioavailability (apple juice-fexofenadine, atenolol, aliskiren; orange juice-aliskiren, atenolol, celiprolol, montelukast, fluoroquinolones, alendronate; pomelo juice-sildenafil; grape juice-cyclosporine), increased

Table 6.4 Clinically significant grapefruit juice–drug interactions

Drug class	Examples of drugs with increased bioavailability	Therapeutic drug monitoring routinely performed?
Calcium channel blockers	Felodipine Nicardipine Nifedipine Verapamil	Only verapamil is occasionally monitored
Benzodiazepines	Alprazolam Diazepam Midazolam Triazolam	Sometimes monitored for compliance but mostly in urine
Statins	Atorvastatin Lovastatin Simvastatin	Not monitored
Cardioactive	Amiodarone	Occasionally monitored
Immunosuppressant	Cyclosporine Tacrolimus	Yes Yes
Anticonvulsant	Carbamazepine	Yes
Antibiotic	Clarithromycin Erythromycin	Not monitored Not monitored
Protease inhibitors	Amprenavir Saquinavir	Rarely monitored Rarely monitored
Proton pump inhibitors	Lansoprazole Omeprazole	Not monitored Not monitored
Narcotic analgesic	Methadone Oxycodone	Sometimes monitored for compliance but mostly in urine.
Analgesic	Acetaminophen	Only monitored in suspected overdose

bioavailability (Seville orange juice-felodipine, pomelo juice-cyclosporine, orange-aluminum containing antacids). However, unlike furanocoumarin-rich grapefruit juice, other fruit juices very rarely precipitate severely detrimental food–drug interaction. Physicians, pharmacists and other health professionals should properly educate patients about potential adverse juice–drug interactions and help minimize occurrences of adverse effects due to juice–drug interactions [37].

6.4 Conclusions

When investigating unexpected changes in serum drug concentrations or therapeutic effectiveness, one should ask about the addition or deletion of herbal supplement or a change in food. Clinical laboratory professionals need to be aware of such interactions. In many cases, patients consider herbal supplements to be natural and thus safe, and they may not inform their clinician about self-medication. Therefore when the validity of a TDM result is questioned, and the analytical performance of the test has been validated, the laboratory professional should include these products as potential causes.

References

[1] Choi JG, Eom SM, Kim J, Kim SH, Huh E, Kim H, et al. A comprehensive review of recent studies on herb-drug interaction: a focus on pharmacodynamic interaction. J Altern Complement Med 2016;22:262–79.

[2] Pawar RS, Grundel E. Overview of regulation of dietary supplements in the USA and issues of adulteration with phenethylamines (PEAs). Drug Test Anal 2017;9:500–17.

[3] Kupiec T, Raj V. Fatal seizures due to potential herb-drug interactions with Ginkgo biloba. J Anal Toxicol 2005;29:755–8.

[4] Sood A, Sood R, Brinker FJ, Mann R, et al. Potential for interaction between dietary supplements and prescription medications. Am J Med 2008;121:207–11.

[5] Shi S, Klotz U. Drug interactions with herbal medicines. Clin Pharamcokinet 2012;51:77–104.

[6] Wentworth JM, Agostini M, Love J, Schwabe JW, Chatterjee VK. St John's wort, a herbal antidepressant. Activates the steroid X receptor. J Endocrinol 2000;166:R11–6.

[7] Chrubasik-Hausmann S, Vlachojannis J, McLachlan AJ. Understanding drug interactions with St John's wort (Hypericum perforatum L.): impact of hyperforin content. J Pharm Pharmacol 2019;71:129–38.

[8] Ernst E. St. John's wort supplements endanger the success of organ transplantation. Arch Surg 2002;137:316–9.

[9] Jiang X, Williams KM, Liauw WS, Ammit AJ, et al. Effect of St. John's wort and ginseng on the pharmacokinetics and pharmacodynamics of warfarin in healthy subjects. Br J Clin Pharamcol 2004;57:592–9.

[10] van den Bout-van den Beukel CJ, Koopmans PP, van der Ven AJ, De Smet PA, Burtger DM. Possible drug metabolism interactions of medicinal herbs with antiretroviral agents. Drug Metab Rev 2006;38:477–514.

[11] Piscitelli SC, Burstein AH, Chaitt D, Alfaro RM, Fallon J. Indinavir concentrations and St. John's wort. Lancet 2000;355:547–8.

[12] de Maat MM, Hoetelmans RM, t RA M, Van Gorp EC, et al. Drug interaction between St. John's wort and nevirapine. AIDS 2001;15:420–1.

[13] Johne A, Brockmoller J, Bauer S, Maurer A, et al. Pharmacokinetic interaction of digoxin with an herbal extract from St John's wort (Hypericum perforatum). Clin Pharmacol Ther 1999;66:338–45.

[14] Smith P. The influence of St. John's wort on the pharmacokinetics and protein binding of imatinib mesylate. Pharmacotherapy 2004;24:1508–14.

[15] Mathijssen RH, Verweij J, de Bruijn P, Loos WJ, Sparreboom A. Effects of St. John's wort on irinotecan metabolism. J Natl Cancer Inst 2002;94:1247–9.

[16] Izzo AA, Ernst E. Interaction between herbal medicines and prescribed drugs. Drugs 2009;69:1777–98.

[17] Murphy PA, Kern SE, Stanczyk FZ, Westhoff CL. Interaction of St. John's wort with oral contraceptives: effects on the pharmacokinetics of norethindrone and ethinyl estradiol, ovarian activity and breakthrough bleeding. Contraception 2005;71:4102–408.

[18] Wang LS, Zhou G, Zhu B, Wu J, et al. St. John's wort induces both cytochrome P450 3A4 catalyzed sulfoxidation and 2 C19 dependent hydroxylation of omeprazole. Clin Pharmacol Ther 2004;75:191–7.
[19] Tannergren C, Engman H, Knutson L, Hedeland M, et al. St John's wort decreases the bioavailability of R and S-verapamil through induction of the first pass metabolism. Clin Pharmacol Ther 2004;5:298–309.
[20] Xu H, Williams KM, Liauw WS, Murray M, et al. Effects of St. John's wort and CYP2C9 genotype on the pharmacokinetics and pharmacodynamics of gliclazide. Br J Pharmacol 2008;153:1579–86.
[21] Sugimoto K, Ohmori M, Tsuruoka S, Nishiki K, et al. Different effect of St. John's wort on the pharmacokinetics of simvastatin and pravastatin. Clin Pharmacol Ther 2001;70:518–24.
[22] Alissa E. Medicinal herbs and therapeutic drugs interactions. Ther Drug Monit 2014;36:413–22.
[23] Gurley BJ, Fifer EK, Gardner Z. Pharmacokinetic herb-drug interactions (part-2): drug interactions involving popular botanical dietary supplements and their clinical relevance. Planta Med 2012;78:140–1514.
[24] Singh YN. Potential for interaction of kava and St. John's wort with drugs. J Ethnopharmacol 2005;100:108–13.
[25] Dannawi M. Possible serotonin syndrome after combination of buspirone and St. John's wort. J Psychopharmacol 2002;16:401.
[26] Milić N, Milosević N, Golocorbin Kon S, Bozić T, et al. Warfarin interactions with medicinal herbs. Nat Prod Commun 2014;9:1211–6.
[27] Leite PM, Martins MA, Castilho RO. Review on mechanisms and interactions in concomitant use of herbs and warfarin therapy. Biomed Pharmacother 2016;83:14–21.
[28] Yang CY, Chao PD, Hou YC, Tsai SY, et al. Marked decrease of cyclosporine bioavailability caused by coadministration of ginkgo and onion. Food Chem Toxicol 2006;44:1572–8.
[29] Granger AS. Ginkgo biloba precipitating epileptic seizures. Age Ageing 2001;30:523–5.
[30] Li XZ, Ramzan I. Role of ethanol in kava hepatotoxicity. Phytother Res 2010;24:475–80.
[31] Almedi JC, Grimsley EW. Coma from the health food store: interaction between kava and alprazolam. Ann Intern Med 1996;125:940–1.
[32] Fegan TC, Walle T, Oexmann MJ, Walle UK, et al. Increased clearance of propranolol and theophylline by high-protein compared with high carbohydrate diet. Clin Pharmacol Ther 1987;41:402–6.
[33] Bailey DG, Spence JD, Munoz C, Arnold JM. Interaction of citrus juices with felodipine and nifedipine. Lancet 1991;337:268–9.
[34] Saito M, Hirata-Koizumi M, Matsumoto M, Urano T, Hasegawa R. Undesirable effects of citrus juice on the pharmacokinetics of drugs: focus on recent studies. Drug Saf 2005;28:677–94.
[35] Mouly S, Lloret-Linares C, Sellier PO, Sene D, et al. Is the clinical relevance of drug-food and drug-herb interactions limited to grapefruit juice and Saint-John's wort? Pharmacol Res 2017;118:82–92.
[36] Johnston PE, Milstone A. Probable interaction of bergamottin and cyclosporine in a lung transplant recipient [Letter to the editor]. Transplantation 2005;27:746.
[37] Chen M, Zhou SY, Fabriaga E, Zhang PH, Zhou Q. Food-drug interactions precipitated by fruit juices other than grapefruit juice: an update review. J Food Drug Anal 2018;26(2S):S61–71.

CHAPTER 7

Pharmacogenomics and therapeutic drugs monitoring: Are they complementary?

7.1 Introduction

Pharmacogenomics (also known as *pharmacogenetics*) is the area of medicine that addresses the impact of genetic factors on drug actions [1]. For therapeutic drugs, the goal of pharmacogenomics is to optimize drug efficacy while minimizing or avoiding toxic effects. In principle, genetic variation can influence pharmacodynamics, pharmacokinetics, or both. Many current clinical applications of pharmacogenomics involve drug metabolism, but with continued research into pharmacogenomics involving pharmacodynamics [2,3]. Targeted cancer therapies utilize medications designed to impact specific somatic mutations in neoplasms [4]. Many of the current pharmacogenomic clinical applications focus on the cytochrome P450 (CYP) enzymes, especially CYP2C9, CYP2C19, and CYP2D6 [1]. Current clinical uses of pharmacogenomics are still rather limited, primarily due to limited evidence that such testing is cost effective and provides clinical benefit [5,6]. Nevertheless, pharmacogenomics applications continue to evolve and, for some drugs, provide information that complements therapeutic drug monitoring (TDM).

7.2 Types of genetic variation

The most common type of genetic variation is single-nucleotide polymorphism (SNP), a situation in which some individuals have one nucleotide at a given position while other individuals have another nucleotide (e.g., cytosine vs guanine, C/G). SNPs occurring in the coding region of genes can change the amino acid sequence that results when DNA is transcribed into RNA and the RNA is then translated into protein [1]. Less common types of genetic variation include insertions or deletions (collectively known as

Therapeutic Drug Monitoring Data
https://doi.org/10.1016/B978-0-12-815849-4.00007-4

indels), alterations of gene splicing, partial or total gene deletions, gene duplication or multiplication, and variation in gene promoter affecting gene expression.

Although there is a systematic nomenclature for precisely describing genetic variation (e.g., following the HUGO Gene Nomenclature standards), genetic variants important in pharmacogenomics are often still referenced by a historical naming system based on when variants were first characterized and reported [1,2]. By this historical convention, the normal allele (individual copy of a gene on a chromosome) is defined as *1 (e.g., CYP2C9*1). In order of historical discovery, variant alleles were designated *2, *3, *4, and so on. Unfortunately, this nomenclature does not give any clue to the nature of the genetic variation. For instance, *2 and *5 could represent fairly benign genetic variants (e.g., point mutations with minimal impact on protein function) whereas *6 could signify a variant that results in complete absence of enzyme activity such as gene deletion or a mutation resulting in a frameshift or protein truncation.

7.3 General considerations for pharmacogenomics involving drug metabolism

Although many non-genetic factors influence the effects of medications—including variables such as concomitant drugs or organ failure described in other sections of this manual—there are now many examples in which inter-individual differences in medication response are due to variants in genes encoding drug-metabolizing enzymes, drug transport proteins, and drug receptor targets [1]. Most well-established applications of pharmacogenomics involve drug metabolism.

As an example, genetic variation of CYP2D6 was one of the first well-described pharmacogenomic applications [3]. In the 1970s, a now obsolete antihypertensive drug debrisoquine was being tested in clinical trials. While most individuals tolerated the drug well, some individuals developed prolonged hypotension after receiving the drug. Debrisoquine was later found to be metabolized mainly by CYP2D6, an enzyme predominantly found in the liver that is now known to metabolize approximately 25% of all drugs currently prescribed in the United States. CYP2D6 shows the widest range of genetic variation of the human CYP enzymes, with common genetic variants that include total gene deletion and gene duplication, with rare individuals documented that have over four functional copies of the CYP2D6 gene (instead of the usual two alleles). Genetic variation of CYP2D6, as will be discussed below, has clinical importance for multiple medications. Individuals with low CYP2D6 activity may experience adverse effects to standard doses of certain drugs due to reduced clearance, whereas those with higher-than-average activity may eliminate a drug so rapidly that therapeutic concentrations are not achieved with standard doses.

There is some nomenclature related to pharmacogenomics of drug-metabolizing enzymes that can be confusing [3]. *Poor metabolizers* represent individuals with little or no enzymatic activity, due either to lack of expression of the enzyme (partial or complete gene deletions; intron splice variants; frameshift mutations) or mutations that reduce enzymatic activity (e.g., a variant impacting the catalytic site) in both alleles. *Extensive metabolizers* are considered the "wild-type" situation and generally represent individuals with two normal copies of the enzyme gene on each chromosome. *Intermediate metabolizers* have enzymatic activity roughly 50% that of extensive metabolizers. The most common genetic reason underlying intermediate activity is one copy of the normal gene and one variant copy associated with low activity (i.e., heterozygous for a low activity variant allele). *Ultrarapid metabolizers* have enzymatic activity significantly greater than that of the average population. This often results when an individual has more than the normal two copies of a gene. Gene duplication or multiplication is not seen with many genes but can occur with CYP2D6 as mentioned above. For many drugs, CYP or other drug-metabolizing enzymes inactivate the drug and thus play an important role in clearance. Poor metabolizers may thus experience drug toxicity with standard doses due to slower drug clearance. Ultrarapid metabolizers may see little clinical effect with standard drug doses due to accelerated clearance of the drug.

Special circumstances arise when pharmacogenomics impacts prodrugs [1,3]. For example, two common medications, clopidogrel (anti-platelet agent) and codeine (opiate analgesic), are *prodrugs* that are inactive until converted to active metabolites by CYP enzymes. Clopidogrel is activated by CYP2C19, and codeine is converted to the more active opiate morphine by CYP2D6. For these prodrugs, poor metabolizers of the respective CYP enzyme will be more likely to show lack of efficacy, as less active metabolites will be generated. Tamoxifen, a selective estrogen receptor modulator used in breast cancer treatment, is also functionally a prodrug dependent on CYP2D6 for conversion to metabolites with higher affinity for estrogen receptors. CYP2D6 poor metabolizers will show less response to standard doses of tamoxifen.

7.4 General considerations for pharmacogenomics affecting pharmacodynamics

Understanding the genetic variation involving pharmacodynamics has developed more slowly than that of pharmacokinetics [2]. An exception has been the rapid growth of targeted therapies for somatic mutations in cancers. It is been harder, however, to utilize germline genetic variation for pharmacogenomics. In part, this is because the molecular targets of certain drugs are incompletely understood [1]. An example of pharmacodynamic genetic variation is for the β_2-adrenergic receptor, the target of β-agonists used in asthma therapy such as albuterol and salmeterol. Genetic variation of the receptor influences how well β-agonist therapy works. However, genetic testing of the β_2-adrenergic receptor has not yet had much impact clinically.

One of the best established examples of pharmacogenomics involving pharmacodynamics is the molecular target of warfarin, the vitamin K epoxide reductase complex subunit 1 (VKORC1) protein, which will be discussed in more detail below. The importance of understanding pharmacodynamic variation is that it holds the potential of predicting therapeutic efficacy (or lack thereof). This could be especially valuable for psychiatric disorders where therapies such as anti-depressants or anti-psychotics may take weeks or even months show effectiveness.

Overall, pharmacogenomics may be useful for a number of clinical indications (Table 7.1). Table 7.2 lists pharmacogenomic associations mentioned in package inserts for approved medications. Over time, more pharmacogenomic applications will emerge. The following sections describe some specific applications.

Table 7.1 Potential indications for pharmacogenomic testing

Indication	Examples
Testing required for drug therapy	Oncology, including 'paired diagnostics'
Predict rare but serious toxic effects	Abacavir, carbamazepine Azathioprine, 6-mercaptopurine
Lessen risk of treatment failure	Clopidogrel
Guide initial and/or maintenance dosing	Warfarin
Investigate reason for therapeutic failure	Antidepressants
Evaluate unexpected adverse effects	Many examples, especially drug metabolism-based pharmacogenomics

Table 7.2 Selected pharmacogenomic biomarkers and drugs with package insert data on pharmacogenomics

Biomarker	Function	Specific drugs
CYP2C9	Drug metabolism	Celecoxib Warfarin
CYP2C19	Drug metabolism	Clopidogrel Carisoprodol Diazepam Proton pump inhibitors (e.g., omeprazole)
CYP2D6	Drug metabolism	Antidepressants—SSRIs (e.g., citalopram, fluoxetine) Antidepressants—SNRIs (venlafaxine) Antidepressants—tricyclics (e.g., amitriptyline, nortriptyline) Antipsychotics (e.g., aripiprazole, iloperidone, risperidone) Beta-adrenergic antagonists (e.g., carvedilol, metoprolol) Codeine
HLA-B*1502	Genetic link to hypersensitivity	Carbamazepine, phenytoin
HLA-B*5701	Genetic link to hypersensitivity	Abacavir
TPMT	Drug metabolism	Azathioprine, 6-mercaptopurine
UGT1A1	Drug metabolism	Irinotecan
VKORC1	Pharmacodynamic target	Warfarin

Abbreviations: *CYP*, cytochrome P450; *SNRI*, serotonin-norepinephrine reuptake inhibitor; *SSRI*, selective serotonin reuptake inhibitor; *TPMT*, thiopurine methyltransferase; *UGT*, UDP-glucuronosyltransferase. Data from US Food and Drug Administration: http://www.fda.gov/Drugs/ScienceResearch/ResearchAreas/Pharmacogenetics/ucm083378.htm.

7.5 Drug metabolized by CYP2C19

Clopidogrel is a commonly used anti-platelet drug administered to reduce risk of myocardial infarction, peripheral vascular ischemia, and stroke. Clopidogrel is converted to an active thiol metabolite mainly by CYP2C19. Patients who are CYP2C19 poor metabolizers (most commonly the *2 and *3 alleles) may show poor therapeutic response (sometimes referred to as 'Plavix resistance' using the common tradename for clopidogrel) and have higher risk for cardiovascular events while on clopidogrel [7,8]. The TRITON-TIMI 38 trial demonstrated that patients with CYP2C19 loss-of-function mutations had a relative 53% increased risk of death from cardiovascular causes (myocardial infarction, stroke) and a three-fold increased rate of stent thrombosis compared to patients with normal CYP2C19 genotype [9]. The United States Food and Drug Administration (FDA) issued a label warning update in March 2010 for CYP2C19 poor metabolizers. The 2013 Clinical Pharmacogenetics Implementation Consortium (CPIC) guidelines recommended that patients who have CYP2C19 genotypes associated with poor or intermediate metabolizer status be prescribed alternate anti-platelet therapy [8].

However, CYP2C19 genotyping for patients who are candidates for clopidogrel therapy has not been widely adopted. One important point is that the CYP2C19 function is only one factor of many that can influence clopidogrel response, with drug-drug interactions also of importance. In addition, a functional test (platelet aggregometry) can produce similar clinical information. As a result, multiple task forces, including the Veterans Health Administration Clinical Pharmacogenetics Subcommittee do not recommend routine genetic screening for clopidogrel. CYPC19 genotyping but recognize that such testing may be clinically useful for patients at high risk patients for poor outcomes or recurrent coronary events [7]. The related drug prasugrel is an alternative anti-platelet agent for patients.

Several other categories of drugs are inactivated by CYP2C19 metabolism [3]. These include carisoprodol (muscle relaxant), citalopram (antidepressant), clobazam (anti-epileptic drug), diazepam (anxiolytic), and most of the "proton pump inhibitors" used to treat gastroesophageal reflux (e.g., lansoprazole, omeprazole). CYP2C19 poor metabolizers may experience toxicity to these drugs at standard doses. The frequencies of CYP2C19 inactivating genetic variants are highest in Polynesians (up to 75%) and Asians (~15–25%) and lower in African-American (4%) and Caucasian (2–5%) populations. Genotyping of CYP2C19 has been more widely applied in Japan than in the United States because of the prevalence of CYP2C19-poor metabolism in the Japanese population. CYP2C19 genotyping may be helpful in patients with poor clinical response or unusual toxicity to drugs metabolized by CYP2C19.

7.6 Warfarin

Warfarin remains the most commonly prescribed oral anticoagulant, although the growing use of other oral anticoagulants (e.g., direct thrombin and factor X inhibitors) has reduced warfarin usage. Standard monitoring of warfarin therapy utilizes prothrombin time (PT)/international normalized ratio (INR), with INR target ranges for specific indications (e.g., atrial fibrillation, mechanical heart valve, etc.). Warfarin-related bleeding complications occur in approximately 3% of patients in the first 3 months of therapy. Patients are at highest risk of bleeding events during the first several months of warfarin therapy even with close monitoring of PT/INR. During warfarin maintenance therapy, there is a risk of bleeding complications of 7.6–16.5 per 100 patient-years. The goal of warfarin pharmacogenomics is to help predict both the initial and maintenance doses of warfarin [10]. One limitation of pharmacogenomics testing is that it does not eliminate the need to monitor warfarin therapy by PT/INR.

Two major genes are involved in warfarin pharmacogenomics: CYP2C9 and VKORC1. CYP2C9 metabolizes warfarin to an inactive metabolite [11]. VKORC1 is the molecular target inhibited by warfarin, thus reducing the production of vitamin K-dependent clotting factors. CYP2C9-poor metabolizers are at risk for warfarin toxicity at standard doses because of reduced clearance of the drug and excess concentrations of warfarin. The two most common mutations associated with poor metabolizers are CYP2C9*2 and *3. VKORC1 has several genetic variants that influence warfarin pharmacodynamics. Collectively, genetic variation in CYP2C9 and VKORC1 account for approximately one-third of the observed variation in warfarin dosage that leads to stable anticoagulation. Additional proteins involved in warfarin effect include CYP4F2 and gamma-glutamyl carboxylase (GGCX), with genetic variation in these genes also having an effect on warfarin.

Various algorithms are available to predict optimal warfarin dosage [12]. The most widely used (www.warfarindosage.org) is freely available and open-access. This website allows the clinician to input various factors that influence warfarin response (e.g., age; gender; weight; concomitant medications such as amiodarone, statins, and trimethoprim/sulfamethoxazole; tobacco use; and genotypes for CYP2C9, VKORC1, CYP4F2, and GGCX, if available). The output of the algorithm is an estimated initial and maintenance dose of warfarin.

Large-scale clinical trials evaluating genotype-guided warfarin dosing have produced mixed results, in some cases showing no benefit or even inferiority to standard dosing algorithms [13–15]. This has limited the clinical use of warfarin pharmacogenomics, especially in the United States where warfarin pharmacogenomic testing is generally not covered by health insurance and remains mostly a research tool. One broad limitation of warfarin pharmacogenomics is the presence of multiple other factors impacting drug action, necessitating monitoring of warfarin effect by following PT/INR.

7.7 Drugs metabolized by CYP2D6

As mentioned above, codeine is a prodrug that needs to be converted by CYP2D6 to morphine for therapeutic effect [3]. CYP2D6 poor metabolizers often show lack of efficacy to codeine due to lack of morphine conversion. Such patients more often benefit from other opiates such as hydrocodone, oxycodone, or morphine itself. On the other extreme, patients who are CYP2D6 ultra-rapid metabolizers convert codeine to morphine rapidly and may even experience morphine toxicity. Life-threatening codeine toxicity due to rapid CYP2D6 metabolism has been especially seen in the pediatric population, with some fatalities [16,17]. Codeine toxicity may occur in CYP2D6 ultra-rapid metabolizing children who are administered codeine (e.g., for analgesia for dental procedures) or in breastfeeding infants whose mothers are CYP2D6 ultra-rapid metabolizers prescribed codeine [18]. In either case, the excess of morphine can cause respiratory depression and other symptoms of opiate overdose. The FDA issued separate warnings on use of codeine in breastfeeding mothers and for postsurgical pain management for children in 2007 and 2012, respectively. CYP2D6 pharmacogenomic testing has not become standard of care given the availability of alternate analgesic medications.

Numerous other drugs are metabolized by CYP2D6 [3]. In contrast to the activation of codeine by CYP2D6, many drugs are inactivated by CYP2D6 metabolism, often representing the major route of elimination for these drugs. The major categories of CYP2D6 substrates include antidepressants (including some selective serotonin reuptake inhibitors, serotonin-norepinephrine reuptake inhibitors, and tricyclic antidepressants), antipsychotics, and β-adrenergic receptor antagonists (carvedilol, metoprolol, and propranolol) (see Table 7.2). For drugs inactivated by CYP2D6, poor metabolizers are at risk for drug toxicity at standard doses, due to slow clearance of the drug. Poor metabolizers may require reduced dosage or use of an alternative drug. Ultrarapid metabolizers may show lack of efficacy at standard dosages and need higher doses to compensate for more rapid metabolism.

7.8 Pharmacogenomics impacting cancer therapy

There are increasing pharmacogenomic applications involving cancer therapy. Some examples are related to drug metabolism (germline genetic variation), while others involve somatic mutations in the cancer itself.

Azathioprine and 6-mercaptopurine (6MP) are agents used in the treatment of cancers (e.g., acute lymphoblastic leukemia) and autoimmune disorders such as inflammatory bowel disease and rheumatoid arthritis [19]. Azathioprine is the prodrug of 6MP. One route for inactivation and clearance of 6MP is by the enzyme thiopurine methyltransferase (TPMT). Although azathioprine and 6MP both can produce toxic effects on the bone marrow if used in high doses, approximately 1 in 300 Caucasians (less in most other populations) experiences very profound bone marrow toxicity following standard doses of azathioprine and 6MP.

Many of the patients experiencing severe toxicity in response to azathioprine or 6MP therapy have very low TPMT enzymatic activity (they are "TPMT-poor metabolizers") [19]. Several clinical laboratory tests are available to determine whether individuals will have difficulty metabolizing 6MP and azathioprine. In 2004–2005, the package inserts for azathioprine and 6MP were revised to include warnings on toxicity related to genetic variation of TPMT. TPMT-poor metabolizers can still receive 6MP or azathioprine but need markedly reduced doses. Functional testing for TPMT enzyme activity in the blood (the enzyme is found in erythrocytes) is currently commonly performed prior to initiating therapy with 6MP or azathioprine.

A second oncology application involves the drug irinotecan, a chemotherapeutic agent used in the treatment of colorectal cancer and some other solid tumors [20]. Irinotecan has complicated metabolism but is inactivated by glucuronidation mediated by UDP-glucuronosyltransferase 1A1 (UGT1A1), an enzyme that also carries out conjugation of bilirubin. A variety of rare, severe mutations in UGT1A1 can result in the Crigler-Najjar syndromes, devastating disorders that can be fatal in childhood unless liver transplantation is performed. A milder mutation, designated UGT1A1*28, is the most common cause of *Gilbert syndrome*, a usually benign condition often diagnosed incidentally in the primary care setting following detection of (usually mild) unconjugated hyperbilirubinemia on routine chemistry laboratory studies or in the workup of jaundice. These individuals are, however, at high risk for severe toxicity following irinotecan therapy. With standard doses, such individuals may develop life-threatening neutropenia or diarrhea poorly responsive to therapy. Genetic testing for UGT1A1*28 is FDA-approved and, similar to 6MP and azathioprine, the package insert for irinotecan now includes specific information on UGT1A1 genetic variation. However, genetic screening for UGT1A1 has not been adopted universally.

Pharmacogenomics also finds application in oncology in targeted therapies for certain cancers [4,21]. There are now many examples of this therapeutic. One of the best examples is the use of trastuzumab (Herceptin) for breast cancers overexpressing the HER2 protein. In the pathology workup of breast cancer, determination of HER2 expression status determines whether trastuzumab is a therapeutic option. A more recent trend in oncology is the emergence of *paired diagnostics*, where a drug and its companion diagnostic test are approved simultaneously. An example is vemurafenib (Zelboraf), a drug that targets metastatic melanoma with a specific mutation in *BRAF*. Vemurafenib and the companion test to determine *BRAF* mutation status were approved simultaneously by the FDA. This is an area that continues to develop rapidly.

7.9 Pharmacogenomics and drug hypersensitivity

Drug hypersensitivity is a relatively common problem and may manifest on a spectrum from mild symptoms to more severe presentations such as Stevens-Johnson syndrome and toxic epidermal necrolysis [22]. Genetic variation in the human leukocyte antigen (HLA)-B gene has now been strongly linked to severe hypersensitivity to abacavir (antiviral used to treat human immunodeficiency virus, HIV) and carbamazepine (anticonvulsant) [23].

Approximately 5–8% of HIV patients develop hypersensitivity to abacavir, with presence of the HLA-B*5701 allele associated with high risk of hypersensitivity. If the HLA-B*5701 is absent, the risk of hypersensitivity is very low. Package insert warnings to abacavir hypersensitivity were issued in July 2008.

For carbamazepine, the HLA-B*1502 allele is positively correlated to hypersensitivity. This allele occurs mainly in individuals of Southeast Asian descent and is uncommon in other populations. Label warning for carbamazepine was issued in December 2007. The HLA-B*1502 is also linked to hypersensitivity to phenytoin and its prodrug fosphenytoin. Pharmacogenomic testing for abacavir and carbamazepine hypersensitivity carry strong recommendations from multiple societies and policy recommending groups [7,8].

7.10 Conclusions

The clinical application of pharmacogenomics has developed slowly but continues to evolve. So far, it has been challenging to demonstrate clinical efficacy and cost-effectiveness of pharmacogenomics testing. Consequently, pharmacogenomic therapy is often used retrospectively to try to determine why a particular patient has experienced toxicity or unexplained lack of efficacy to a drug. An educational challenge is to emphasize that genetics may account for only a minor portion of variability in response to a drug and pharmacogenomics may not eliminate need for TDM or close follow-up for patients on pharmacotherapy. Other factors such as concomitant medications or organ failure may be more important in some circumstances. Clinical experience and ongoing research trials should better define useful applications of pharmacogenomics so that testing will become more common, with consensus guidelines emerging to guide clinical practice. Evidence demonstrating clinical and economic benefit of pharmacogenomics testing is an important future goal.

References

[1] Wang L, McLeod HL, Weinshilboum RM. Genomics and drug response. N Engl J Med 2012;364:1144–53.

[2] St Sauver JL, Bielinski SJ, Olson JE, Bell EJ, et al. Integrating pharmacogenomics into clinical practice: promise vs reality. Am J Med 2016;129:1093–9. e1091.

[3] Ma JD, Lee KC, Kuo GM. Clinical application of pharmacogenomics. J Pharm Pract 2012;25:417–27.

[4] Duffy MJ, O'Donovan N, Crown J. Use of molecular markers for predicting therapy response in cancer patients. Cancer Treat Rev 2011;37:151–9.

[5] Janzic A, Locatelli I, Kos M. The value of evidence in the decision-making process for reimbursement of pharmacogenetic dosing of warfarin. Am J Cardiovasc Drugs 2017;17:399–408.

[6] Verhoef TI, Redekop WK, Langenskiold S, Kamali F, et al. Cost-effectiveness of pharmacogenetic-guided dosing of warfarin in the United Kingdom and Sweden. Pharm J 2016;16:478–84.

[7] Vassy JL, Stone A, Callaghan JT, Mendes M, et al. Pharmacogenetic testing in the veterans health administration (VHA): policy recommendations from the VHA clinical pharmacogenetics subcommittee. Genet Med 2018;21:382–90.

[8] Caudle KE, Klein TE, Hoffman JM, Muller DJ, et al. Incorporation of pharmacogenomics into routine clinical practice: the clinical pharmacogenetics implementation consortium (CPIC) guideline development process. Curr Drug Metab 2014;15: 209–17.

[9] Mega JL, Close SL, Wiviott SD, Shen L, et al. Cytochrome p-450 polymorphisms and response to clopidogrel. N Engl J Med 2009;360:354–62.

[10] Lesko LJ. The critical path of warfarin dosing: finding an optimal dosing strategy using pharmacogenetics. Clin Pharmacol Ther 2008;84:301–3.

[11] Schwarz UI, Ritchie MD, Bradford Y, Li C, et al. Genetic determinants of response to warfarin during initial anticoagulation. N Engl J Med 2008;358:999–1008.

[12] Klein TE, Altman RB, Eriksson N, Gage BF, et al. Estimation of the warfarin dose with clinical and pharmacogenetic data. N Engl J Med 2009;360:753–64.

[13] Kimmel SE, French B, Kasner SE, Johnson JA, et al. A pharmacogenetic versus a clinical algorithm for warfarin dosing. N Engl J Med 2013;369:2283–93.

[14] Pirmohamed M, Burnside G, Eriksson N, Jorgensen AL, et al. A randomized trial of genotype-guided dosing of warfarin. N Engl J Med 2013;369:2294–303.
[15] Gage BF, Bass AR, Lin H, Woller SC, et al. Effect of genotype-guided warfarin dosing on clinical events and anticoagulation control among patients undergoing hip or knee arthroplasty: the GIFT randomized clinical trial. JAMA 2017;318:1115–24.
[16] Agrawal YP, Rennert H. Pharmacogenomics and the future of toxicology testing. Clin Lab Med 2012;32:509–23.
[17] Niesters M, Overdyk F, Smith T, Aarts L, et al. Opioid-induced respiratory depression in paediatrics: a review of case reports. Br J Anaesth 2013;110:175–82.
[18] Willmann S, Edginton AN, Coboeken K, Ahr G, et al. Risk to the breast-fed neonate from codeine treatment to the mother: a quantitative mechanistic modeling study. Clin Pharmacol Ther 2009;86:634–43.
[19] Asadov C, Aliyeva G, Mustafayeva K. Thiopurine S-methyltransferase as a pharmacogenetic biomarker: significance of testing and review of major methods. Cardiovasc Hematol Agents Med Chem 2017;15:23–30.
[20] de Man FM, Goey AKL, van Schaik RHN, Mathijssen RHJ, et al. Individualization of irinotecan treatment: a review of pharmacokinetics, pharmacodynamics, and pharmacogenetics. Clin Pharmacokinet 2018;57:1229–54.
[21] Duffy MJ, Crown J. Companion biomarkers: paving the pathway to personalized treatment for cancer. Clin Chem 2013;59:1447–56.
[22] Garon SL, Pavlos RK, White KD, Brown NJ, et al. Pharmacogenomics of off-target adverse drug reactions. Br J Clin Pharmacol 2017;83:1896–911.
[23] Fricke-Galindo I, A LL, Lopez-Lopez M. An update on HLA alleles associated with adverse drug reactions. Drug Metab Pers Ther 2017;32:73–87.

CHAPTER 8

Therapeutic drug monitoring using alternative specimens

8.1 Introduction

Assays for drugs or drug metabolites for therapeutic drug monitoring (TDM) typically utilize plasma, serum or whole blood obtained by venous blood sampling (see Chapter 2). Most commercialized assays for TDM (e.g., immunoassays run on automated clinical chemistry analyzers) have been approved for plasma, serum, or whole blood. Consequently, reference ranges have typically been established for these matrices and reported in assay package inserts and/or published literature [1]. Venipuncture-obtained samples for TDM work well in settings that have established phlebotomy and specimen transport resources, allowing for specimens to stay within acceptable parameters prior to analysis. However, there are some disadvantages to obtaining specimens by venipuncture, including the need for phlebotomy and time-sensitive sample processing (including centrifugation and temperature control) [2]. These drawbacks may be especially an issue for certain patient populations (especially young children and the elderly) and in remote or resource-limited settings [1]. In addition, plasma, serum, or whole blood may not adequately correlate with the pharmacodynamics for some drugs; in these cases, alternative body fluids or tissue may hold advantages for TDM.

8.2 Overview of alternative specimens

Theoretically, a variety of specimens could be used for TDM; however, some sample types such as tissue from internal organs are impractical except for very limited applications. Dried blood spots (DBS), saliva (oral fluid), and urine have received the most attention and research as alternative TDM specimens, with each having potential advantages over venipuncture-obtained specimens for some drugs [2–4]. DBS and saliva particularly offer advantages in ease of specimen collection and transport. Other alternative specimens include cerebrospinal fluid (CSF), hair, tears, nails, and tissue [2].

Therapeutic Drug Monitoring Data
https://doi.org/10.1016/B978-0-12-815849-4.00008-6

This chapter will focus in most detail on DBS and saliva but will cover other specimen types briefly.

8.3 Dried blood spots for therapeutic drug monitoring

Dried blood spot (DBS) sampling has emerged as a viable specimen for TDM for some drugs, with potential advantages over conventional venipuncture [4–6]. DBS samples have been used for decades for newborn metabolic testing using heel sticks; the experience with newborn samples provides a wealth of background for the strengths and limitations of this specimen type [7]. For DBS sampling for therapeutic drug monitoring (TDM), capillary blood is typically obtained by a finger prick using an automatic lancet (e.g., autoretracting 1.8–2.4 mm needle) [5, 6]. DBS sampling for TDM in newborns and infants may use specimens obtained by heel stick. The first blood drop is typically discarded to minimize contamination with tissue fluids. The next blood drops are used to fill premarked circles on filter paper, with one drop in each designated circle. The DBS dries on the filter paper at ambient temperature and then is packaged for transport to the clinical laboratory. Upon specimen receipt, the laboratory assesses the homogeneity of the blood spots and punches out discs of defined diameter. The analyte(s) of interest are extracted from the disc and then measured.

There are potential benefits of DBS for TDM, especially in terms of patient convenience and specimen stability/transport [5, 6]. The simplicity of DBS sampling makes it feasible for patients or family members to collect specimens at home or other convenient setting. Alternatively, healthcare personnel unskilled in venipuncture can be quickly trained to perform DBS sampling. DBS sampling can occur at times where it may be infeasible to travel to a healthcare facility. This can make possible TDM sampling protocols that would otherwise be impractical, including area under the concentration-time curve measurements. Use of capillary blood also limits iatrogenic blood loss, an advantage for young children and anemic patients. Perhaps the major advantages of DBS sampling relate to specimen storage and transport. Many drugs and drug metabolites are as stable or more stable in DBS compared to specimens collected by venipuncture and then frozen after processing. DBS specimens can be transported with very low infectious risk by routine postal systems without specialized biohazard shipping containers [4]. These characteristics make DBS sampling attractive for specimens collected in remote and/or resource-limited settings.

Table 8.1 Comparison of alternative matrices with serum/plasma

Variable	Serum/plasma	Dried blood spots	Saliva (oral fluid)
Need for skilled collector	Yes	No	No
Suitability for remote or low resource location	Poor	Excellent	Good
Transport by regular mail	No	Yes	Yes
Volume of sample required	High	Low	Low
Immunoassay analysis feasible	Yes	No	No
Stability at ambient temperature	Low	High	Relatively stable
Impact of varying hematocrit	Minimal	Yes	No
Established reference ranges	Yes	Generally no	Generally no
Need for extensive validation	Generally no	Yes	Yes

There are some disadvantages to DBS sampling for TDM [5, 6]. The small amount of sample in DBS necessitates use of sensitive analytical methods (especially chromatography/mass spectrometry techniques) and generally precludes use of automated immunoassays commonly used for serum and plasma. In addition, there is limited ability to perform repeated analysis or more than one type of analysis on a blood spot. There are also pre-analytical issues with DBS sampling. Contamination can occur if the person performing the sample collection also handles the dosing of the drug and transfers pharmaceutical material to the filter paper. Improper application of blood to the filter paper may lead to spot inhomogeneity. Analytically, use of DBS sampling for TDM requires extensive validation that should include evaluation of factors such as hemolysis, sample homogeneity, extraction recovery, analyte stability, and impact of varying hematocrits. Hematocrit variation represents perhaps the biggest challenge to drug analysis using DBS [5, 6]. Reference ranges for specimens obtained by DBS sampling need to be determined by detailed validation and correlated with those for plasma, serum, or whole blood. Table 8.1 summarizes some of the features of DBS as a specimen for TDM compared to serum/plasma.

8.4 Saliva (oral fluid) for therapeutic drug monitoring

Interest in saliva (oral fluid) as an alternative specimen for TDM dates back to the 1970s, with early investigations focusing on first-generation anti-epileptic drugs (AEDs) such as carbamazepine, phenobarbital, and phenytoin [3].

Over the last decade, saliva has been investigated as a specimen for other classes of drugs [2, 8]. Saliva is an ultrafiltrate of blood, and salivary drug levels are typically reflective of free drug levels in plasma for certain drugs that are not ionizable within the salivary pH range. For some drugs with high degrees of protein binding, saliva offers an alternative to unbound/free drug analysis using blood.

Similar to DBS, simplicity of specimen collection is a major advantage of saliva as a specimen for TDM [2, 8]. Specimens can be obtained without any invasive procedure and in a variety of settings without need for trained personnel. This allows for flexibility and timing of specimen collection along with ease in collecting multiple specimens, if indicated. The relative non-invasiveness of saliva collection can be especially advantageous in infants or those with skin or vascular conditions that impede collection by finger stick or venipuncture.

There are some pre-analytical challenges with saliva collection [2, 8]. Salivary flow rates vary significantly both between individuals and under different conditions; the use of stimulated saliva has advantages over resting saliva in this regard. Salivary flow rate, pH, sampling condition, and pathophysiological factors may influence the concentration of a particular drug in saliva. Spurious results can occur with drug contamination in the mouth. This is more likely to happen if saliva sampling occurs shortly after drug ingestion or in cases where drug residues or fragments are retained in the oral cavity (e.g., patients with severe gingivitis and pockets of inflamed gum tissue). The lowest risk of specimen contamination is with trough collections shortly before oral ingestion of drug. Specimen volume is also important when using saliva for TDM. Low volume of saliva collected can be suboptimal for laboratory analysis. This can occur with inexperienced collectors or if clear directions regarding specimen volume are not followed. Alternatively, some patients have medical conditions that limit saliva production and/or secretion such as HIV/AIDS, Sjögren's syndrome, diabetes mellitus, and Parkinson's disease.

Drugs can also limit saliva production [9]. Common examples includes medications with antimuscarinic effects (e.g., tricyclic antidepressants, diphenhydramine, scopolamine). Low specimen volume can sometimes be overcome with dilution of the specimen by the laboratory with distilled water or other appropriate diluent, assuming that the concentration of the diluted specimen is still above the lower limit of quantitation for the analytical method. Lastly, saliva may be viscous and difficult to pipette, possibly leading to need for recollection. Table 8.1 summarizes some of the features of saliva as a specimen for TDM compared to serum/plasma.

AEDs have received significant attention with respect to saliva TDM [3, 10]. The majority of older and newer generation AEDs can be monitored using saliva. There are some AEDs for which saliva either does not appear to be a good specimen for TDM or there is insufficient data to support use of saliva. Those with insufficient evidence include clonazepam, pregabalin, stiripentol, tiagabine, and vigabatrin [10]. Early data suggested that valproic acid was not a good candidate for TDM using saliva due to erratic distribution [3]; however, more recent data has challenged this [11]. This will require further investigations.

A growing literature has also developed on the applicability of TDM using saliva in neonates and infants, a population where there is obvious benefit in limiting blood loss and invasive specimen collection [2]. A comprehensive review of the subject published in 2018 found literature support for salivary monitoring of neutral and acidic compounds including busulfan, fluconazole, oxcarbazepine, phenytoin, primidone, and theophylline [8]. These drugs exhibited very good correlation between blood and salivary concentrations. For studies in neonates and infants, lamotrigine was the only basic compound to exhibit tight correlation between blood and salivary concentrations. Overall, saliva shows promise as a specimen for TDM of selected drugs in neonates and infants.

8.5 Urine for therapeutic drug monitoring

Urine has traditionally been used for qualitative drug analysis, especially in the realm of drugs of abuse and toxicology testing [12–15]. Urine accumulates many drugs and metabolites and has the advantage of low-cost and ready availability. In most individuals, adequate specimen volumes for drug analysis can be easily obtained. Numerous immunoassay and chromatographic assays exist for drug detection in urine.

Quantitative drug analysis in urine has some applications in TDM including analysis of opioids and benzodiazepines to confirm treatment compliance and to identify non-medical drug use or suspected diversion of medications [13, 16]. This type of testing may be a condition for medication contracts used in patients prescribed controlled substances. One challenge with quantitation of drugs and metabolites in urine is that a wide range of concentrations can be observed even between individuals with the same drug therapy. Patient hydration status, renal function, urinary pH, and use of diuretic medications are just some of the factors that can influence drug disposition in urine. When quantitative urine testing is used for assessing adherence to medication therapy, interpretation of data should be cautious.

There is also the possibility that patients adulterate the urine specimen to limit detection of non-medical drug use or substitute something else in place of their own urine specimen (e.g., someone else's urine, animal urine, or synthetic urine) [17, 18].

8.6 Cerebrospinal fluid for therapeutic drug monitoring

Cerebrospinal fluid (CSF) is uncommonly used for TDM due to the invasiveness of sampling (typically by lumbar puncture) and relatively small sample volumes collected [2]. Nevertheless, CSF does have a potential advantage as a pharmacological "sanctuary" fluid providing insight into concentrations of compounds that have crossed the blood–brain barrier into the central nervous system (CNS) [19]. Drugs that do not adequately penetrate the blood–brain barrier may achieve ineffective concentrations in the CNS despite concentrations in plasma or serum in the therapeutic range. One logical application for use of CSF for TDM is for antimicrobial or antineoplastic therapy targeting the CNS [20–22]. To this end, chromatographic/mass spectrometry methods have been reported for quantitation of vancomycin and antifungal drugs (amphotericin B, fluconazole, fluorocytosine, and caspofungin) in CSF as a means to assess penetration of these drugs to treat cerebral bacterial and fungal infections. Other work has focused on antiretroviral therapies such as abacavir and efavirenz.

8.7 Uncommon matrices for TDM

A variety of other specimens have some potential for TDM including sweat, tears, hair, and tissue [2]. These currently have very limited applications, but these may grow as analytical techniques develop. Each of these specimens will be discussed briefly.

Sweat has limited value for quantitative drug analysis but some potential for assessing adherence to medication therapy [23]. An advantage of sweat as a specimen is non-invasive collection, typically using specialized patches. Surface and environmental contamination present practical challenges. Analysis of sweat generally requires chromatographic/mass spectrometry techniques.

Tears can be easily sampled and are less prone to environmental contamination as compared with saliva or sweat [2]. Tears represent the gold standard matrix for TDM of topical ophthalmologic compounds; however, this is rarely assessed in clinical practice. Disadvantages of tears include low specimen volume and variability in chemical composition from stimulated and unstimulated tears [24]. This specimen type does have potential applications in small animal research.

Hair has emerged as a specimen for assessing longer-term exposure to drugs, most commonly used to assess drugs of abuse [25]. In contrast to blood and urine, hair provides a much wider window for detection of drugs up to several months. The most common applications of hair testing are in the drugs of abuse/toxicology testing realm (e.g., assessing drug abstinence in someone in a substance abuse treatment program). Challenges with hair include technical difficulties in analysis of this matrix and multiple factors that can impact drug incorporation into the hair matrix. Important variables include rate of hair growth, use of hair cleaning products, artificial hair coloring, and surface contamination from sweat and the environment. Hair currently has limited clinical utility for TDM, but possible applications include assessment of long-term adherence to treatment such as antiretroviral therapy [25].

8.8 Conclusions

While serum and plasma remain the most common specimens for TDM, use of alternative matrices has progressed considerably in the last decade. Advances in chromatography/mass spectrometry techniques will undoubtedly allow easier analysis of specimens such as DBS or saliva that require sensitive analytical techniques. There are still challenges in validating these matrices to allow them to be used more widely in clinical settings.

References

[1] Zhao W, Jacqz-Aigrain E. Principles of therapeutic drug monitoring. Handb Exp Pharmacol 2011;205:77–90.
[2] Avataneo V, D'Avolio A, Cusato J, Cantu M, De Nicolo A. LC-MS application for therapeutic drug monitoring in alternative matrices. J Pharm Biomed Anal 2018;166:40–51.
[3] Patsalos PN, Berry DJ. Therapeutic drug monitoring of antiepileptic drugs by use of saliva. Ther Drug Monit 2013;35:4–29.
[4] Wilhelm AJ, den Burger JC, Swart EL. Therapeutic drug monitoring by dried blood spot: progress to date and future directions. Clin Pharmacokinet 2014;53:961–73.
[5] Antunes MV, Charao MF, Linden R. Dried blood spots analysis with mass spectrometry: potentials and pitfalls in therapeutic drug monitoring. Clin Biochem 2016;49:1035–46.
[6] Sharma A, Jaiswal S, Shukla M, Lal J. Dried blood spots: concepts, present status, and future perspectives in bioanalysis. Drug Test Anal 2014;6:399–414.
[7] Wilcken B, Wiley V. Newborn screening. Pathology 2008;40:104–15.

[8] Hutchinson L, Sinclair M, Reid B, Burnett K, Callan B. A descriptive systematic review of salivary therapeutic drug monitoring in neonates and infants. Br J Clin Pharmacol 2018;84:1089–108.
[9] Drummer OH. Introduction and review of collection techniques and applications of drug testing of oral fluid. Ther Drug Monit 2008;30:203–6.
[10] Patsalos PN, Spencer EP, Berry DJ. Therapeutic drug monitoring of antiepileptic drugs in epilepsy: a 2018 update. Ther Drug Monit 2018;40:526–48.
[11] Dwivedi R, Gupta YK, Singh M, Joshi R, Tiwari P, Kaleekal T, Tripathi M. Correlation of saliva and serum free valproic acid concentrations in persons with epilepsy. Seizure 2015;25:187–90.
[12] Hammett-Stabler CA, Pesce AJ, Cannon DJ. Urine drug screening in the medical setting. Clin Chim Acta 2002;315:125–35.
[13] Kwong TC, Magnani B, Moore C. Urine and oral fluid drug testing in support of pain management. Crit Rev Clin Lab Sci 2017;54:433–45.
[14] McMillin GA, Marin SJ, Johnson-Davis KL, Lawlor BG, Strathmann FG. A hybrid approach to urine drug testing using high-resolution mass spectrometry and select immunoassays. Am J Clin Pathol 2015;143:234–40.
[15] Tenore PL. Advanced urine toxicology testing. J Addict Dis 2010;29:436–48.
[16] Mahajan G. Role of urine drug testing in the current opioid epidemic. Anesth Analg 2017;125:2094–104.
[17] Dasgupta A. The effects of adulterants and selected ingested compounds on drugs-of-abuse testing in urine. Am J Clin Pathol 2007;128:491–503.
[18] Fu S. Adulterants in urine drug testing. Adv Clin Chem 2016;76:123–63.
[19] Cory TJ, Schacker TW, Stevenson M, Fletcher CV. Overcoming pharmacologic sanctuaries. Curr Opin HIV AIDS 2013;8:190–5.
[20] Calcagno A, Simiele M, Alberione MC, Bracchi M, Marinaro L, Ecclesia S, Di Perri G, D'Avolio A, Bonora S. Cerebrospinal fluid inhibitory quotients of antiretroviral drugs in HIV-infected patients are associated with compartmental viral control. Clin Infect Dis 2015;60:311–7.
[21] Nwogu JN, Ma Q, Babalola CP, Adedeji WA, Morse GD, Taiwo B. Pharmacokinetic, pharmacogenetic, and other factors influencing CNS penetration of antiretrovirals. AIDS Res Treat 2016;2016:2587094.
[22] Qu L, Qian J, Ma P, Yin Z. Utilizing online-dual-SPE-LC with HRMS for the simultaneous quantification of amphotericin B, fluconazole, and fluorocytosine in human plasma and cerebrospinal fluid. Talanta 2017;165:449–57.
[23] Fucci N, De Giovanni N, Pascali VL. The sweat matrix: a new perspective for drugs analysis. Skin Res Technol 2015;21:129–30.
[24] Stuchell RN, Feldman JJ, Farris RL, Mandel ID. The effect of collection technique on tear composition. Invest Ophthalmol Vis Sci 1984;25:374–7.
[25] Pragst F, Balikova MA. State of the art in hair analysis for detection of drug and alcohol abuse. Clin Chim Acta 2006;370:17–49.

CHAPTER 9
Antiepileptic drugs

9.1
Introduction

The primary treatment of epilepsy is the prophylactic use of anticonvulsant drugs, also known as antiepileptic drugs (AEDs). The goal is prevention of all seizures with minimization of negative effects on general well-being, cognition, mood, and endocrine function. Following determination that recurrent seizures are probable, successful treatment with AEDs begins with proper determination of the seizure type and diagnosis of epileptic syndrome. Seizure type is typically determined based on clinical and electroencephalogram (EEG) manifestations. There are several ways in which seizures and epilepsy syndromes may be classified. The International League Against Epilepsy (ILAE) released an updated seizure-type classification in 2017. Both basic and extended versions of the classification are available [1]. Key signs and symptoms of seizures are used for classification and are categorized based on whether seizures are focal or generalized from onset or with unknown onset. Any focal seizure can be further categorized based on whether awareness is retained or impaired. Impaired awareness during any segment of the seizure should be considered as a focal impaired awareness seizure. Focal seizures are further characterized by motor onset signs and symptoms: atonic, automatisms, clonic, epileptic spasms, or hyperkinetic, myoclonic, or tonic activity. Seizures of non-motor-onset (absence seizures) can manifest as autonomic, behavior arrest, cognitive, emotional, or sensory dysfunction. The earliest prominent manifestation defines the seizure type, which might then progress to other signs and symptoms. Focal seizures can become bilateral tonic–clonic. Generalized seizures engage bilateral networks from onset. Generalized motor seizure characteristics comprise atonic, clonic, epileptic spasms, myoclonic, myoclonic–atonic, myoclonic–tonic–clonic, tonic, or tonic–clonic. Absence seizures are typical or atypical, or seizures that present prominent myoclonic activity or eyelid myoclonia.

Therapeutic Drug Monitoring Data
https://doi.org/10.1016/B978-0-12-815849-4.00009-8

Seizures of unknown onset may have features that can still be classified as motor, absence, tonic–clonic, epileptic spasms, or behavior arrest.

Therapeutic drug monitoring (TDM) using blood levels of AEDs began in the mid-1960s and replaced the former practice of dosing to pharmacodynamic effectiveness. TDM became widespread in the late-1970s with the introduction of immunoassays that made drug monitoring widely accessible [2]. Routine monitoring of the first generation ("traditional" or "classical") AEDs (phenobarbital, phenytoin, carbamazepine, primidone, ethosuximide, and valproic acid) is considered an integral tool in the treatment of epileptic patients. Since 1990, second and third generation AEDs have been introduced including brivaracetam, clobazam, eslicarbazepine acetate, ezogabine, felbamate, gabapentin, lacosamide, lamotrigine, levetiracetam, oxcarbazepine, perampanel, pregabalin, rufinamide, stiripentol, tiagabine, topiramate, vigabatrin and zonisamide. In general, many of the newer generation AEDs (including vigabatrin, felbamate, gabapentin, lamotrigine, tiagabine, topiramate, levetiracetam, oxcarbazepine, zonisamide, pregabalin, rufinamide, and lacosamide) enjoy both improved tolerability and safety compared with older agents such as phenobarbital, phenytoin, carbamazepine, and valproic acid [3]. As a result, some of these newer AEDs do not require TDM. However, TDM may be useful for gabapentin, lamotrigine, levetiracetam, oxcarbazepine, tiagabine, topiramate, and zonisamide. In this chapter, TDM of classical anticonvulsants as well as monitoring of certain newer AEDs is discussed.

Protein binding of AEDs (both classical and newer generation drugs) vary widely. Gabapentin and pregabalin are not protein bound. Clobazam, clonazepam, perampanel, phenytoin, retigabine, stiripentol, tiagabine and valproic acid are strongly protein bound. Therapeutic ranges of AEDs also vary widely [4]. Protein binding and therapeutic ranges of AEDs are listed in Table 9.1.

Several factors should be considered when choosing an AED, but efficacy based on evidence-based studies is a primary concern. Summaries of studies and recommendations of specific AEDs for particular seizure type or epilepsy syndrome have been developed [5–11]. Treatment guidelines involving newer AEDs have been recently published [12, 13]. Other factors to be considered include drug-specific (adverse effects, toxicity, and drug interactions), patient-specific (gender, age, contraception, genetics) and possibly practical variables (availability, cost) [10, 11]. As expected, newer generation AEDs demonstrate less drug–drug interaction and better safety profile in treating pregnant women compared to first generation AEDs [13].

Table 9.1 Protein binding and therapeutic ranges of antiepileptic drugs

Antiepileptic drug	Therapeutic range (μg/mL)	Protein binding
Carbamazepine	4–12	80%
Clobazam	0.03-0.3	90%
Clonazepam	0.02-0.07	91%
Gabapentin	2–20	Negligible
Eslicarbazepine	3–35	44%
Ethosuximide	40–100	22%
Felbamate	30–60	48%
Lacosamide	10–20	14%
Lamotrigine	3–15	55%
Levetiracetam	12–46	<5%
Perampanel	200–1000	98%
Phenobarbital	10–40	40%
Phenytoin	10–20	90%
Pregabalin	2–8	No protein binding
Primidone	5–10	34%
Rufinamide	30–40	28%
Stiripentol	4–22	96%
Tiagabine	0.02-0.2	98%
Topiramate	5–20	20%
Valproic acid	50–100	80–95% (Concentration dependent binding)
Vigabatrin	2–36	17%
Zonisamide	10–40	44%

For therapy with classical AEDs, once a drug is chosen, the appropriate dosage should be given with slow titration to desired clinical response, utilizing appropriate serum or plasma levels to monitor achievement of therapeutic levels and avoidance of toxic concentrations. If the initial monotherapy is ineffective or poorly tolerated, switching to another drug is recommended. However, if the first drug is well tolerated but only partially effective, addition of another AED should be considered. Often more than one AED is needed for proper seizure control. Addition of one or more drugs for combination therapy increases the probability of drug–drug interactions impacting pharmacokinetics. Specific interactions will be mentioned in the discussion of each drug, but some general concepts can be stated. Several first-generation AEDs induce hepatic enzymes and can thus cause dramatic reductions of serum concentrations of other drugs which are substrates of these enzymes. For example, the plasma concentrations of valproic acid or lamotrigine can be reduced over 50% in patients comedicated with enzyme inducers. More specifically, carbamazepine, phenytoin, phenobarbital and primidone stimulate a variety of cytochrome P450 (CYP) enzymes as well as glucuronyl transferases and epoxide hydrolyase; in contrast, the newer AEDs generally do not share the broad spectrum enzyme-inducing activity of these drugs [14].

In addition to affecting concentrations of comedicated AEDs, significant effects on concentrations of other medications can occur. Of particular concern are interactions of AEDs with oral contraceptives and anticoagulants. The efficacy of estrogen/progesterone containing contraceptives may be reduced due to induction of metabolizing enzymes by the following AEDs: carbamazepine, felbamate, oxcarbazepine, lamotrigine, phenobarbital, phenytoin, primidone and topiramate. This can result in unplanned pregnancy if concentrations of ethinyl estradiol or other components of oral contraceptives become sub-therapeutic [14]. In contrast, studies suggest that valproic acid, gabapentin, levetiracetam, zonisamide, and lacosamide do not interact with oral contraceptives and may therefore be regarded as safe with respect to possible contraceptive failure. [15]. Reciprocal interactions may also occur and administration of the combined contraceptive pill to patients taking some AEDs can cause a reduction in serum levels of AEDs resulting in loss of seizure control. Markedly decreased levels of lamotrigine and valproic acid during the 21 days of intake of the contraceptive pills have been reported [14].

Interaction between warfarin and AEDs may have serious clinical consequence. The enzyme-inducing AEDs stimulate metabolism of warfarin and other coumarin dugs, thereby increasing risk of treatment failure unless the dosage of anticoagulant drug is increased. Clark et al. reported that warfarin dose and dose/INR ratio (a measure of the impact of warfarin on prothrombin time and the internationalized normalized ratio) was significantly increased after administration of carbamazepine, while oxcarbazepine, phenobarbital, and phenytoin initiation did not significantly affect warfarin dosing. The authors concluded that interaction between carbamazepine and warfarin is clinically significant. Frequent INR monitoring and warfarin dose escalation are recommended in this setting [16]. Bruun commented that elderly patients with newly diagnosed epilepsy are at high risk of clinically relevant pharmacokinetic interactions with other drugs, especially if exposed to carbamazepine. Patients on dihydropyridine calcium-channel blockers, statins, warfarin, and risperidone face the highest risk of drug–drug interactions [17].

The majority of the new AEDs gained initial United States Food and Drug Administration (FDA) approval based on clinical trials in which the new drug was used as adjunctive treatment, typically for patients with refractory partial-onset seizures. However, these drugs are often prescribed for off-label use, particularly for monotherapy and sometimes for purposes other than seizure control [18]. Although some studies indicate advantages of the new AEDs over the first-generation drugs, there is currently no consensus that classical AEDS are no longer relevant in treating epilepsy. Moreover, newer AEDs are not free from potential adverse effects. Cahill reported that neuropathy appears to be associated with the length of exposure to new AEDs which may be related to the effects of new AEDs on vitamin B_{12} and folate metabolism. The authors recommended monitoring for neuropathy and vitamin B_{12} and folate levels in patients on long-term treatment with new AEDs [19].

References

[1] Fisher RS, Cross JH, D'Souza C, French JA, et al. Instruction manual for the ILAE 2017 operational classification of seizure types. Epilepsia 2017;58:531–42.

[2] Pippenger CE. Therapeutic drug monitoring assay development to improve efficacy and safety. Epilepsy Res 2006;68:60–3.

[3] French JA, Gazzola DM. New generation antiepileptic drugs: what do they offer in terms of improved tolerability and safety? Ther Adv Drug Saf 2011;2:141–58.

[4] Patsalos PN, Zugman M, Lake C, James A, et al. Serum protein binding of 25 antiepileptic drugs in a routine clinical setting: a comparison of free non-protein-bound concentrations. Epilepsia 2017;58:1234–43.

[5] Bergey GK. Evidence-based treatment of idiopathic generalized epilepsies with new antiepileptic drugs. Epilepsia 2005;46(S9):161–8.

[6] Walker M. Status epilepticus: an evidence based guide. BMJ 2005;331:673–7.

[7] French JA, Kanner AM, Bautista J, Abou-Khalil B, et al. Efficacy and tolerability of the new antiepileptic drugs I: treatment of new onset epilepsy. Neurology 2004;62:1252–60.

[8] French JA, Kanner AM, Bautista J, Abou-Khalil B, et al. Efficacy and tolerability of the new antiepileptic drugs II: treatment of refractory epilepsy. Neurology 2004;62:1261–73.

[9] Glauser T, Ben-Menachem E, Bourgeois B, Cnaan A, et al. ILAE treatment guidelines: evidence-based analysis of antiepileptic drug efficacy and effectiveness as initial monotherapy for epileptic seizures and syndromes. Epilepsia 2006;47(7):1094–120.

[10] Beyenburg S, Bauer J, Reuber M. New drugs for the treatment of epilepsy: a practical approach. Postgrad Med J 2004;80:581–7.

[11] Sander JW. The use of antiepileptic drugs—principles and practice. Epilepsia 2004;45(6):28–34.

[12] Eskioglou E, Perrenoud MP, Ryvlin P, Novy J. Novel treatment and new drugs in epilepsy treatment. Curr Pharm Des 2017;23:6389–98.

[13] Hanaya R, Arita K. The new antiepileptic drugs: their neuropharmacology and clinical indications. Neurol Med Chir (Tokyo) 2016;56(5):205–20.

[14] Perucca E. Clinically relevant drug interactions with antiepileptic drugs. Br J Clin Pharmacol 2005;61(3):246–65.

[15] Reimers A, Brodtkorb E, Sabers A. Interactions between hormonal contraception and antiepileptic drugs: clinical and mechanistic considerations. Seizure 2015;28:66–70.

[16] Clark NP, Hoang K, Delate T, Horn JR, Witt DM. Warfarin interaction with hepatic cytochrome P-450 enzyme-inducing anticonvulsants. Clin Appl Thromb Hemost 2018;24(1):172–8.

[17] Bruun E, Virta LJ, Kälviäinen R, Keränen T. Co-morbidity and clinically significant interactions between antiepileptic drugs and other drugs in elderly patients with newly diagnosed epilepsy. Epilepsy Behav 2017;73:71–6.
[18] Golden AS, Haut SR, Moshe SL. Nonepileptic uses of antiepileptic drugs in children and adolescents. Pediatr Neurol 2006;34:421–32.
[19] Cahill V, McCorry D, Soryal I, Rajabally YA. Newer anti-epileptic drugs, vitamin status and neuropathy: a cross-sectional analysis. Rev Neurol (Paris) 2017;173:62–6.

9.2

Carbamazepine

Chemical properties	
Solubility in H_2O	Insoluble
Molecular weight	236.26
pKa	n/a
Melting point	204–206 °C
Dosing	
Recommended dose, adult	Initial dosage: 200 mg twice a day, maintenance dose: 800–1200 mg/day, maximum dosage: 1600 mg/day
Recommended dose, child	15–30 mg/kg/d
Monitoring	
Sample	Serum, plasma
Effective concentrations	4–12 μg/mL
Toxic concentrations	>20 μg/mL
Methods	Immunoassay (most common), HPLC, GC
Pharmacokinetic properties	
Oral dose absorbed (bioavailability)	78%
Time to peak concentration	4–8 h
Protein bound	65–80%, albumin
Volume of distribution	Child: 1.9 L/kg Adult: 0.59–2.00 L/kg
Half-life, adult	10–20 h
Time to steady state, adult	2–6 d
Half-life, child	8–19 h
Time to steady state, child	2–4 d

Excretion, urine			
	% Excreted	Active	Detected in blood
Unchanged	0.5–1.0	Yes	Yes
Carbamazepine-10,11-epoxide	<2	Yes	Yes
10,11 diol-carbamazepine, free and conjugated	10–20	No	Unknown

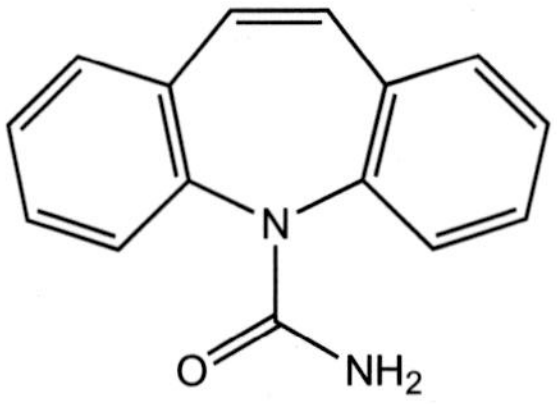

Carbamazepine

Carbamazepine (Tegretol) is an antiepileptic drug (AED) approved for use in the United States in 1974. This drug comes with a United States Food and Drug Administration (FDA) black box warning that it may cause life-threatening allergic reactions called Stevens-Johnson syndrome and toxic epidermal necrolysis. However, the alert states that carbamazepine-induced skin reactions are significantly more common in patients with the human leukocyte antigen (HLA)-B*1502 allele almost exclusively found in people whose ancestry is from 'broad areas of Asia', including South Asian Indians [1].

Although structurally related to the tricyclic antidepressants, it has none of the pharmacologic properties of that class of drugs. Carbamazepine was first synthesized in the 1950s. Currently, carbamazepine is a first-line treatment for partial seizures (psychomotor or temporal lobe) and is indicated for bipolar disorder and trigeminal neuralgia. Carbamazepine is slowly and erratically absorbed following oral administration. The oral bioavailability is approximately 0.78 [2]. Peak plasma levels are usually reached in 4–8 h; however, administration of a large dose may delay peak levels for up to 24 h. Carbamazepine is also significantly protein bound (65–80%) to a combination of albumin and alpha-1-acid glycoprotein and exhibits an initial low clearance that increases two to three fold due to autoinduction of cytochrome P450 (CYP) enzymes expressed in the liver. Carbamazepine has an active metabolite, carbamazepine-10,11-epoxide (carbamazepine-E) that possesses anticonvulsant activity and central nervous system (CNS) toxicity similar to the parent compound. This metabolite is detoxified by epoxide hydrolase [3].

Carbamazepine is primarily metabolized by CYP3A4 and to a lesser degree by CYP3A5. Coadministration of drugs known to inhibit CYP3A4 activity (such as azole antifungal drugs, clarithromycin, erythromycin, and isoniazid) would be expected to cause increased serum concentrations of carbamazepine. However, carbamazepine itself induces the activity of a variety of CYP enzymes including CYP1A2, CYP2C9, CYP2C19, and CYP3A4. Hence, carbamazepine can reduce the serum concentration of AEDs and many other types of drugs when administered concurrently.

Affected drugs include valproic acid, ethosuximide, lamotrigine, topiramate, oxcarbazepine, antidepressants, antimicrobials, antineoplastics, antipsychotics, immunosuppressants, steroids (including oral contraceptives), and cardiovascular drugs. Conversely, discontinuance of carbamazepine may be followed by an increase to toxic concentrations of concurrently administered AEDs. Phenytoin, phenobarbital, and felbamate enhance the metabolism of carbamazepine by induction of CYP enzyme expression.

The half-life ($t_{½}$) of carbamazepine can vary significantly depending on factors such as length of therapy and co-administered drugs. In patients on chronic therapy, the $t_{½}$ is 10–20 h. One study reported that $t_{½}$ was significantly longer in men than women as well as in African Americans when compared to Caucasians [2]. The $t_{½}$ of carbamazepine is reduced to 9–10 h in patients who are receiving other enzyme-inducing drugs. The $t_{½}$ of carbamazepine-E is slightly less than the parent compound.

Carbamazepine-E represents approximately 15–20% of the parent drug concentration in normal individuals. However, in patients with renal failure, concentration of carbamazepine-E is significantly increased. Therapeutic drug monitoring (TDM) of carbamazepine is strongly recommended. Effective levels of carbamazepine are 4–12 μg/mL, although considerable variability and toxic effects (especially CNS related) can be seen at these levels. The toxic level is considered as >20 μg/mL. TDM of carbamazepine-E is not routinely performed mainly due to lack of commercially available immunoassays. However, measuring the epoxide metabolite level may be beneficial in patients with renal failure due to potential accumulation of carbamazepine-E. Although the therapeutic range of carbamazepine-E has not been clearly established, serum concentrations exceeding 9 μg/mL are associated with toxicity. Desirable level of carbamazepine-E is 0.4–4 μg/mL [4]. Monitoring carbamazepine-E concentration in patients overdosed with carbamazepine is also recommended, although these levels may not be available fast enough to aid with clinical decision-making [5]. Carbamazepine-E concentration may also increase in patients receiving both valproic acid and carbamazepine due to inhibition of epoxide hydrolase by valproic acid.

Acute overdose with carbamazepine may result in stupor, coma, hyperirritability, convulsions, and respiratory depression. Adverse effects seen with chronic administration include drowsiness, vertigo, ataxia, and blurred vision. The frequency of seizures may increase with chronic overdosage. Other toxic effects include nausea, vomiting, serious hematological effects, hypersensitivity reactions, and water retention. Carbamazepine has been also been reported to cause hyponatremia and decreased thyroid function tests.

Accidental overdose of carbamazepine may cause coma. A 9-year-old girl presented with alleged history of deliberate ingestion of 4 g of carbamazepine, following which she became comatose. A bedside electroencephalogram (EEG) was performed to rule out nonconvulsive status epilepticus, which showed presence of spindle coma (SC) not reactive to noxious stimuli. Following hemodialysis, a second EEG showed absence of SC and complete clinical recovery [6]. Fluconazole may cause carbamazepine toxicity presumably by inhibiting the CYP3A4 isoenzyme [7]. Less well recognized is the possibility that influenza vaccination may significantly increase carbamazepine blood levels causing carbamazepine toxicity [8].

Immunoassay kits are commercially available for TDM of carbamazepine. These assays can be easily adopted on automated analyzers for rapid determination of carbamazepine concentrations in serum or plasma. However, carbamazepine-E cross-reacts with certain carbamazepine assays. Although carbamazepine-E cross-reactivity on commercially available immunoassays typically varies from 0% to 22%, PETINIA (particle-enhanced turbidimetric inhibition immunoassay) carbamazepine assay marketed by Siemens Diagnostics shows a high cross-reactivity of approximately 90%. Parant et al. reported a case where a patient overdosed with carbamazepine showed a carbamazepine level of 42.5 μg/mL using the PETINIA assay but only 29.5 μg/mL with the EMIT 2000 assay. The carbamazepine level measured by using high performance liquid chromatography was 26.2 μg/mL, while the carbamazepine-E concentration was 18.2 μg/mL. The authors concluded that significantly elevated carbamazepine concentration measured by the PETINIA assay was due to significant cross-reactivity of carbamazepine-E with the assay antibody [9]. Hydroxyzine and cetirizine, if present at very high concentrations (e.g., overdose), can falsely increase serum carbamazepine level using the PETINIA assay [10]. However, other immunoassays are not affected [11].

References

[1] Payne PW. Ancestry-based pharmacogenomics, adverse reactions and carbamazepine: is the FDA warning correct? Pharm J 2014;14:473–80.
[2] Marino SE, Birnbaum AK, Leppik IE, Conway JM, et al. Steady-state carbamazepine pharmacokinetics following oral and stable-labeled intravenous administration in epilepsy patients: effects of race and sex. Clin Pharmacol Ther 2012;91:483–8.
[3] Rosa M, Bonnaillie P, Chanteux H. Prediction of drug–drug interactions with carbamazepine-10,11-epoxide using a new in vitro assay for epoxide hydrolase inhibition. Xenobiotica 2016;46:1076–84.
[4] Burianová I, Bořecká K. Routine therapeutic monitoring of the active metabolite of carbamazepine: is it really necessary? Clin Biochem 2015;48:866–9.

[5] Mittag N, Meister S, Berg AM, Walther UI. A case report of a carbamazepine overdose with focus on pharmacokinetic aspects. Pharmacopsychiatry 2016;49:76–8.
[6] Chauhan B, Patanvadiya A, Dash GK. Carbamazepine toxicity-induced spindle coma: a novel case report. Clin Neuropharmacol 2017;40:100–2.
[7] Nair DR, Morris HH. Potential fluconazole-induced carbamazepine toxicity. Ann Pharmacother 1999;33:790–2.
[8] Robertson Jr. WC. Carbamazepine toxicity after influenza vaccination. Pediatr Neurol 2002;26:61–3.
[9] Parant F, Bossu H, Gagnieu MC, Lardet G, Moulsma M. Cross-reactivity assessment of carbamazepine-10,11-epoxide, oxcarbazepine, and 10-hydroxy-carbazepine in two automated carbamazepine immunoassays: PETINIA and EMIT 2000. Ther Drug Monit 2003;25:41–5.
[10] Parant F, Moulsma M, Gagnieu MC, Lardet G. Hydroxyzine and metabolites as a source of interference in carbamazepine particle-enhanced turbidimetric inhibition immunoassay (PETINIA). Ther Drug Monit 2005;27:457–62.
[11] Dasgupta A, Tso G, Johnson M, Chow L. Hydroxyzine and cetirizine interfere with the PENTINA carbamazepine assay but not with the ADVIA CENTEUR carbamazepine assay. Ther Drug Monit 2010;32:112–5.

9.3

Ethosuximide

Chemical properties	
Solubility in H_2O	190 mg/mL
Molecular weight	141.17
pKa	9.3
Melting point	64–65 °C
Dosing	
Recommended dose, adult	250 mg tablet twice a day
Recommended dose, child	Below 6 years of age, 250 mg/day
Monitoring	
Sample	Serum, plasma
Effective concentrations	40–100 μg/mL
Toxic concentrations	>150 μg/mL
Methods	Immunoassay (most common), HPLC, GC
Pharmacokinetic properties	
Oral dose absorbed	100%
Time to peak concentration	1–2 h
Protein bound	22%
Volume of distribution	0.7 L/kg
Half-life, adult	50–60 h
Time to steady state, adult	8–12 d
Half-life, child	30 d
Time to steady state, child	6–10 d

Excretion, urine

	% Excreted	Active	Detected in blood
Unchanged	20	Yes	Yes
2-Hydroxyethyl-2-methylsuccinimide	14	No	No

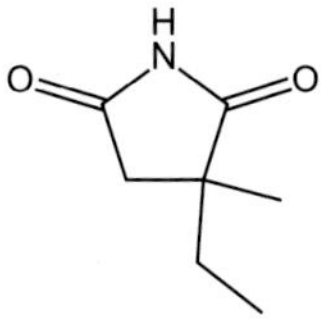

Ethosuximide

Succinimides are antiepileptic drugs (AEDS) used in the treatment of absence seizures. The most effective succinimide is ethosuximide (Zarontin). It has been clearly established that treatment with ethosuximide can significantly reduce the number of absence seizures in children. Methsuximide (Celontin) has been reported to be effective not only in absence seizures but also in certain types of focal and other refractory seizures and as an "add on" drug in intractable epilepsies. Phensuximide (Milontin) was originally thought to possess antiepileptic properties. However, this drug is so rapidly metabolized that it is not an effective AED.

Ethosuximide is a chiral molecule with high bioavailability and low degree of binding to plasma proteins. Ethosuximide has a long elimination half-life (between 40 and 60 h in adults, 30 and 40 h in children). The therapeutic range is established at 40–100 μg/mL (283–708 μmol/L), but the upper limit is probably underestimated, with some patients experiencing effective seizure control with minimal side effects at higher concentrations [1]. Ethosuximide is effective only in the treatment of absence-type seizures and is not effective in the treatment of major motor seizures. Moreover, with regards to both efficacy and tolerability, ethosuximide represents the optimal initial empirical monotherapy for children and adolescents with absence seizures. However, if absence and generalized tonic–clonic seizures coexist, valproic acid should be preferred, as ethosuximide has minimal or no efficacy for tonic–clonic seizures [2].

Ethosuximide is metabolized primarily by hepatic cytochrome P450 (CYP) 3A4 enzyme isoform to the inactive metabolite, 2-hydroxyethyl-2-methylsuccinimide. CYP3A4 inducers such as carbamazepine, phenobarbital and phenytoin decrease the levels and therapeutic efficacy of ethosuximide. The enhancement of ethosuximide clearance in patients comedicated with these enzyme-inducing anticonvulsants is clinically significant, requiring higher ethosuximide dosages in order to achieve therapeutic concentrations in these patients [3]. Conversely, CYP3A4 inhibitors such as azole antifungals, isoniazid, erythromycin, ciprofloxacin, protease inhibitors, quinidine, and verapamil may increase the effects and serum levels of ethosuximide [4]. The addition of valproic acid to ethosuximide therapy for epilepsy caused increased serum concentration of ethosuximide in four of five patients observed. Ethosuximide levels increased from 73 to 112 μg/mL (53% higher), with concomitant toxicity. Both were reversed by reduction of the ethosuximide dose from 27.4 to 20.4 mg/kg. Serum concentrations of these two drugs should be monitored closely when they are given together [5].

The most common side effects involve the gastrointestinal (nausea, vomiting, and anorexia) and central nervous systems (drowsiness, lethargy, euphoria, dizziness, headache, and hiccough). These seem to be dose related and often some tolerance develops during chronic therapy. Patients with prior history of psychiatric disorders have been reported to experience behavioral effects such as restlessness, agitation, anxiety, aggressiveness, and inability to concentrate [6]. Therapeutic drug monitoring of ethosuximide is recommended, with immunoassays being the most common analytical method [7].

References

[1] Bentué-Ferrer D, Tribut O, Verdier MC. Therapeutic drug monitoring of ethosuximide. Therapie 2012;67:391–6. (Article in French).

[2] Brigo F, Igwe SC. Ethosuximide, sodium valproate or lamotrigine for absence seizures in children and adolescents. Cochrane Database Syst Rev 2017;14(2):CD003032.

[3] Giaccone M, Bartoli A, Gatti G, Marchiselli R, et al. Effect of enzyme inducing anticonvulsants on ethosuximide pharmacokinetics in epileptic patients. Br J Clin Pharmacol 1996;41:575–9.

[4] Perucca E. Clinically relevant drug interactions with antiepileptic drugs. Br J Clin Pharmacol 2005;61(3):246–65.

[5] Mattson RH, Cramer JA. Valproic acid and ethosuximide interaction. Ann Neurol 1980;7:583–4.

[6] Mattson RH. Efficacy and adverse effects of established and new antiepileptic drugs. Epilepsia 1995;36(Suppl. 2):S13–26.

[7] Patsalos PN, Spencer EP, Berry DJ. Therapeutic drug monitoring of antiepileptic drugs in epilepsy: a 2018 update. Ther Drug Monit 2018;40:526–48.

9.4
Gabapentin

Chemical properties	
Solubility in H_2O	4.49 mg/mL
Molecular weight	171.2
pKa	N/A
Melting point	162–166 °C
Dosing	
Recommended dose, adult	800–1800 mg/d
Recommended dose, child	25–35 mg/kg/d
Monitoring	
Sample	Serum
Effective concentrations	2–20 μg/mL
Toxic concentrations	>20 μg/mL
Methods	HPLC, LC–MS, GC, GCMS, immunoassay
Pharmacokinetic properties	
Oral dose absorbed	50–60%
Time to peak concentration	1.5–4 h
Protein bound	<3%
Volume of distribution	
1.0 Adult	0.85 L/kg
1.0 Child 4–6	3.12 L/kg
1.0 Child 7–12	1.53 L/kg
Half-life, adult	5–9 h
Time to steady state, adult	1–2 d
Half-life, child	4–6 h
Time to steady state, child	1–2 d

Excretion, urine

	% Excreted	Active	Detected in blood
Unchanged	100	Yes	Yes

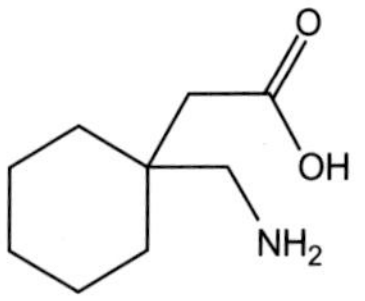

Gabapentin (1-[aminomethyl] cyclohexaneacetic acid; Neurontin) is a structural analog of the naturally occurring neurotransmitter gamma-aminobutyric acid (GABA). Gabapentin was approved in 1993 by the United States Food and Drug Administration (FDA) initially as adjunct therapy for treatment of partial seizures in pediatric patients 3 years of age and older and to control partial seizures with or without secondary generalization in patients older than 12 years of age. It is also used for monotherapy of newly diagnosed focal epilepsy. In 2004, gabapentin was also approved as an analgesic for post-herpetic neuralgia. The European Medicines Agency approved gabapentin in 2006 for epilepsy and certain types of neuropathic pain and the United Kingdom National Institute for Clinical Excellence (NICE) recommends gabapentin as a first-line treatment for all neuropathic pain. Off-label use of gabapentin is estimated to be 83–95% as it is widely used to treat various disorders including insomnia, neuropathic pain conditions, drug and alcohol addiction, anxiety, bipolar disorder, borderline personality disorder, menopausal conditions, vertigo, pruritic disorders, and migraines. Interestingly, gabapentin does not bind to $GABA_A$ receptors (receptor target for barbiturates and benzodiazepines) despite structural similarity to GABA; however, it can increase GABA concentrations in the brain and can decrease glutamate concentration. Gabapentin also does not bind opioid or cannabinoid receptors. A proposed molecular target for gabapentin is binding to the alpha-2-delta type 1 subunit of voltage gated calcium channels, selectively inhibiting Ca^{2+} influx through these voltage-operated Ca^{2+} channels. This molecular action would reduce postsynaptic excitability and decrease the release of excitatory neurotransmitters.

Gabapentin is safely tolerated over a very broad range of doses from approximately 800–1800 mg/day (although package inserts suggest that some patients may safely tolerate doses as high as 3600 mg/day). In clinical practice, dosing is typically titrated starting from lower doses (i.e., <400 mg/day) and moving rapidly upward. However, most guidelines recommend not exceeding maximum dosage of 1800 mg/day. Although gabapentin was initially viewed as a drug with low likelihood for abuse, recent studies indicate that gabapentin has abuse potential [1].

Gabapentin is rapidly absorbed in part by the saturable L-amino acid transport system that impacts bioavailability across clinically used dosing. For example, bioavailability is approximately 60% for a 900 mg dose but only 27% for a 4800 mg dose per day. Coadministration of antacids causes a 20% decrease in bioavailability of gabapentin. The histamine-2 receptor blockers decrease renal clearance of gabapentin [2]. Frequency of dosage is sometimes increased from 2 to 3 doses per day to 4 doses per day in order to compensate for saturable bioavailability. Peak plasma levels are usually reached within 4 h. The volume of distribution and clearance rate of gabapentin are age-dependent, with younger children (4–6 years) having a larger volume of distribution (3.12 vs 1.53 L/kg) and higher clearance rate (9.37 vs 5.92 L/kg/day) than children aged 7–12 years [2]. Therefore, gabapentin requires age-adjusted dosing [3]. Half-life is prolonged in patients with renal failure. Dose adjustment is mandatory in patients with creatinine clearance <60 mL/min and also in the elderly due to reduced renal function [4]. Gabapentin does not bind to plasma proteins.

Gabapentin does not undergo hepatic metabolism and is excreted unchanged in urine which serves as an advantage in patients with liver disorders [5]. Another pharmacokinetic advantage is that gabapentin does not affect plasma concentrations of other drugs including AEDs and oral contraceptives. It is also well tolerated and may be rapidly titrated [6]. Adverse experiences most commonly noted during gabapentin therapy include somnolence, dizziness, and ataxia. Weight gain was sometimes reported with higher doses of gabapentin, and pediatric reports cite occasional patients showing prominent behavioral changes including hyperactivity, irritability, and agitation [7].

Although not routinely monitored, pharmacokinetic variability of gabapentin is extensive, requiring a need for individualization of therapy regardless of indication. Therapeutic drug monitoring (TDM) is most common in patients with epilepsy [8]. The therapeutic range is 2–20 μg/mL [9]. Acute gabapentin overdose may be fatal. A 47-year-old female was found dead at work with her daughter's bottle of gabapentin (600 mg) where 26 tablets were missing. Her postmortem peripheral blood concentration of gabapentin was 37 μg/mL while central blood gabapentin concentration was 32 μg/mL. In addition her liver gabapentin level was 26 mg/kg. The authors concluded that fatality appeared to be associated with isolated and acute gabapentin ingestion [10].

An immunoassay for measurement of gabapentin in plasma/serum recently became available [11]. Chromatographic methods such as gas chromatography/electron ionization mass spectrometry or liquid chromatography combined with mass spectrometry or tandem mass spectrometry can also be used for TDM of gabapentin [12].

References

[1] Smith RV, Havens JR, Walsh SL. Gabapentin misuse, abuse and diversion: a systematic review. Addiction 2016;111:1160–74.

[2] McLean MJ. Clinical pharmacokinetics of gabapentin. Neurology 1994;44(Suppl. 5): S17–22.

[3] McLean MJ. Gabapentin. Epilepsia 1995;36(Suppl 2):S73–86.

[4] Armijo JA, Pena MA, Adin J, Vega-Gil N. Association between patient age and gabapentin serum concentration to dose-ratio: a preliminary multivariant analysis. Ther Drug Monit 2004;26:633–7.

[5] Mason BJ, Quello S, Shadan F. Gabapentin for the treatment of alcohol use disorder. Expert Opin Investig Drugs 2018;27:113–24.

[6] Perucca E. Clinically relevant drug interactions with antiepileptic drugs. Br J Clin Pharmacol 2005;61(3):246–65.

[7] Morris GL. Gabapentin. Epilepsia 1999;40(Suppl 5):S63–70.

[8] Johannessen Landmark C, Beiske G, Baftiu A, Burns ML, Johannessen SI. Experience from therapeutic drug monitoring and gender aspects of gabapentin and pregabalin in clinical practice. Seizure 2015;28:88–91.

[9] Patsalos PN, Spencer EP, Berry DJ. Therapeutic drug monitoring of antiepileptic drugs in epilepsy: a 2018 update. Ther Drug Monit 2018;40:526–48.

[10] Cantrell FL, Mena O, Gary RD, McIntyre IM. An acute gabapentin fatality: a case report with postmortem concentrations. Int J Legal Med 2015;129:771–5.

[11] Juenke JM, Wienhoff KA, Anderson BL, McMillin GA, Johnson-Davis KL. Performance characteristics of the ARK diagnostics gabapentin immunoassay. Ther Drug Monit 2011;33:398–401.

[12] Jacob S, Nair AB. An updated overview on therapeutic drug monitoring of recent antiepileptic drugs. Drugs RD 2016;16:303–16.

9.5

Lamotrigine

Chemical properties	
Solubility in H_2O	0.17 mg/mL
Molecular weight	256.091
pKa	5.7
Melting point	216–218 °C
Dosing	
Recommended dose, adult	200–400 mg/d
Recommended dose, child	0.6–1.2 mg/kg/d
Monitoring	
Sample	Serum, plasma
Effective concentrations	3–15 μg/mL
Toxic concentrations	>20 μg/mL
Methods	Immunoassay, HPLC, LCMS
Pharmacokinetic properties	
Oral dose absorbed	Close to 100%
Time to peak concentration	1–3 h
% Protein bound	55%
Volume of distribution	Approximately 0.87 L/kg
Half-life, adult	24–34 h
Time to steady state, adult	3–10 d
Half-life, child	~30 h
Time to steady state, child	3–10 d

Excretion, urine

	% Excreted	Active	Detected in blood
Unchanged	10	Yes	Yes
N-2 glucuronide	70	No	No
N-5 glucuronide	10	No	No

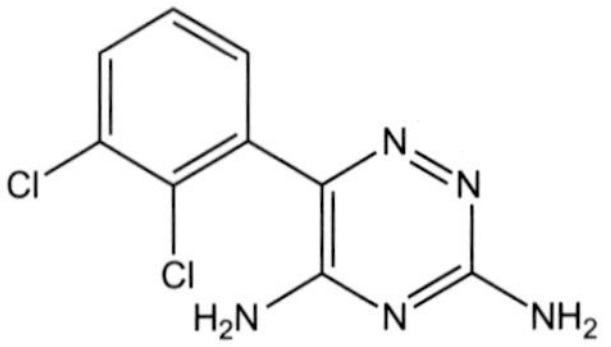

Lamotrigine

Lamotrigine (3,5-diamino-6-[2,3-dichlorophenyl]-1,2,4-triazine; Lamictal) was approved for use by the United States Food and Drug Administration (FDA) in 1994 and is effective as monotherapy or adjunctive therapy in partial seizures, primary and secondarily generalized tonic–clonic seizures, absence seizures, and attacks associated with Lennox–Gastaut syndrome. In 2003, the FDA approved lamotrigine for treating bipolar disorder. Lamotrigine stabilizes presynaptic neuronal membranes by acting at voltage-sensitive sodium channels and modulating presynaptic transmitter release of excitatory neurotransmitters such as aspartate and glutamate [1]. In contrast to some other medications used to treat seizure disorders that have clear associations with birth defects or other adverse effects on the developing fetus (e.g., phenytoin, valproic acid), lamotrigine has an excellent safety record in pregnancy and is a drug of choice for management of pregnant women with epilepsy [2].

After oral administration, lamotrigine is well absorbed with bioavailability close to 100%. Peak plasma levels are reached within 1–3 h. Lamotrigine is approximately 55% bound to serum proteins. Lamotrigine has a volume of distribution of ~0.87 L/kg in healthy adults. The half-life in healthy individuals is 24–34 h. The drug is highly metabolized in the liver, predominately by UDP-glucuronosyltransferase. Most of the drug is ultimately excreted in urine as glucuronic acid conjugates [1].

Many drug–drug interactions have been reported with lamotrigine. Drugs which are inducers of hepatic drug-metabolizing enzyme expression (e.g. phenytoin, phenobarbital, primidone, oxcarbazepine, and olanzapine) increase metabolism of lamotrigine, causing reduced serum levels and elimination half-life (8–20 h in the presence of enzyme inducing antiepileptic drugs). In contrast, inhibitors of liver drug-metabolizing enzymes such as valproic acid, fluoxetine, and sertraline inhibit metabolism of lamotrigine, causing elevated serum levels of lamotrigine. Valproic acid and sertraline may cause approximately twofold increases in serum lamotrigine levels. Lamotrigine half-life may be increased to 60 h when combined with valproic acid [3–5]. Hence, lamotrigine dosage should be decreased in the presence of valproic acid use. Oral contraceptives can cause serum lamotrigine decreases of up to 50%. Conversely, a rebound increase in serum lamotrigine levels may occur when the contraceptive pill is discontinued. For oral contraceptive regimens that include a hormone-free week, cyclic effects on the lamotrigine metabolism can occur. Cessation of oral contraceptives can lead to an 84% increase in the concentration of lamotrigine due to altered lamotrigine glucuronidation. The change in lamotrigine concentrations could be observed within 1 week of the shift of treatment [6].

Introduction of lamotrigine requires slow titration to reduce risk of severe dermatological reactions including Stevens-Johnson syndrome, an adverse side effect that has warranted a black box warning by the FDA for lamotrigine. Skin rashes occur in 5–10% of patients on lamotrigine, with severe dermatological reactions in approximately 0.1% of patients. Severe reactions typically occur within the first 2–8 weeks of therapy. Slow titration of the drug allows for recognition of early signs of a rash and then immediate discontinuation of the drug. Recommended dosage schedule includes low doses (25 mg) every other day for 2 weeks progressing to once a day weekly and then gradually increasing dosage as needed up to no >400 mg per day [7]. In addition to development of a rash, other adverse effects of lamotrigine toxicity include blurred vision, double vision, clumsiness, dizziness and drowsiness [1].

Therapeutic drug monitoring (TDM) of lamotrigine is beneficial for optimizing dosage of lamotrigine. TDM-based dosage adjustment is especially beneficial for patients undergoing hemodialysis or in patients with severe liver impairment. Lamotrigine serum levels have been shown to increase by 90% from pre-pregnancy baseline to third trimester. Therefore, TDM of lamotrigine is essential in pregnant women receiving lamotrigine. Dosage adjustment should be made based on trough lamotrigine level at least once a month. Lamotrigine clearance is higher in children; therefore, TDM is recommended for pediatric patients. The therapeutic range for treatment of epilepsy is considered to be 2.5–15 μg/mL. Immunoassays are commercially available for TDM of lamotrigine. In addition, high performance liquid chromatography, liquid chromatography combined with mass spectrometry, or gas chromatography based methods may also be used for TDM of lamotrigine [4].

References

[1] Yasam VR, Jakki SL, Senthil V, Eswaramoorthy M, et al. A pharmacological overview of lamotrigine for the treatment of epilepsy. Expert Rev Clin Pharmacol 2016;9: 1533–46.

[2] Pariente G, Leibson T, Shulman T, Adams-Webber T, et al. Pregnancy outcomes following in utero exposure to lamotrigine: a systematic review and meta-analysis. CNS Drugs 2017;31:439–50. [Erratum in: CNS Drugs. 2017;31:451].

[3] Johannessen LC, Patsalos PN. Drug interactions involving the new second- and third-generation antiepileptic drugs. Expert Rev Neurother 2010;10:119–40.
[4] Jacob S, Nair AB. An updated overview on therapeutic drug monitoring of recent antiepileptic drugs. Drugs RD 2016;16:303–16.
[5] Biton V. Pharmacokinetics, toxicology and safety of lamotrigine in epilepsy. Expert Opin Drug Metab Toxicol 2006;2:1009–18.
[6] Christensen J, Petrenaite V, Altterman J, Sidenius P, et al. Oral contraceptives induce lamotrigine metabolism: evidence from a double blind placebo controlled trial. Epilepsia 2007;48:484–9.
[7] Frey N, Bodmer M, Bircher A, Rüegg S, et al. The risk of Stevens-Johnson syndrome and toxic epidermal necrolysis in new users of antiepileptic drugs. Epilepsia 2017;58:2178–85.

9.6

Levetiracetam

Chemical properties	
Solubility in H_2O	1040 mg/mL
Molecular weight	170.209
pKa	NA
Melting point	117 °C
Dosing	
Recommended dose, adult	1000–3000 mg/d
Recommended dose, child	10–20 mg/kg/d
Monitoring	
Sample	Serum, plasma
Effective concentrations	10–60 μg/mL
Toxic concentrations	Not established
Methods	HPLC, GC, LCMS
Pharmacokinetic properties	
Oral dose absorbed	95–100%
Time to peak concentration	1 h
Protein bound	<5%
Volume of distribution	0.5–0.7 L/kg
Half-life, adult	6–8 h
Time to steady state, adult	2 d
Half-life, child	5–7 h
Time to steady state, child	1–2 d

Excretion, urine

	% Excreted	Active	Detected in blood
Unchanged	64	Yes	Yes
2-Pyrrolidone-*N*-butyric acid	27	No	No
Hydroxyl-2-oxopyrrolidine	2	No	No
2-Oxopyrrolodine	1	No	No

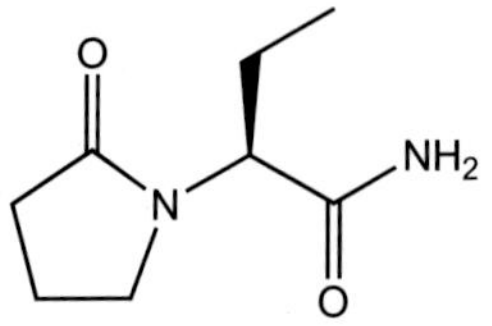

Levetiracetam

Levetiracetam (2-(2-oxopyrrolidin-1-yl) butanamide; Keppra) was approved for use in 1999 by the United State Food and Drug Administration (FDA) as monotherapy as well as for adjunctive treatment of adults with partial-onset seizures and has been effective in reducing seizures for both partial and primary generalized epilepsy. It is also used for the management of juvenile myoclonic and generalized tonic clonic seizure in patients 12 years of age and older [1]. Its mechanism of action is unknown but seems to be different from other available antiepileptic drugs (AEDs).

After oral ingestion, levetiracetam is rapidly absorbed, with peak concentration occurring after 1.3 h, and its bioavailability is >95%. Levetiracetam is not bound to plasma proteins and has a volume of distribution of 0.5–0.7 L/kg. Plasma concentrations increase in proportion to dose over the clinically relevant dosage range. Steady-state blood concentrations are achieved within 24–48 h. The elimination half-life in both adult volunteers and adults with epilepsy varies from 6 to 8 h. The half-life in children is 5–7 h, and in elderly patients half-life may be prolonged to 10 to 11 h. Approximately 34% of a levetiracetam dose is metabolized and 66% is excreted in urine unchanged. Interestingly, levetiracetam metabolism is not hepatic but occurs primarily in blood by hydrolysis to 2-pyrrolidone-*N*-butyric acid (also known as L057). Autoinduction of metabolism is not seen with levetiracetam. The predominant renal excretion of levetiracetam means that clearance correlates with creatinine clearance. Therefore, dosage adjustments are necessary for patients with moderate to severe renal impairment. To date, no clinically relevant pharmacokinetic interactions between AEDs and levetiracetam have been identified, including lack of interaction with digoxin, warfarin, and oral contraceptives. However, adverse pharmacodynamic interactions with carbamazepine and topiramate have been reported [2].

Oral levetiracetam 1000, 2000 and 3000 mg/day as adjunctive therapy is generally well tolerated with relatively few adverse effects. The most commonly reported adverse effects during clinical trials were central nervous system-related and included somnolence, asthenia, headache and dizziness. There have been reported psychotic episodes associated with levetiracetam treatment, and caution is recommended when considering its use in patents who may be predisposed to psychiatric disorders [3, 4].

Therapeutic drug monitoring (TDM) of levetiracetam may be useful to monitor compliance, adjust dosage, and avoid overdose. For a daily dosage of 1000–3000 mg, a target therapeutic range of 12–46 μg/mL has been suggested, although some patients may respond well to higher serum concentrations without adverse effects. Serum levetiracetam levels may be reduced by approximately 60% during pregnancy. Immunoassays for measurement of levetiracetam concentrations in plasma/serum recently became available [5]. Chromatographic methods such as high performance liquid chromatography, gas chromatography or gas chromatography combined with mass spectrometry can also be used for measuring levetiracetam serum or plasma levels [6]. More recently, serum levetiracetam determination using liquid chromatography combined with tandem mass spectrometry has been described [7].

References

[1] Ito S, Yano I, Hashi S, Tsuda M, et al. Population pharmacokinetic modeling of levetiracetam in pediatric and adult patients with epilepsy by using routinely monitored data. Ther Drug Monit 2016;38:371–8.

[2] Patsalos PN. Clinical pharmacokinetics of levetiracetam. Clin Pharmacokinet 2004; 43(11):707–24.

[3] Dooley M, Plosker GL. Levetiracetam. A review of its adjunctive use in the management of partial onset seizures. Drugs 2000;60:871–93.

[4] Welty TE, Gidal BE, Ficker DM, Privitera MD. Levetiracetam: a different approach to the pharmacotherapy of epilepsy. Ann Pharmacother 2002;36:296–304.

[5] Bianchi V, Arfini C, Vidali M. Therapeutic drug monitoring of levetiracetam: comparison of a novel immunoassay with an HPLC method. Ther Drug Monit 2014;36:681–5.

[6] Jacob S, Nair AB. An updated overview on therapeutic drug monitoring of recent antiepileptic drugs. Drugs RD 2016;16:303–16.

[7] Van Matre ET, Mueller SW, Fish DN, MacLaren R, et al. Levetiracetam pharmacokinetics in a patient with intracranial hemorrhage undergoing continuous veno-venous hemofiltration. Am J Case Rep 2017;18:458–62.

9.7

Oxcarbazepine

Chemical properties	
Solubility in H_2O	0.308 mg/mL
Molecular weight	252.268
pKa	NA
Melting point	215.5 °C
Dosing	
Recommended dose, adult	1200–2400 mg/d
Recommended dose, child	10–30 mg/kg/d
Monitoring	
Sample	Serum, plasma
Effective concentrations of MHD metabolite	3–35 μg/mL
Toxic concentrations	Not established
Methods	HPLC, GC, LCMS, GCMS
Pharmacokinetic properties	
Oral dose absorbed	99%
Time to peak concentration MHD	3–5 h
Protein bound of MHD	40%, albumin
Volume of distribution of MHD	7.8–12.5 L/kg
Half-life, MHD adult	8–15 h
Time to steady state, MHD adult	2 d
Half-life, MHD child <6 year	5–9 h
Time to steady state, child <6 year	1 d

Excretion, urine

	% Excreted	Active	Detected in blood
Unchanged	<1	Yes	No
10-Monohydroxy carbamazepine (MHD)	27	Yes	Yes
MHD glucuronide	49	Yes	Yes
10,11 Dihyroxy metabolite (DHD)	3	No	No
Oxcarbazepine and MHD conjugates	13	No	No

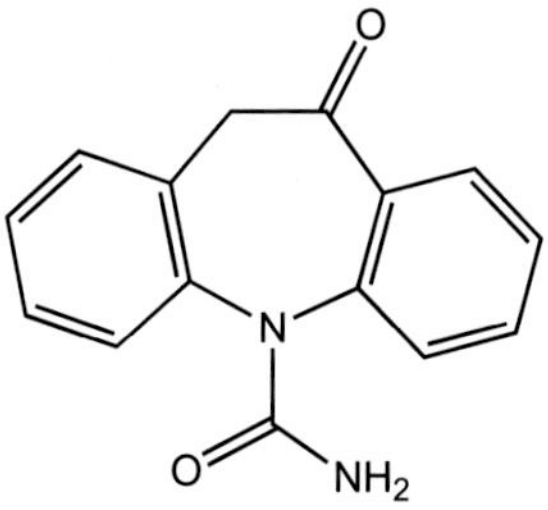

Oxacarbazepine

Oxcarbazepine (10, 11-dihydro-10-oxo-carbamazepine; Trileptal), a 10-keto analog of carbamazepine, is used for monotherapy and adjunctive treatment for partial seizures and generalized tonic–clonic seizures in children and adults. The slight structural difference between oxcarbazepine and carbamazepine was introduced to avoid metabolism to the carbamazepine-10,11-epoxide metabolite and thus reduce adverse effects while retaining efficacy. The exact mechanism of action of oxcarbazepine is unknown, but it is thought to have the same or similar mechanism as carbamazepine involving sodium channel inhibition but at much lower concentrations. Oxcarbazepine has been shown to be equivalent to carbamazepine and phenytoin in efficacy and superior in individually determined dose-related tolerability [1].

Oxcarbazepine is a prodrug which undergoes rapid pre-systemic reduction metabolism to its active monohydroxy metabolite 10-hydroxycarbazepine (MHD). Therapeutic drug monitoring (TDM) of oxcarbazepine focuses on MHD. The bioavailability of oxcarbazepine assessed from MHD plasma concentration data is 99%, and its apparent volume of distribution is 7.8–12.5 L/kg in epileptic patients. MHD concentrations peak at 3–5 h, and protein binding of MHD is approximately 40% (primarily albumin). However, the protein binding is approximately 59% for oxcarbazepine. Most of the administered dose of oxcarbazepine (79%) is eventually excreted through the kidneys as either unchanged MHD or glucuronide conjugates of MHD. <1% is excreted as unchanged oxcarbazepine, and 9% as inactive glucuronide conjugates of oxcarbazepine. Approximately 4% of MHD is further oxidized with formation of the inactive metabolite 10,11-dihydro-10,11-trans-dihydroxycarbazepine (DHD) [1, 2].

The half-life of MHD in adults is 8–15 h, and this is significantly reduced in children. Children aged 2–5 years have a clearance rate approximately twice that of adults and the rate for children 6–12 years old is between that of adults and younger children. Body weight-normalized daily doses for children younger than 6 years of age should be approximately 50% higher than doses administered to adults and older children in order to achieve comparable plasma MHD concentrations [3].

Unlike carbamazepine, phenobarbital, phenytoin, and primidone, oxcarbazepine metabolism is not induced or inhibited by the cytochrome P450 (CYP) system. Fewer adverse interactions with other concomitantly given medications occur because of oxcarbazepine's route of metabolism. [4]. However, high doses of oxcarbazepine and MHD do inhibit CYP2C19 which can cause increases in concentrations of drugs such as phenobarbital and phenytoin that are metabolized by this enzyme. Oxcarbazepine and MHD induce CYP3A4 and CYP3A5 which are responsible for metabolism of dihydropyridine calcium antagonists and oral contraceptives, resulting in lower plasma concentrations of these drugs. Therefore, concurrent use of oxcarbazepine with hormonal oral contraceptives may render them less effective. Unlike carbamazepine, oxcarbazepine does not induce its own metabolism [5, 6]. Common adverse effects of oxcarbazepine included somnolence, headache, dizziness, rash, weight gain, gastrointestinal disturbances, and hyponatremia. Hyponatremia occurs with greater frequency in elderly patients [1].

Because oxcarbazepine is rapidly converted into MHD after oral administration, monitoring MHD concentrations reflect the pharmacological effects of oxcarbazepine. Based on the available evidence, TDM of MHD is not routinely warranted but may be beneficial in optimizing seizure control at the extremes of age, during pregnancy, in renal insufficiency, or to determine the significance of potential drug interactions or rule out noncompliance [7]. Immunoassays for MHD have recently become commercially available; one from Sekisui (https://www.sekisuidiagnostics.com/products-all/oxcarbazepine-metabolite-assay/) and another from ARK (http://arktdm.com/products/epilepsy/oxcarbazepine/pdfs/ARK_Oxcarbazepine_Metabolite_Assay_Rev03_February_2017.pdf).

Li et al. developed a high performance liquid chromatography combined with ultraviolet detection at 254 nm for analysis of MHD in both serum and saliva. The authors observed good correlation between plasma and saliva MHD levels in Chinese children, indicating that salivary monitoring of MHD is a feasible approach for TDM of oxcarbazepine [8].

References

[1] May TW, Korn-Merker E, Rambeck B. Clinical pharmacokinetics of oxcarbazepine. Clin Pharmacokinet 2003;42(12):1023–42.

[2] Antunes NJ, van Dijkman SC, Lanchote VL, Wichert-Ana L, et al. Population pharmacokinetics of oxcarbazepine and its metabolite 10-hydroxycarbazepine in healthy subjects. Eur J Pharm Sci 2017;109S:S116–23.

[3] Rey E, Bulteau C, Motte J, Tran A, et al. Oxcarbazepine pharmacokinetics and tolerability in children with inadequately controlled epilepsy. J Clin Pharmacol 2004;44:1290–300.
[4] Johannessen Landmark C, Svendsen T, Dinarevic J, Kufaas RF, et al. The impact of pharmacokinetic interactions with eslicarbazepine acetate versus oxcarbazepine and carbamazepine in clinical practice. Ther Drug Monit 2016;38:499–505.
[5] Perucca E. Clinically relevant drug interactions with antiepileptic drugs. Br J Clin Pharmacol 2005;61(3):246–65.
[6] Bialer M. The pharmacokinetics and interactions of new antiepileptic drugs: an overview. Ther Drug Monit 2005;27:722–6.
[7] Bring P, Ensom MH. Does oxcarbazepine warrant therapeutic drug monitoring? A critical review. Clin Pharmacokinet 2008;47:767–78.
[8] Li RR, Sheng XY, Ma LY, Yao HX, et al. Saliva and plasma monohydroxycarbamazepine concentrations in pediatric patients with epilepsy. Ther Drug Monit 2016;38:365–70.

9.8

Phenobarbital

Chemical properties	
Solubility in H_2O	1.0 mg/mL
Molecular weight	230.23
pKa	7.3
Melting point	174–178 °C

Dosing	
Recommended dose, adult	60–200 mg
Recommended dose, child	3–6 mg/kg/d

Monitoring	
Sample	Serum, plasma
Effective concentrations	15–40 μg/mL
Toxic concentrations	>40 μg/mL
Methods:	Immunoassay, HPLC, GC

Pharmacokinetic properties	
Oral dose absorbed	80–100%
Time to peak concentration	2–18 h
Protein bound	40%, albumin
Volume of distribution	0.7 L/kg
Half-life, adult	81–117 h
Time to steady state, adult	17–24 d
Half-life, child	40–70 h (increased in neonates)
Time to steady state, child	8–15 d

Excretion, urine

	% Excreted	Active	Detected in blood
Unchanged	20–30	Yes	Yes
p-Hydroxyphenobarbital and glucuronide	25–50	No	Yes

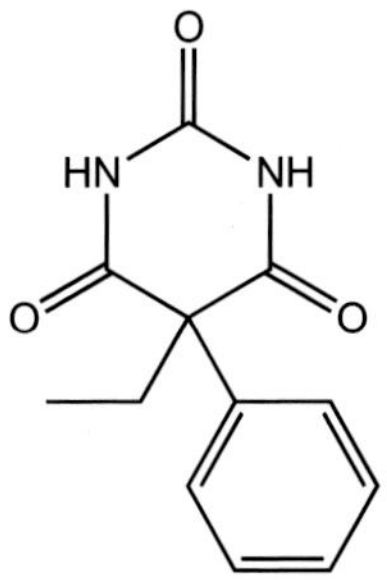

Phenobarbital

Phenobarbital, derived from barbituric acid, has been prescribed as an antiepileptic drug (AED) since 1912. Although a spectrum of new AEDs are available, phenobarbital remains in use as it is inexpensive, often effective, and has relatively low toxicity. Therefore, >100 years after initial introduction into clinical use, phenobarbital is still the most widely prescribed AED in the developing world and remains a cost-effective choice for people with epilepsy in some industrialized countries [1].

Phenobarbital is used to treat generalized tonic–clonic, partial, and febrile seizures. Other agents such as carbamazepine and phenytoin given alone or in combination are often used as the first-line drug(s) of choice with phenobarbital as an alternative if the initial agents fail. Previous studies have demonstrated that phenobarbital is as effective in monotherapy as phenytoin and carbamazepine. Sedation is the most common adverse effect of phenobarbital therapy, although tolerance may develop with chronic treatment [2]. Nystagmus, ataxia, rash, and hypoprothrombinemia have been reported in the newborn of mothers treated with phenobarbital. Other reported adverse effects of phenobarbital include megaloblastic anemia, agitation and confusion in the elderly, and hyperactivity and irritability in children. Phenobarbital was associated with a higher rate of drug withdrawal compared to other AEDs, although there was no evidence to suggest that phenobarbital caused more adverse events compared to carbamazepine, valproic acid or phenytoin. However, in the case of pregnant women, it is important for clinicians to evaluate the benefits and risks of phenobarbital administration before making a final recommendation [3]. Overdose of phenobarbital (a long-acting barbiturate) can cause prolonged central nervous system and cardiovascular depression. Seizure aggravation can be induced by phenobarbital toxicity. An electroencephalographic evaluation seems reasonable in cases of phenobarbital poisoning to detect underlying seizure and help direct further management [4]. Acute fatalities may occur after ingestion of as little as 6 g of phenobarbital where blood concentration was 65 μg/mL. Death may result from cardiorespiratory arrest, pulmonary and cerebral edema or pneumonia [5]. An uncommon side effect of phenobarbital therapy is the development of secondary pellagra, a deficiency of niacin (vitamin B_3) resulting from deficiency of the precursor tryptophan. Pellagra can manifest as dermatologic, gastrointestinal, neurological and psychiatric signs and symptoms. Carbamazepine and phenytoin have also been associated with secondary pellagra. The mechanism by which carbamazepine, phenobarbital, and phenytoin produce secondary pellagra may be interference with metabolism pathways involving tryptophan. There is a case report where a 29-year-old female patient with a 3 year history of discoloration of hands and feet diagnosed with pellagra due to treatment with phenobarbital. The woman eventually died [6].

Phenobarbital metabolism includes oxidation to form the para-hydroxy inactive metabolite and excretion as the glucuronide. The main enzyme systems involved cytochrome P450 (CYP) 2C9 and CYP2C19. Phenobarbital (like carbamazepine, phenytoin, and primidone) stimulates the expression of a variety of CYP enzymes including CYP1A2, CYP2C9, CYP2C19, and CYP3A4. Hence, phenobarbital therapy can reduce the serum concentration of other AEDs and drugs metabolized by CYP enzymes. AEDs potentially impacted by the CYP enzyme induction of phenobarbital include carbamazepine, clonazepam, ethosuximide, lamotrigine, oxcarbazepine, topiramate, and valproic acid. In addition, the metabolism of antimicrobials, antipsychotics, bilirubin, cardiovascular drugs, immunosuppressants, vitamins D, vitamin K, warfarin, and tricyclic antidepressants may be impacted by phenobarbital therapy. Conversely, discontinuation of phenobarbital may be followed by an increase to toxic concentrations of concurrently administered AEDs and other drugs. Phenobarbital concentrations are in turn affected by other drugs, most significantly other AEDs. Interactions between phenytoin and phenobarbital are variable. Phenobarbital may increase phenytoin metabolism (by induction of the liver microsomal enzyme system), decrease metabolism (by competitive inhibition of enzymes), and decrease absorption of oral phenytoin. Concurrent administration of phenytoin has been reported to increase phenobarbital concentrations. Valproic acid treatment causes a consistent and significant (up to 40%) increase in phenobarbital concentrations due to decreased metabolism and phenobarbital dosage may need to be reduced by as much as 80% [7–9].

Therapeutic drug monitoring (TDM) of phenobarbital is recommended for the following indications: (1) to establish that the drug has had an adequate clinical trial before discontinuing therapy because of incomplete seizure control; (2) if compliance or toxicity is questioned; (3) if a drug adjustment is made, or (4) when a potentially interacting drug is given. Therapeutic range of phenobarbital is 15–40 μg/mL and plasma protein binding is approximately 40%, mostly to albumin. Close monitoring is important in the neonate and during puberty, as phenobarbital clearance changes during development. The long elimination half-life of phenobarbital (approximately 2–3 days) means that drug levels may be obtained before steady state is achieved. Before increasing a dose, it is important the relevant serum concentration be measured during steady state [10]. Immunoassays are commercially available for TDM of phenobarbital.

References

[1] Savica R, Beghi E, Mazzaglia G, Innocenti F, et al. Prescribing patterns of antiepileptic drugs in Italy: a nationwide population-based study in the years 2000–2005. Eur J Neurol 2007;14:1317–21.
[2] Brodie MJ, Kwan P. Current position of phenobarbital in epilepsy and its future. Epilepsia 2012;53(Suppl 8):40–6.
[3] Zhang LL, Zeng LN, Li YP. Side effects of phenobarbital in epilepsy: a systematic review. Epileptic Disord 2011;13:349–65.
[4] Hassanian-Moghaddam H, Ghadiri F, Shojaei M, Zamani N. Phenobarbital overdose presenting with status epilepticus: a case report. Seizure 2016;40:57–8.
[5] Bruce AM, Smith H. The investigation of phenobarbitone, phenytoin, primidone in the death of epileptics. Med Sci Law 1977;17:195–9.
[6] Pancar Yuksel E, Sen S, Aydin F, Senturk N, Sen N, et al. Phenobarbital-induced pellagra resulted in death. Cutan Ocul Toxicol 2014;33:76–8.
[7] Sander JW. The use of antiepileptic drugs—principles and practice. Epilepsia 2004;45(6):28–34.
[8] Perucca E. Clinically relevant drug interactions with antiepileptic drugs. Br J Clin Pharmacol 2005;61(3):246–65.
[9] Bourgeois BFD. Pharmacokinetic properties of current antiepileptic drugs: what improvements are needed? Neurology 2000;55(11):S11–6. [Suppl 3].
[10] Bentué-Ferrer D, Verdier MC, Tribut O. Therapeutic drug monitoring of primidone and phenobarbital. Therapie 2012;67:381–90.

Further reading

[11] Neels HM, Sierens AC, Naelaerts K, Scharpé SL, et al. Therapeutic drug monitoring of old and newer anti-epileptic drugs. Clin Chem Lab Med 2004;42(11):1228–55.

9.9

Phenytoin

Chemical properties			
Solubility in H_2O	Limited		
Molecular weight	352.26		
pKa	8.3		
Melting point	295–298 °C		
Dosing			
Recommended dose, adult	Up to 300 mg/day		
Recommended dose, child	5–10 mg/kg/d		
Monitoring			
Sample	Serum, plasma		
Effective concentrations	10–20 μg/mL		
Toxic concentrations	>20 μg/mL		
Methods	Immunoassay, HPLC, GLC		
Pharmacokinetic properties			
Oral dose absorbed	70–90%		
Time to peak concentration	1.5–3.0 h slow release formulation 4–12 h		
Protein bound	87–93%, albumin		
Volume of distribution	0.5–0.8 L/kg		
Half-life, adult (h)	12–36 h		
Time to steady state, adult	5–7 d		
Half-life, child	7–29 h		
Time to steady state, child	2–5 d (dose dependent; children average 14 h)		
Excretion, urine			
	% Excreted	Active	Detected in blood
Unchanged	1–5	Yes	Yes
Parahydroxyphenylhydantoin (HPPH) (free and conjugated)	60–80	No	Yes

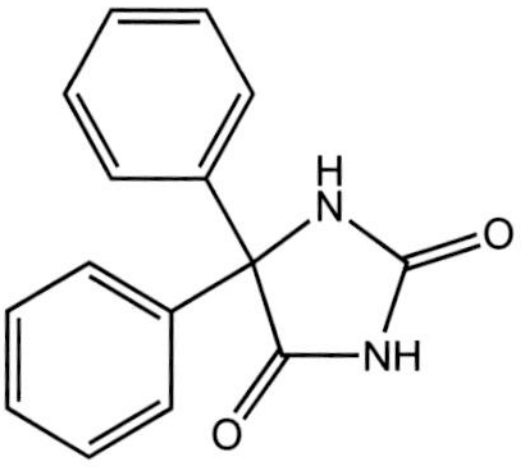

Phenytoin

Phenytoin (Dilantin) is a hydantoin that has effects on Na^+, K^+ and Ca^{2+} ion conductance and membrane potentials at therapeutic concentration. Phenytoin blocks repetitive firing of neurons and maintains Na^+ channels in the inactivated state. Phenytoin is a primary drug used for all types of seizures except absence seizures. It was the first antiepileptic drug (AED) which, in ordinary doses, does not cause general central nervous system (CNS) depression. Phenytoin is available in both oral (regular and extended release) and intravenous (IV) formulations. The absorption of oral phenytoin is slow and variable but with overall high bioavailability. Antacid preparations containing calcium may interfere with the absorption of phenytoin, and ingestion times of such preparations and phenytoin should be staggered. Differences in the bioavailability of different pharmaceutical preparations of phenytoin have been documented. Both regular and slow release formulations are available; peak plasma levels are reached in 1.5–3 h (regular) and 4–12 h (slow release). The half-life of phenytoin is 12–36 h. It takes 5–7 days for phenytoin to reach steady state. Although giving an IV loading dose achieves therapeutic serum levels faster than an oral loading dose, it may be associated with significant adverse effects. The most serious are those affecting the cardiovascular system, particularly significant hypotension, bradyarrhythmias, and cardiac arrest. Local reactions at the infusion site might occur; these are usually limited to pain and burning, but significant adverse effects can occur from extravasation injuries and the rare 'purple glove' syndrome. Oral loading doses are free from these adverse effects but may be associated with nausea and vomiting. Both routes of administration may result in dose-related neurotoxicity such as nystagmus and ataxia [1].

Phenytoin is extensively (>95% of a dose) metabolized by the liver microsomal enzymes, especially cytochrome P450 (CYP) 2C9 and CYP2C19. The major metabolite, parahydroxy-phenylhydantoin (HPPH), is inactive and accounts for up to 80% of a single dose. Other minor metabolites have been identified and are also inactive. At plasma levels below 10 μg/mL, elimination is first order. At higher plasma concentrations, elimination becomes dose-dependent (zero order) due to saturation of liver metabolism. As a consequence, the plasma half-life increases with drug dose and concentration. Since pediatric metabolism is faster than adult metabolism, higher doses are necessary to maintain plasma levels. Because CYP2C9 may be involved in 80–90% of metabolism of phenytoin, polymorphisms in CYP2C9 may result in significant reduction in the metabolism of phenytoin and an elevation of serum phenytoin levels.

Two common polymorphisms of the *CYP2C9* gene impacting phenytoin metabolism have been described (CYP2C9*2 and CYP2C9*3). Both result in diminished enzyme activity, with CYP2C9*3 having the more dramatic effect. In Caucasians, the most common variant of CYP2C9 is CYP2C9*2 (10–13% of the population), whereas the frequency of CYP2C9*3 varies from 5 to 9%. In Asians and Africans, these two alleles appear at lower frequency than seen in the Caucasians. In Chinese and Japanese, the CYP2C9*2 allele has not been detected. The frequencies of CYP2C9*2 in Indian population have found to be between 3 and 5% whereas CYP2C9*3 ranges from 4 to 8%. Individuals with reduced enzymatic activity of CYP2C9 due to genetic polymorphism are at higher risk of phenytoin toxicity [2].

Drug interactions of phenytoin with a variety of other drugs have been documented. Phenytoin (like carbamazepine, phenobarbital, and primidone) stimulates the activity of a variety of CYP enzymes including CYP1A2, CYP2C9, CYP2C19, and CYP3A4. AEDs potentially impacted by the CYP enzyme induction of phenytoin include carbamazepine, clonazepam, ethosuximide, lamotrigine, oxcarbazepine, topiramate, and valproic acid. In addition, the metabolism of antimicrobials, antipsychotics, bilirubin, cardiovascular drugs, immunosuppressants, vitamins D, vitamin K, warfarin, and tricyclic antidepressants may be impacted by phenytoin therapy. Conversely, discontinuation of phenytoin may be followed by an increase to toxic concentrations of concurrently administered AEDs and other drugs. Phenobarbital may also reduce oral absorption of phenytoin [3–5].

Interaction between valproic acid and phenytoin is complex. Valproic acid increased the unbound fraction of phenytoin in both single- and multiple-dose studies by 15% and 41%, respectively. Single-dose valproic acid increased the total area under the curve (AUC) of phenytoin by 11%. Multiple-dose valproic acid decreased the total AUC by 7%. Single- and multiple-dose valproic acid increased the unbound AUC by 25% and 18%, respectively, probably due to the inhibition on the metabolizing enzymes. The authors concluded that there are at least two mechanisms involved in valproic acid-phenytoin interaction. Valproic acid, being a strongly protein bound drug, displaces phenytoin from albumin binding sites and thus increases free phenytoin concentration. Increased free phenytoin results in increased clearance by liver enzymes. On the other hand, enzyme inhibition by valproic acid increased both the total and unbound concentration of phenytoin. The two conflicting mechanisms may result in different effects on the total plasma concentration of phenytoin. Therapeutic drug monitoring (TDM) based on the total concentration of phenytoin may be misleading when valproic acid is co-administered. Proper dosage adjustment of phenytoin should ideally be made based on measurement of free (unbound) phenytoin concentrations [6].

TDM is useful for individualization of phenytoin therapy, because there is usually a good correlation between plasma levels of phenytoin and clinical effect. Effective levels are 10–20 μg/mL. Levels >20 μg/mL are generally associated with the appearance of toxic symptoms; nystagmus (>20 μg/mL), ataxia (>30 μg/mL) and lethargy (>40 μg/mL). However, some patients may respond well even when plasma concentrations exceed 20 μg/mL. The most common side effect of long-term phenytoin therapy is gingival hyperplasia, a cosmetically undesirable effect that may require dental surgery.

The high protein binding of phenytoin makes determination of free levels potentially valuable. The accepted therapeutic range for free phenytoin is 0.8–2.1 μg/mL [1]. The free phenytoin concentration correlates better with clinical outcome. Categories of patients for which free levels may be helpful include renal failure, liver disease, pregnancy, elderly, and those with hypoalbuminemia [1, 7]. Burt et al. commented that monitoring total phenytoin is not as reliable as free phenytoin as a clinical indicator for therapeutic and nontherapeutic concentrations. Thus, authors recommend that TDM should use free phenytoin concentrations only [8]. Free phenytoin can be determined in the protein free ultrafiltrate (prepared by centrifuging serum for at least 20 min using Centrifree Micropartition Filter with molecular cut-off of 30,000 Da). Historically, the Abbott TDx method has been widely used for the measurement of free phenytoin but this method is no longer available. Williams et al. reported that Beckman-Coulter DxC800 based free phenytoin method compared well with the TDx free phenytoin method [9]. The free phenytoin assay on the Cobas e601 analyzer also correlated well with the TDx phenytoin assay [10]. An alternative approach is to calculate free phenytoin based on total albumin concentration and total phenytoin concentration in serum using an equation such as Winter-Tozer. However, in one study the authors concluded that in patients with end stage renal disease receiving hemodialysis, Winter-Tozer equation was not accurate in predicting free phenytoin concentration. However, a revised equation was found to be significantly more accurate [11].

9.9.1 Fosphenytoin

Fosphenytoin is a phosphate ester prodrug developed as an alternative to intravenous phenytoin for acute treatment of seizures. Advantages include more convenient and rapid intravenous administration, availability for intramuscular injection, and low potential for adverse local reactions at injection sites. Drawbacks include the occurrence of transient paraesthesias and pruritus at rapid infusion rates, and cost. Fosphenytoin, like phenytoin, is strongly protein bound (93–98%). Fosphenytoin is entirely eliminated through metabolism to phenytoin by blood and tissue phosphatases. The bioavailability of the derived phenytoin relative to intravenous phenytoin is approximately 100% following intravenous or intramuscular administration. The half-life for conversion of fosphenytoin to phenytoin ranges from 7 to 15 min. The rapid achievement of effective concentrations permits the use of fosphenytoin in emergency situations, such as status epilepticus. Following intramuscular administration, therapeutic phenytoin plasma concentrations are observed within 30 min and maximum plasma concentrations occur at approximately 30 min for fosphenytoin and at 2–4 h for derived phenytoin [12]. Purple glove syndrome (PGS) is a poorly understood severe adverse drug reaction that is typically associated with intravenous phenytoin administration. Although fosphenytoin is thought to circumvent this risk of PGS, Newman et al. reported a rare case of PGS in a patient treated with fosphenytoin therapy [13]. Because fosphenytoin is a prodrug, TDM of phenytoin or preferably free phenytoin is recommended after administration of fosphenytoin [14].

References

[1] Gallop K. Review article: phenytoin use and efficacy in the ED. Emerg Med Australas 2010;22:108–18.
[2] Chaudhary N, Kabra M, Gulati S, Gupta YK, et al. Frequencies of CYP2C9 polymorphisms in North Indian population and their association with drug levels in children on phenytoin monotherapy. BMC Pediatr 2016;16:66.
[3] Sander JW. The use of antiepileptic drugs—principles and practice. Epilepsia 2004;45(6):28–34.
[4] Perucca E. Clinically relevant drug interactions with antiepileptic drugs. Br J Clin Pharmacol 2005;61(3):246–65.

[5] Bourgeois BFD. Pharmacokinetic properties of current antiepileptic drugs: what improvements are needed? Neurology 2000;55(11):S11–6. suppl 3.
[6] Lai ML, Huang JD. Dual effect of valproic acid on the pharmacokinetics of phenytoin. Biopharm Drug Dispos 1993;14:365–70.
[7] Dasgupta A. Usefulness of monitoring free (unbound) concentrations of therapeutic drugs in patient management. Clin Chim Acta 2007;377:1–13.
[8] Burt M, Anderson DC, Kloss J, Apple FS. Evidence-based implementation of free phenytoin therapeutic drug monitoring. Clin Chem 2000;46:1132–5.
[9] Williams C, Jones R, Akl P, Blick K. An automated real-time free phenytoin assay to replace the obsolete Abbott TDx method. Lab Med 2014;45:48–51.
[10] Dasgupta A, Davis B, Chow L. Validation of a free phenytoin assay on Cobas c501 analyzer using calibrators from Cobas Integra free phenytoin assay by comparing its performance with fluorescence polarization immunoassay for free phenytoin on the TDx analyzer. J Clin Lab Anal 2013;27:1–4.
[11] Soriano VV, Tesoro EP, Kane SP. Characterization of free phenytoin concentrations in end-stage renal disease using the Winter-Tozer equation. Ann Pharmacother 2017;51:669–74.
[12] Fischer JH, Patel TV, Fischer PA. Fosphenytoin: clinical pharmacokinetics and comparative advantages in the acute treatment of seizures. Clin Pharmacokinet 2003;42:33–58.
[13] Newman JW, Blunck JR, Fields RK, Croom JE. Fosphenytoin-induced purple glove syndrome: a case report. Clin Neurol Neurosurg 2017;160:50–3.
[14] DasGupta R, Alaniz C, Burghardt D. Evaluation of intravenous phenytoin and fosphenytoin loading doses: influence of obesity and sex. Ann Pharmacother Ann Pharmacother 2019;53:458–63.

9.10

Primidone

Chemical properties	
Solubility in H_2O	0.6 mg/mL
Molecular weight	218.25
pKa	n/a
Melting point	281–282 °C
Dosing	
Recommended dose, adult	Day 1–3, 125 mg/day daily which can be gradually increased to 125 mg twice or thrice daily and finally to 250 mg 3–4 times a day after 10 days. Maximum dosage 2 g per day
Recommended dose, child	15–30 mg/kg/d
Monitoring	
Sample	Serum, plasma
Effective concentrations	5–10 μg/mL
Toxic concentrations	>15 μg/mL
Methods	Immunoassay, HPLC, GLC
Pharmacokinetic properties	
Oral dose absorbed	80–90%
Time to peak concentration	1.5–3 h
Protein bound	20–30%, albumin
Volume of distribution	2–3 L/kg
Half-life, adult	6–12 h
Time to steady state, adult	30–40 h
Half-life, child	4–6 h
Time to steady state, child	20–30 h

Excretion, urine

	% Excreted	Active	Detected in blood
Unchanged	30–50	Yes	Yes
Phenobarbital	15–25	Yes	Yes
Phenylethylmalonamide (PEMA)	20	Yes	Yes

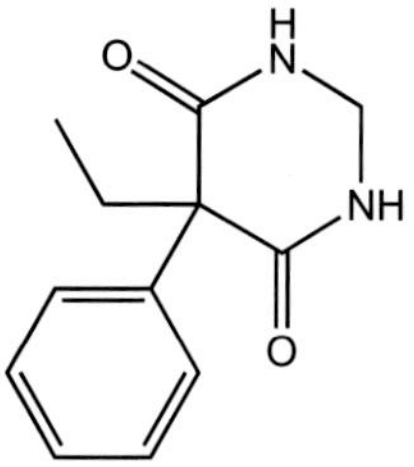

Primidone

Primidone (Mysoline) is an antiepileptic drug (AED) effective primarily against tonic–clonic seizures, simple partial seizures, and complex partial seizures. It is also effective in controlling seizures in neonates who have very immature drug-metabolizing enzyme systems. Primidone is ineffective in treating absence seizures. Primidone is commonly administered concurrently with phenytoin and/or carbamazepine; however, monotherapy is preferred. Primidone is also effective in treating essential tremor in elderly patients [1].

Primidone dosage must be individualized for each patient. A major concern is drug–drug interactions, including with other AEDs. Another concern is the administration of generic forms of primidone with different bioavailability. If a patient switches from a preparation with complete absorption to one with less absorption, the plasma concentration could decrease and the patient could have increased seizure activity. If the new preparation has higher bioavailability, then the plasma concentration could increase and the patient could suffer adverse reactions [2].

Primidone should be given to the patient in low dosages when initiating therapy. The recommended initial dose is 100–125 mg per day for adults and 50–100 mg per day for children. During the next several days to few weeks, the dosage is gradually increased until the maintenance dose regimen is reached. For adults, the maximum dosage it is up to 250 mg three times a day. This protocol is necessary to prevent sedation and gastrointestinal disturbances which are often seen when high doses of primidone are administered at the start of treatment. However, for treating essential tremor in elderly, low dose primidone (250 mg/day) is equally effective as high dose primidone (750 mg/day) [3].

The essential mechanism of primidone's action is not fully understood. Its mode of action has been described by the term "neuronal membrane stabilizing effect." This term implies prevention rather than suppression of electrical excitation. Evidence suggests that primidone prevents the spread of neural excitation from an epileptogenic focus; that is, it impedes the conduction of electrical impulses across synapses. This impairment of electrical transmission across consecutive synapses may prevent a potential epileptic discharge from occurring. Primidone is also a γ-aminobutyric acid (GABA) agonist. After oral administration, primidone is converted into phenobarbital due to oxidation of the second carbon on primidone by hepatic cytochrome P450 (CYP) 2C19. Alternatively, ring cleavage of the drug at the second carbon position converts primidone to phenylethylmalonamide (PEMA). Both primidone and phenobarbital are hydroxylated at the para position of the phenyl ring and undergo conjugation. Although the active metabolite phenobarbital accounts for some of the anticonvulsant effect of primidone, the parent drug also has antiepileptic activities [4].

Based on the study using human subjects, the concentration of primidone reached a peak concentration 1.5–3 h and then declined monoexponentially in most subjects. The major metabolites of primidone, phenobarbital and PEMA have longer half-lives than parent drug primidone. Approximately 74% of dosage was recovered in urine as primidone, phenobarbital and PEMA. Therefore, when adjusting primidone dosage, it is necessary to consider the plasma concentrations of both primidone and phenobarbital and the time it takes for both drugs to reach steady state. The parent drug primidone rapidly reaches steady state (30–40 h), while phenobarbital takes longer (17–24 days). Both primidone and phenobarbital exhibit a linear relationship between the daily dose of the drug administered and the plasma concentration produced [1, 5, 6].

Due to its wide volume of distribution and its low protein binding (only 33.5% protein bound), primidone crosses the placenta and enters the breast milk of epileptic mothers. Neonates who are exposed to primidone in utero often develop marked withdrawal symptoms (hyperexcitability, tremor, nervousness, unmotivated crying, vomiting, and restlessness). Nursing infants also have adverse reactions such as sedation, suckling problems, and possible withdrawal symptoms [7].

Primidone (like carbamazepine, phenytoin, and phenobarbital) induces the expression of a variety of CYP enzymes including CYP1A2, CYP2C8/9, CYP2B6, and CYP3A4. Hence, primidone reduces the serum concentration of some other AEDs and many other types of drugs when administered concurrently. AEDs potentially impacted by the CYP enzyme induction of primidone include carbamazepine, clonazepam, ethosuximide, lamotrigine, oxcarbazepine, topiramate, and valproic acid. In addition, the metabolism of antimicrobials, antipsychotics, bilirubin, cardiovascular drugs, immunosuppressants, vitamins D, vitamin K, warfarin, and tricyclic antidepressants may be impacted by primidone therapy. Conversely, discontinuance of primidone may be followed by an increase to toxic concentrations of concurrently administered AEDs [8].

Gastrointestinal disturbances are commonly seen in patients during early stages of primidone treatment. These symptoms generally disappear if the dose is decreased. Gastrointestinal manifestations such as nausea, vomiting, and diarrhea can be alleviated if the drug is taken with food. Baseline liver function tests should be performed before initiating primidone therapy and then monitored several times a year. Renal function tests, complete blood counts, and urinalysis should also be monitored regularly. The other adverse effects of primidone are due to elevated levels of primidone, as well as its metabolites phenobarbital and PEMA. The most common adverse reactions are associated with primidone's depressive effect on the central nervous system (CNS). Toxic overdoses of primidone primarily cause neurological dysfunctions such as ataxia, confusion, vertigo, dizziness, sedation, fatigue, slurred speech, and stupor which can lead to coma. Visual disturbances such as nystagmus, diplopia, and blurred vision are common. Primidone overdoses can also cause life-threatening respiratory depression. Lin and Chung commented that the toxic effects, such as CNS depression and dysequilibrium, are due mostly to primidone itself, rather than its metabolite phenobarbital [9]. Life threatening overdose due to primidone has also been reported. A case report described a patient who was in danger of dying from a massive primidone overdose. The woman was comatose, hypotensive and suffering from acute renal failure with crystalluria. Her plasma primidone level was 209 μg/mL. However, her life was saved by hemoperfusion [10].

The rate of conversion of primidone to phenobarbital is highly variable according to the subject. Therefore, when initiating therapeutic drug monitoring (TDM), both primidone and the active metabolite phenobarbital should be monitored. Immunoassays are commercially available for measuring serum or plasma concentration of both primidone and phenobarbital. The recommended therapeutic range of primidone is 5–10 μg/mL. Generally accepted therapeutic range for phenobarbital is between 15 and 40 μg/mL. TDM of primidone alone has low clinical utility.

References

[1] Martines C, Gatti G, Sasso E, Calzetti S, Perucca E. The disposition of primidone in elderly patients. Br J Clin Pharmacol 1990;30:607–11.

[2] Holtkamp M, Theodore WH. Generic antiepileptic drugs-safe or harmful in patients with epilepsy? Epilepsia 2018;59:1273–81.

[3] Serrano-Duenas M. Use of primidone in low doses (250 mg/day) versus high doses (750 mg/day) in the management of essential tremor. Double-blind comparative study with one-year follow-up. Parkinsonism Relat Disord 2003;10:29–33.
[4] El-Masri HA, Portier CJ. Physiologically based pharmacokinetics model of primidone and its metabolites phenobarbital and phenylethylmalonamide in humans, rats, and mice. Drug Metab Dispos 1998;26:585–94.
[5] Sander JW. The use of antiepileptic drugs—principles and practice. Epilepsia 2004;45(6):28–34.
[6] Bourgeois BFD. Pharmacokinetic properties of current antiepileptic drugs: what improvements are needed? Neurology 2000;55(11):S11–6. Suppl 3.
[7] Kuhnz W, Koch S, Helge H, Nau H. Primidone and phenobarbital during lactation period in epileptic women: total and free drug serum levels in the nursed infants and their effects on neonatal behavior. Dev Pharmacol Ther 1988;11:147–54.
[8] Perucca E. Clinically relevant drug interactions with antiepileptic drugs. Br J Clin Pharmacol 2005;61(3):246–65.
[9] Lin SL, Chung MY. Acute primidone intoxication: report of a case. Taiwan Yi Xue Hui Za Zhi 1989;88:1053–5.
[10] van Heijst AN, de Jong W, Seldenrijk R, van Dijk A. Coma and crystalluria: a massive primidone intoxication treated with haemoperfusion. J Toxicol Clin Toxicol 1983;20:307–18.

Further reading

[11] Bentué-Ferrer D, Verdier MC, Tribut O. Therapeutic drug monitoring of primidone and phenobarbital. Therapie 2012;67:381–90. [Article in French].

9.11

Tiagabine

Chemical properties	
Solubility in H_2O	0.0211 mg/mL
Molecular weight	375.55
pKa	4.14
Melting point	189–194 °C
Dosing	
Recommended dose	32–56 mg/d (adult)
Monitoring	
Sample	Serum, plasma
Effective concentrations	0.02–0.2 μg/mL
Toxic concentrations	Not established
Methods	GC/MS, HPLC, LC/MS
Pharmacokinetic properties	
Oral dose absorbed	>90%
Time to peak concentration	1–2 h
Protein bound	96%, albumin, alpha-1 acid glycoprotein
Volume of distribution	1 L/kg (adults)
Half-life, adult	5–9 h
Time to steady state, adult	1–2 day
Half-life, child	3–6 h
Time to steady state, child	1–1.5 day

Excretion, urine

	% Excreted	Active	Detected in blood
Unchanged	2	Yes	Yes
Other	25	No	No

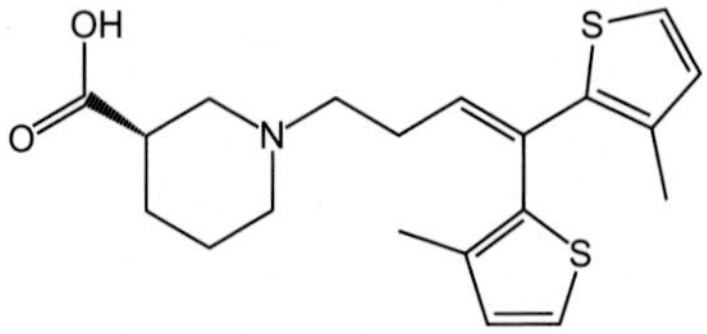

Tiagabine

Tiagabine (Gabitril) is an antiepileptic drug (AED) approved in the United States in 1997 for adjunctive therapy of partial seizures [1]. The mechanism of action of tiagabine has not been completely resolved but is thought to involve at least in part inhibition of reuptake of the inhibitory neurotransmitter γ-aminobutyric acid (GABA). Although effective as an AED, the therapeutic use of tiagabine in Europe and the United States has been limited by a propensity of tiagabine to trigger seizures and, rarely, to result in life-threatening, non-convulsive status epilepticus [2, 3]. With a steady influx of other newer AEDs as alternatives for management of epilepsy, tiagabine has assumed a more limited role, generally reserved for seizure disorders refractory to other AEDs.

Tiagabine is rapidly absorbed after oral ingestion with excellent bioavailability [4]. Tiagabine is highly bound (>95%) to plasma proteins. The high degree of binding to plasma proteins can result in drug–drug interactions with valproic acid, which can displace tiagabine from protein binding sites and thereby increase the free fraction of tiagabine. Tiagabine is extensively metabolized by the liver, with only 2% of the unchanged drug excreted in urine. Liver metabolism is the major route for elimination. None of the metabolites of tiagabine are known to be therapeutically active. Drugs that induce hepatic liver metabolism (including carbamazepine, phenobarbital, and phenytoin) increase the metabolism of tiagabine and accelerate its clearance. The elimination half-life of tiagabine is 5–9 h for those not on concomitant liver enzyme inducers. The half-life decreases to 2–4 h if the patient is on another medication capable of inducing hepatic enzyme metabolism. Children clear tiagabine faster than adults [5]. Severe liver failure can increase half-life to 12–16 h [6]. Renal insufficiency has little impact on the clearance of tiagabine [7].

The variable metabolism of tiagabine, including potential for drug–drug interactions, makes therapeutic drug monitoring (TDM) potentially valuable [8, 9]. A wide therapeutic range of 20–200 ng/mL (0.02–0.2 μg/mL) has been proposed based on a multi-center study, although individualized determination of effective concentrations is important [10]. Analytical methodologies for measuring tiagabine in serum or plasma including gas chromatography/mass spectrometry, high-performance liquid chromatography, and liquid chromatography coupled to tandem mass spectrometry [11]. The high degree of plasma protein binding suggests that measurement of free (unbound) tiagabine concentrations may be clinically valuable [12]; however, analytical methods to accurately determine the very low free tiagabine concentrations have yet to be reported.

References

[1] LaRoche SM, Helmers SL. The new antiepileptic drugs: clinical applications. JAMA 2004;291:615–20.
[2] Balslev T, Uldall P, Buchholt J. Provocation of non-convulsive status epilepticus by tiagabine in three adolescent patients. Eur J Paediatr Neurol 2000;4:169–70.
[3] Kellinghaus C, Dziewas R, Ludemann P. Tiagabine-related non-convulsive status epilepticus in partial epilepsy: three case reports and a review of the literature. Seizure 2002;11:243–9.
[4] Gustavson LE, Mengel HB. Pharmacokinetics of tiagabine, a γ-aminobutyric acid-uptake inhibitor, in healthy subjects after single and multiple doses. Epilepsia 1995;36: 605–11.
[5] Gustavson LE, Boellner SW, Granneman GR, Qian JX, Guenther HJ, el-Shourbagy T, Sommerville KW. A single-dose study to define tiagabine pharmacokinetics in pediatric patients with complex partial seizures. Neurology 1997;48:1032–7.
[6] Lau AH, Gustavson LE, Sperelakis R, Lam NP, El-Shourbagy T, Qian JX, Layden T. Pharmacokinetics and safety of tiagabine in subjects with various degrees of hepatic function. Epilepsia 1997;38:445–51.
[7] Cato 3rd A, Gustavson LE, Qian J, El-Shourbagy T, Kelly EA. Effect of renal impairment on the pharmacokinetics and tolerability of tiagabine. Epilepsia 1998;39:43–7.
[8] Jacob S, Nair AB. An updated overview on therapeutic drug monitoring of recent antiepileptic drugs. Drugs RD 2016;16:303–16.
[9] Patsalos PN, Spencer EP, Berry DJ. Therapeutic drug monitoring of antiepileptic drugs in epilepsy: a 2018 update. Ther Drug Monit 2018;40:526–48.
[10] Uthman BM, Rowan AJ, Ahmann PA, Leppik IE, Schachter SC, Sommerville KW, Shu V. Tiagabine for complex partial seizures: a randomized, add-on, dose–response trial. Arch Neurol 1998;55:56–62.
[11] Krasowski MD, McMillin GA. Advances in anti-epileptic drug testing. Clin Chim Acta 2014;436:224–36.
[12] Dasgupta A. Usefulness of monitoring free (unbound) concentrations of therapeutic drugs in patient management. Clin Chim Acta 2007;377:1–13.

9.12

Topiramate

Chemical properties	
Solubility in H_2O	9.8 mg/mL
Molecular weight	339.36
pKa	8.61
Melting point	125–126 °C

Dosing	
Recommended dose	25–50 mg/day (adult)

Monitoring	
Sample	Serum, plasma
Effective concentrations	5–20 μg/mL
Toxic concentrations	Not established
Methods	Immunoassay, HPLC, GC, LC–MS/MS

Pharmacokinetic properties	
Oral dose absorbed	>80%
Time to peak concentration	2–4 h
Protein bound	15%
Volume of distribution	0.6–0.8 L/kg
Half-life, adult	20–30 h
Time to steady state, adult	3–5 days
Half-life, child	12–24 h
Time to steady state, child	2–4 d

Excretion, urine

	% Excreted	Active	Detected in blood
Unchanged	70	Yes	Yes
Glucuronide	<5	No	No
Other	<5	No	No

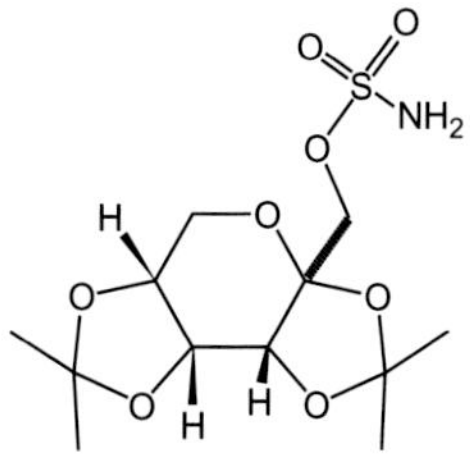

Topiramate

Topiramate (Topamax) is an antiepileptic drug (AED) first approved in the United States in 1996 for the treatment of epilepsy [1]. Topiramate has since gained approvals for other clinical indications including treatment of migraine headaches and, in combination with phentermine, for weight loss. Topiramate is generally well-tolerated. Common adverse effects include dizziness, paresthesias, somnolence, fatigue, and gastrointestinal complaints. The United States Food and Drug Administration notified prescribers in 2017 that topiramate use was associated with acute myopia and secondary angle closure glaucoma in a small subset of patients [2]. Topiramate also inhibits carbonic anhydrase and has the potential to cause metabolic acidosis and increase the risk of kidney stone formation [3].

Topiramate has excellent absorption and high bioavailability following oral absorption, with low binding to plasma proteins [4]. An extended release version of topiramate is available. Hepatic metabolism plays a major role in the excretion of topiramate, with approximately 50% of the absorbed dose metabolized by liver enzymes. Drugs that induce hepatic liver metabolism (including carbamazepine, phenobarbital, and phenytoin) increase the metabolism of topiramate and accelerate its clearance [5, 6]. In adult patients not taking any inducers of liver metabolism, topiramate has an average serum half-life approximately 20–30 h. Concomitant use of medications that induce liver metabolism can reduce the half-life of topiramate to 12 h. Children typically eliminate topiramate faster than adults [7, 8].

The main value of therapeutic drug monitoring (TDM) for topiramate lies in the potential for variable liver metabolism, including drug–drug interactions [9, 10]. Establishment of an individualized reference range for topiramate (e.g., for control of seizures) is valuable. Levels can then be followed when there are changes that may impact topiramate metabolism, especially introduction or withdrawal of concomitant medications that alter liver metabolism.

A therapeutic range of 5–20 μg/mL has been proposed for epilepsy therapy; the target range for other clinical uses of topiramate (e.g., migraine headaches, weight loss) has not been established [11]. The availability of immunoassays for clinical chemistry analyzers allows for rapid determination of topiramate plasma/serum concentrations. Other analytical methodologies include gas chromatography, high-performance liquid chromatography, and liquid chromatography coupled to tandem mass spectrometry [12]. Salivary topiramate levels correlate well with plasma/serum concentrations, making saliva a viable specimen for TDM [13]. Determination of salivary levels requires chromatographic methods.

References

[1] LaRoche SM, Helmers SL. The new antiepileptic drugs: clinical applications. JAMA 2004;291:615–20.

[2] Haque S, Shaffi M, Tang KC. Topiramate associated non-glaucomatous visual field defects. J Clin Neurosci 2016;31:210–3.

[3] Ture H, Keskin O, Cakir U, Aykut Bingol C, Ture U. The frequency and severity of metabolic acidosis related to topiramate. J Int Med Res 2016;44:1376–80.

[4] Easterling DE, Zakszewski T, Moyer MD, Margul BL, Marriott TB, Nayak RK. Plasma pharmacokinetics of topiramate, a new anticonvulsants in humans. Epilepsia 1988;29:662.

[5] Britzi MP, E., Soback S, Levy RH, Fattore C, Crema F, Gatti G, Doose DR, Maryanoff BE, Bialer M. Pharmacokinetic and metabolic investigation of topiramate disposition in healthy subjects in the absence and in the presence of enzyme induction by carbamazepine. Epilepsia 2005;46:378–384.

[6] Mimrod D, Specchio LM, Britzi M, Perucca E, Specchio N, La Neve A, Soback S, Levy RH, Gatti G, Doose DR, et al. A comparative study of the effect of carbamazepine and valproic acid on the pharmacokinetics and metabolic profile of topiramate at steady state in patients with epilepsy. Epilepsia 2005;46:1046–54.

[7] Perucca E. Clinical pharmacokinetics of new-generation antiepileptic drugs at the extremes of age. Clin Pharmacokinet 2006;45:351–64.

[8] Rosenfeld WE, Doose DR, Walker SA, Baldassarre JS, Reifer RA. A study of topiramate pharmacokinetics and tolerability in children with epilepsy. Pediatr Neurol 1999;20:339–44.

[9] Jacob S, Nair AB. An updated overview on therapeutic drug monitoring of recent antiepileptic drugs. Drugs R D 2016;16:303–16.

[10] Patsalos PN, Spencer EP, Berry DJ. Therapeutic drug monitoring of antiepileptic drugs in epilepsy: a 2018 update. Ther Drug Monit 2018;40:526–48.

[11] Johannessen SI, Battino D, Berry DJ, Bialer M, Kramer G, Tomson T, Patsalos PN. Therapeutic drug monitoring of the newer antiepileptic drugs. Ther Drug Monit 2003;25:347–63.

[12] Krasowski MD, McMillin GA. Advances in anti-epileptic drug testing. Clin Chim Acta 2014;436:224–36.

[13] Miles MV, Tang PH, Glauser TA, Ryan MA, Grim SA, Strawsburg RH, deGrauw TJ, Baumann RJ. Topiramate concentration in saliva: an alternative to serum monitoring. Pediatr Neurol 2003;29:143–7.

9.13

Valproic acid

Chemical properties	
Solubility in H_2O	1.3 mg/mL
Molecular weight	144.21
pKa	4.95
Melting point	128–130 °C
Dosing	
Recommended dose	30–60 mg/kg/day (adult)
Monitoring	
Sample	Serum, plasma
Effective concentrations	50–100 μg/mL
Toxic concentrations	>100 μg/mL
Methods	Immunoassay, HPLC, GC
Pharmacokinetic properties	
Oral dose absorbed	100%
Time to peak concentration	1–4 h
Protein bound	90%, albumin; concentration dependent
Volume of distribution	0.1–0.4 L/kg
Half-life, adult	8–15 h
Time to steady state, adult	2–4 days
Half-life, child	8–11 h
Time to steady state, child	2–4 days

Excretion, urine

	% Excreted	Active	Detected in blood
Unchanged	3–7	Yes	Yes
Glucuronide	15–70	No	Yes
Other	7–50	Yes	Yes

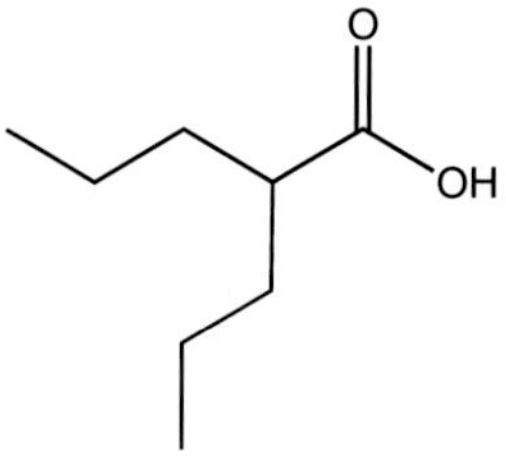

Valproic acid

Valproic acid (2-n-propylpentanoic acid; Depakene, Depakote) is an antiepileptic drug (AED) structurally unrelated to other AEDs. It was first synthesized in 1881, but not approved for use in the United States as an AED until 1978. Valproic acid (VPA) is a widely used AED that is highly effective both in adults and children. It is one of the first line drugs for the treatment of epilepsy. The drug is used in the treatment of simple and complex absence seizures, particularly those refractory to ethosuximide. It has also been shown to be helpful in tonic–clonic seizures and is often used in combination with other AEDs in patients with multiple seizure types. VPA is also used to treat migraine headaches, schizophrenia, and bipolar disorder.

The most frequent adverse effects of VPA include somnolence, weight gain, fatigue and headache. More serious adverse effects include hepatotoxicity and pancreatitis, both of which can be fatal. Risk factors for severe toxicity include age < 3 years old, developmental delay, and polypharmacy. Scheffner described the death of 16 children in Germany due to liver failure as a result of therapy with VPA [1]. Star et al. commented that hepatotoxicity remains a considerable problem of therapy with VPA. The risk appears to be greatest in young children (6 years and below) but can occur at any age. Polytherapy is commonly reported and seems to be a risk factor for hepatotoxicity, pancreatitis and other serious adverse drug reactions with VPA. Pancreatitis due to treatment with VPA may also be fatal [2]. The U.S. Food and Drug Administration (FDA) is advising health care professionals and women that the various formulations of VPA (e.g, valproate sodium, divalproex sodium) are contraindicated and should not be taken by pregnant women for the prevention of migraine headaches.

The mechanism of the antiepileptic action of VPA involves the neurotransmitter γ-aminobutyric acid (GABA). VPA administration is associated with increased GABA level in the brain, thus potentiating GABAergic transmission in specific brain regions. In addition, VPA may exert a direct effect on the potassium channels of the neuronal membrane. However, the neurochemical and neurophysiological effects of VPA are not completely characterized [3].

VPA is rapidly absorbed after oral administration (bioavailability ≥80%). VPA is extensively bound to serum protein (>90%, mostly to albumin) but the extent of binding decreases with increasing drug concentration, resulting in an increase of the free fraction of the drug at the upper end of therapeutic range. Only the unbound (free fraction) of VPA is pharmacologically active. VPA has a very short half-life, which leads to fluctuating plasma levels. Doses are generally administered 3 to 4 times daily [4].

There are at least three routes of VPA metabolism in humans: glucuronidation (~50% of dose), β-oxidation in the mitochondria (~40% of dose), and cytochrome P450 (CYP)-mediated oxidation (minor route ~10% of dose). Valproate glucuronide is the major urinary metabolite of VPA. As a fatty acid, VPA can be metabolized through endogenous pathways in the mitochondria. However, some of the mitochondrial metabolites of VPA showed hepatotoxicity. The CYP-mediated branch of the VPA pathway generates the metabolite 4-ene-valproic acid by CYP2C9, CYP2A6, and, to a lesser extent, by CYP2B6. In addition, these metabolizing enzymes also mediate the metabolism of VPA to the inactive 4-OH-valproic acid and 5-OH-valproic acid. CYP2A6 also contributes partially to the formation of 3-OH-valproic acid [4, 5]. In summary, VPA is a minor substrate of several systems (CYP2A6, CYP2B6, CYP2C8/9, CYP2C19, and CYP2E1) and is also a weak inhibitor of several systems (CYP2C8/9, CYP2C19, CYP2D6, and CYP3A4). The metabolism of VPA is increased by enzyme inducers such as carbamazepine, phenobarbital, phenytoin, and primidone; concentrations of VPA can be reduced by 50–75% in patients comedicated with enzyme inducers. Discontinuation of an enzyme inducer can lead to toxic concentrations of VPA due to a rebound effect. VPA inhibits renal clearance of phenobarbital and is also thought to decrease hydroxylation of the parent drug to the inactive *p*-hydroxyphenobarbital. The result is an increase in phenobarbital concentration of up to 40%. VPA inhibits metabolism of lamotrigine and can cause a twofold increase in lamotrigine concentration [6, 7].

The most common side effects (15–20% of patients) seen with VPA use are gastrointestinal (e.g., anorexia, nausea, and vomiting) and are usually transient. Central nervous system (CNS) side effects include sedation, ataxia, and tremor. These symptoms usually respond to a decrease in dosage. Liver transaminases may be elevated slightly during the first few months of therapy.

Development of fulminate hepatitis is rare and is thought to be the result of formation of a toxic metabolite. Severe toxicity and even fatality from

VPA use in children under age of 3 has been discussed earlier in the monograph. VPA may potentiate the action of CNS depressants such as alcohol and benzodiazepines [1].

In women, VPA is associated with hyperandrogenism, polycystic ovarian syndrome, menstrual disorders, and ovulatory failure. Men on VPA therapy show abnormalities in androgen blood levels, sperm motility, and erectile dysfunctions. Studies have shown that VPA negatively affects the release of luteinizing hormone, follicle stimulating hormone, and prolactin. VPA may also interfere with peripheral endocrine hormones. Its broad inhibitory action on cytochrome and glucuronidation systems can lead to high serum concentration of testosterone, androstenedione and dehydroepiandrosterone sulfate. VPA dependent obesity and hyperinsulinemia can further contribute to an increase in sexual dysfunctions. Because VPA is a first line AED both in children and adult with epilepsy and long-term medication with this drug is sometimes necessary, it is very important for physicians to implement strict monitoring of patients taking VPA in order to identify these kinds of side effects at an early stage [8].

Therapeutic drug monitoring of VPA is recommended. The therapeutic range is 50–100 μg/mL [9]. However, VPA is strongly bound to serum proteins. Monitoring free VPA (in the protein free ultrafiltrate) is recommended for some patient populations, it feasible. Categories of patients for which free VPA levels may be helpful include renal failure, liver disease, pregnancy, elderly, and those with hypoalbuminemia As VPA is very tightly bound to albumin, concentrations of other drugs such as phenytoin, carbamazepine, and salicylate may be altered with concurrent use. For example, VPA can displace protein bound phenytoin, causing therapeutic and toxic effects at phenytoin concentrations lower than those required to produce equivalent effects in patients not taking VPA [10].

References

[1] Scheffner D, Konig S, Rauterberg-Ruland I, Kochen W, et al. Fatal liver failure in 16 children with valproate therapy. Epilepsia 1988;29:530–42.
[2] Star K, Edwards IR, Choonara I. Valproic acid and fatalities in children: a review of individual case safety reports in VigiBase. PLoS One 2014;9(10):e108970.
[3] Loscher W, Bohme G, Schafer H, Kochen W. Effect of metabolites of valproic acid on the metabolism of GABA in brain and brain nerve endings. Neuropharmacology 1981;20:1187–92.
[4] Silva MF, Aires CC, Luis PB, Ruiter JP, et al. Valproic acid metabolism and its effects on mitochondrial fatty acid oxidation: a review. J Inherit Metab Dis 2008;31:205–16.
[5] Ghodke-Puranik Y, Thorn CF, Lamba JK, Leeder JS, et al. Pharmacogenet Genomics 2013;23:236–41.

[6] Perucca E. Clinically relevant drug interactions with antiepileptic drugs. Br J Clin Pharmacol 2005;61(3):246–65.
[7] Bourgeois BFD. Pharmacokinetic properties of current antiepileptic drugs: what improvements are needed? Neurology 2000;55(11):S11–6. [Suppl 3].
[8] Verrotti A, Mencaroni E, Cofini M, Castagnino M, Leo A, et al. Valproic acid metabolism and its consequences on sexual functions. Curr Drug Metab 2016;17:573–81.
[9] Patsalos PN, Zugman M, Lake C, James A, et al. Serum protein binding of 25 antiepileptic drugs in a routine clinical setting: a comparison of free non-protein-bound concentrations. Epilepsia 2017;58:1234–43.
[10] Dasgupta A. Usefulness of monitoring free (unbound) concentrations of therapeutic drugs in patient management. Clin Chim Acta 2007;377:1–13.

9.14

Zonisamide

Chemical properties	
Solubility in H_2O	0.8 mg/mL
Molecular weight	212.23
pKa	10.2
Melting point	161–163 °C
Dosing	
Recommended dose	100–400 mg/day (adult)
Monitoring	
Sample	Serum, plasma
Effective concentrations	10–40 μg/mL
Toxic concentrations	>80 μg/mL
Methods	Immunoassay, HPLC, LC/MS
Pharmacokinetic properties	
Oral dose absorbed	>65%
Time to peak concentration	2–6 h
Protein bound	50%
Volume of distribution	1.5 L/kg
Half-life, adult	50–70 h
Time to steady state, adult	8–12 days
Half-life, child	40–60 h
Time to steady state, child	6–10 days

Excretion, urine

	% Excreted	Active	Detected in blood
Unchanged	15	Yes	Yes
Glucuronide	21	No	Yes
Other	6	No	No

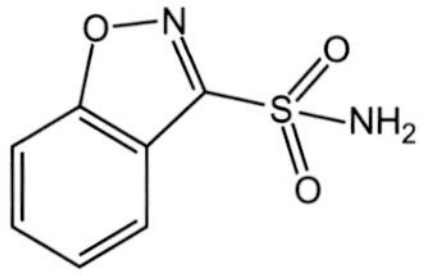

Zonisamide

Zonisamide (Zonegran) is an antiepileptic drug (AED) approved in the United States, Europe, Japan, and Australia for partial seizures [1]. The medication is also used off-label for other clinical indications such as bipolar disorder and migraine headaches. Zonisamide is generally well-tolerated, with common adverse effects that include anorexia, dizziness, irritability, diplopia, and memory impairment. Less common but more severe adverse effects include hypersensitivity reactions and ecchymosis (bruising and hemorrhage). Similar to topiramate, zonisamide inhibits carbonic anhydrase has the potential to create a metabolic acidosis and also increase risk of renal stone formation.

Zonisamide has rapid absorption and excellent bioavailability following oral administration, with only 50% bound to plasma proteins. The major pharmacokinetic variability with zonisamide is in its metabolism [2]. Zonisamide is extensively metabolized by acetylation, oxidation, and other pathways. Oxidation of zonisamide is catalyzed in part by cytochrome P450 (CYP) 3A4, an enzyme isoform commonly linked to drug–drug interactions. The metabolism and clearance of zonisamide is significantly altered by inducers (e.g., carbamazepine, phenobarbital, phenytoin) and inhibitors (e.g., cimetidine, valproic acid) of CYP3A4. The serum half-life of life of zonisamide when used as monotherapy is approximately 60 h; this may be prolonged significantly when zonisamide is used together with valproic acid or another CYP inhibitor [3]. Conversely, inducers of liver enzymes such as phenytoin can decrease zonisamide serum half-life to 25–35 h. Zonisamide clearance is significantly higher in children compared to adults [4]. Children thus require higher doses by weight. Hemodialysis can effectively clear zonisamide from the circulation [5].

The main value of therapeutic drug monitoring (TDM) for zonisamide lies in its variable metabolism, including the potential for drug–drug interactions [6, 7]. A plasma/serum reference range of 10–40 μg/mL has been proposed for use of zonisamide in epilepsy [8]; therapeutic ranges for other indications of zonisamide (e.g., bipolar disorder) have yet to be established. Toxic effects are uncommon at concentrations below 30 mg/L. The recent availability of immunoassays for clinical chemistry analyzers allows for rapid determination of zonisamide plasma/serum concentrations [9]. Other analytical methodologies for measurement of zonisamide in plasma/serum include high-performance liquid chromatography and liquid chromatography coupled to tandem mass spectrometry. TDM using saliva is feasible for zonisamide, although there is limited published data [10, 11].

References

[1] Mimaki T. Clinical pharmacology and therapeutic drug monitoring of zonisamide. Ther Drug Monit 1998;20:593–7.

[2] Buchanan R, Bockbrader HN, Chang T, Sedman AJ. Single- and multiple-dose pharmacokinetics of zonisamide. Epilepsia 1996;37:172.

[3] Perucca E, Bialer M. The clinical pharmacokinetics of the newer antiepileptic drugs. Focus on topiramate, zonisamide and tiagabine. Clin Pharmacokinet 1996;31:29–46.

[4] Perucca E. Clinical pharmacokinetics of new-generation antiepileptic drugs at the extremes of age. Clin Pharmacokinet 2006;45:351–64.

[5] Ijiri Y, Inoue T, Fukuda F, Suzuki K, Kobayashi T, Shibahara N, Takenaka H, Tanaka K. Dialyzability of the antiepileptic drug zonisamide in patients undergoing hemodialysis. Epilepsia 2004;45:924–7.

[6] Jacob S, Nair AB. An updated overview on therapeutic drug monitoring of recent antiepileptic drugs. Drugs RD 2016;16:303–16.

[7] Patsalos PN, Spencer EP, Berry DJ. Therapeutic drug monitoring of antiepileptic drugs in epilepsy: a 2018 update. Ther Drug Monit 2018;40:526–48.

[8] Berent S, Sackellares JC, Giordani B, Wagner JG, Donofrio PD, Abou-Khalil B. Zonisamide (CI-912) and cognition: results from preliminary study. Epilepsia 1987;28:61–7.

[9] Krasowski MD, McMillin GA. Advances in anti-epileptic drug testing. Clin Chim Acta 2014;436:224–36.

[10] Kumagai N, Seki T, Yamada T, Takuma Y, Hirai K. Concentrations of zonisamide in serum, free fraction, mixed saliva and cerebrospinal fluid in epileptic children treated with monotherapy. Jpn J Psychiatry Neurol 1993;47:291–2.

[11] Patsalos PN, Berry DJ. Therapeutic drug monitoring of antiepileptic drugs by use of saliva. Ther Drug Monit 2013;35:4–29.

9.15

Therapeutic drug monitoring of other antiepileptic drugs

The antiepileptic drugs (AEDs) discussed in detail above have the strongest rationale for therapeutic drug monitoring (TDM). For other AEDs, TDM may be occasionally useful, have limited evidence for support, or lack utility altogether. In this section, some of these agents will be discussed with respect to TDM. For AEDs not discussed here, the interested reader is directed towards a recent review [1].

Clobazam is a benzodiazepine that shares a similar mechanism of action to other benzodiazepines like clonazepam and diazepam [2]. Clobazam has predictable pharmacokinetics and has an active metabolite that is generated by cytochrome P450 (CYP) 3A4. This *N*-desmethyl metabolite is primarily eliminated by CYP2C9. The main factors favoring TDM for clobazam are factors that can impact activity of CYP3A4 and CYP2C9, including genetic polymorphisms of CYP2C9 [3]. Chromatographic methods are required for measurement of clobazam in plasma/serum [4].

Felbamate is an AED whose therapeutic use has been greatly limited by association with aplastic anemia (approximately 1 in 4000) and severe hepatic failure (approximately 1 in 30,000), each with mortality approaching 40%. These rare adverse events have led to dramatically restricted use of felbamate [5]. Multiple pharmacokinetic factors favor TDM for felbamate, including significant inter-individual variability in metabolism, potential for drug–drug interactions, and developmental changes in drug clearance [1]. Unfortunately, monitoring of felbamate levels does not predict the rare adverse events.

Lacosamide is a novel AED that structurally is a functionalized amino acid. TDM has limited utility for lacosamide due to predictable pharmacokinetics and low potential for drug–drug interactions [4]. Monitoring of levels may be useful for establishing individualized reference ranges or in patients with liver and/or kidney failure [1].

Rufinamide is an AED that shows nonlinear pharmacokinetics above daily doses of approximately 1600 mg, leading to significant inter-individual differences in plasma concentrations [1]. Potential for drug–drug interactions also exist with the liver metabolism of rufinamide. These factors make TDM potentially useful for rufinamide, although a wide therapeutic range is associated with clinical efficacy [6].

Stiripentol is a newer AED who complicated pharmacokinetics, including saturable metabolism leading to nonlinear pharmacokinetics, resemble those of the first generation AED phenytoin. These factors make stiripentol an excellent candidate for TDM [1]. Stiripentol has had limited clinical use and was only approved by United States Food and Administration in 2018, having been an orphan drug available under compassionate use prior to then.

References

[1] Patsalos PN, Spencer EP, Berry DJ. Therapeutic drug monitoring of antiepileptic drugs in epilepsy: a 2018 update. Ther Drug Monit 2018;40:526–48.

[2] Wheless JW, Phelps SJ. Clobazam: a newly approved but well-established drug for the treatment of intractable epilepsy syndromes. J Child Neurol 2013;28:219–29.

[3] Yamamoto Y, Takahashi Y, Imai K, Miyakawa K, Nishimura S, Kasai R, Ikeda H, Takayama R, Mogami Y, Yamaguchi T, et al. Influence of CYP2C19 polymorphism and concomitant antiepileptic drugs on serum clobazam and *N*-desmethyl clobazam concentrations in patients with epilepsy. Ther Drug Monit 2013;35:305–12.

[4] Krasowski MD, McMillin GA. Advances in anti-epileptic drug testing. Clin Chim Acta 2014;436:224–36.

[5] Pellock JM, Faught E, Leppik IE, Shinnar S, Zupanc ML. Felbamate: consensus of current clinical experience. Epilepsy Res 2006;71:89–101.

[6] Perucca E, Cloyd J, Critchley D, Fuseau E. Rufinamide: clinical pharmacokinetics and concentration-response relationships in patients with epilepsy. Epilepsia 2008;49:1123–41.

CHAPTER 10

Therapeutic drug monitoring of antimicrobial, antifungal and antiviral agents

Appropriate antimicrobial therapy requires a working knowledge of both the patient and the infecting pathogen. The process begins with the recognition that a bacterial infection is indeed the cause of the patient's symptoms and an assessment of other existing conditions such as renal dysfunction or immune suppression [1]. The identification of the pathogen in specimens obtained from the site of infection is paramount for appropriate drug selection and is done using a combination of several techniques. Molecular diagnostics and mass spectrometry have had the biggest impact on organism identification; and today, many laboratories have the ability to identify organisms using these techniques.

The word antibiotic comes from Greek words "anti" means against and "bios" meaning life. The original word coined by Selman Waksman represented compounds which were derived from microorganisms. Today, many antibiotics are synthetic and semi-synthetic drugs. The first effective antibiotic discovered was penicillin. After the discovery of penicillin >100 antibiotics have been developed and currently used in medicine. Antibiotics can be classified based on their chemical structures or on the basis of mechanism [2, 3]. There are five major mechanisms by which an antibiotic exerts its pharmacological action. These include:

Inhibition of cell wall synthesis
Inhibition of bacterial protein synthesis
Alteration of bacterial cell wall
Inhibition of bacterial nucleic acid synthesis
Antimetabolite activities

Inhibition of bacterial cell wall formation and interference with bacterial protein synthesis are probably the most common mechanism by which an antibiotic kills bacteria or inhibits bacterial growth. Antibiotics which

Therapeutic Drug Monitoring Data
https://doi.org/10.1016/B978-0-12-815849-4.00010-4

interfere with cell wall synthesis include beta-lactam antibiotics (penicillins, cephalosporins, carbapenems, monobactams) and vancomycin. Clindamycin, chloramphenicol, lincomycin and macrolide antibiotics interfere with protein synthesis of bacteria by binding to 50S ribosomal subunit. Tetracycline and aminoglycosides interfere with the bacterial 30S ribosomal subunit.

Other antibiotics work by diverse mechanisms. Sulfonamides and trimethoprim kills bacteria by inhibiting folate synthesis. Antibacterial effect of metronidazole, quinolones and novobiocin are due to their capability of interfering with bacterial DNA synthesis, while rifampin interferes with bacterial RNA synthesis. Both polymyxin B and gramicidin kill bacterial by interfering with cell membrane function. However, drug resistance to antimicrobial agents is a big challenge in clinical practice [4].

Typically, minimum inhibitory concentration (MIC) and minimum bactericidal concentration are used to measure in vitro antimicrobial activity and potency. As such, antibiotics can be further characterized as either concentration-dependent (for which achieving a large post-dose concentration to MIC ratio appears important) or concentration-independent/time-dependent (where efficacy is related to maintaining the overall concentration above the MIC). Antibiotics such as beta-lactams (penicillins, cephalosporins, carbapenems, and monobactams), clindamycin, macrolide (erythromycin, clarithromycin), and linezolid can be effective in eradicating bacteria, because these antibiotics can bind to microorganisms for a long time. These antibiotics act as time-dependent antibiotics, with inhibitory effects observed if drug concentration exceeds MIC. For antibiotics which are involved in concentration-dependent killing (e.g., aminoglycosides and quinolones), the peak/MIC ratio is crucial for eradication of the bacteria.

Therapeutic drug monitoring (TDM) plays an important role in the therapy with some antibiotics. Overall, most antimicrobials are well tolerated and do not require routine monitoring using plasma/serum drug concentrations. Aminoglycosides and vancomycin are exceptions where routine TDM is essential due to toxicity (ototoxicity and nephrotoxicity) of these drugs [5]. There is a long history of using TDM for aminoglycosides and vancomycin. Chloramphenicol is a toxic drug. Although this drug is not widely prescribed, TDM is important. TDM is also emerging for other antimicrobials. Some antituberculosis drugs such as isoniazid and rifampin may benefit from TDM. Monitoring certain antifungal agents and antiretrovirals may also be useful.

References

[1] Pinder M, Bellomo R, Lipman J. Pharmacological principles of antibiotic prescription in the critically ill. Anaesth Intensive Care 2002;30:134–44.
[2] Spanu T, Santangelo R, Andreotti F, Cascio GL, et al. Antibiotic therapy for severe bacterial infections: correlation between the inhibitory quotient and outcome. Int J Antimicrob Agents 2004;23:120–8.
[3] Rhee KY, Gardiner DF. Clinical relevance of bacteriostatic versus bactericidal activity in the treatment of gram-positive bacterial infections. Clin Infect Dis 2004;39:755–6.
[4] Munita JM, Arias CA. Mechanisms of antibiotic resistance. Microbiol Spectr 2016;4(2). http://dx.doi.org/10.1128.
[5] Wong G, Sime FB, Lipman J, Roberts JA. How do we use therapeutic drug monitoring to improve outcomes from severe infections in critically ill patients? BMC Infect Dis 2014;14:288.

10.1

Aminoglycosides (amikacin, gentamicin, tobramycin)

10.1 Amikacin

Chemical properties	
Solubility in H_2O	Sparingly soluble
Molecular weight	586.6
pKa	n/a
Melting point	201–204 °C
Dosing	
Recommended dose, adult	10–15 mg/kg/day
Recommended dose, child	Same
Monitoring	
Sample	Serum, plasma (both peak and trough)
Effective concentrations	< 5 μg/mL (trough) 15–25 μg/mL (peak) with conventional dosing
Toxic concentrations	Peak concentration > 32 μg/mL
Methods	Immunoassay
Pharmacokinetic properties	
Oral dose absorbed	Administered by IV only
Time to peak concentration	Variable
Protein bound	<10%
Volume of distribution	0.27 L/kg
Half-life, adult	2.0–3.0 h
Time to steady state, adult	10–15 h
Half-life, child	Same as adult
Time to steady state, child	Same as adult

Excretion, urine

	% Excreted	Active	Detected in blood
Parent	98%	Yes	Yes

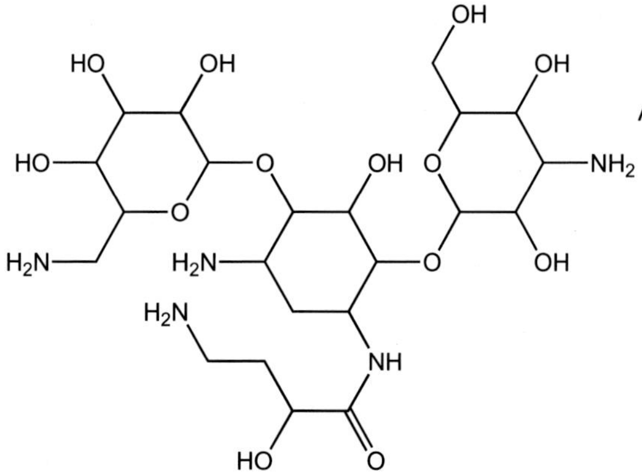

Amikacin

10.2 Gentamicin

Chemical properties	
Solubility in H_2O	100 mg/mL
Molecular weight	477.6 (variable due to other forms)
pKa	8.2
Melting point	102–108 °C, sulfate salt: 218–237 °C
Dosing	
Recommended dose, adult	3.0–5.0 mg/kg/day
Recommended dose, child	3.0–7.5 mg/kg/day
Monitoring	
Sample	Serum, plasma
Effective concentrations	Peak: 5.0–10.0 μg/mL Trough: < 2.0 μg/mL
Toxic concentrations	Peak: >12.0 μg/ml Trough: > 2 μg/mL
Methods	Immunoassay
Pharmacokinetic properties	
Oral dose absorbed	Administered by IV or IM
Time to peak concentration	Varies depending on mode of administration
Protein bound	0–30%
Volume of distribution	0.31 L/kg
Half-life, adult	2.0–3.0 h
Time to steady state, adult	10–15 h
Half-life, child	Same as adult
Time to steady state, child	Same as adult

Excretion, urine

	% Excreted	Active	Detected in blood
Parent	80–90	Yes	Yes

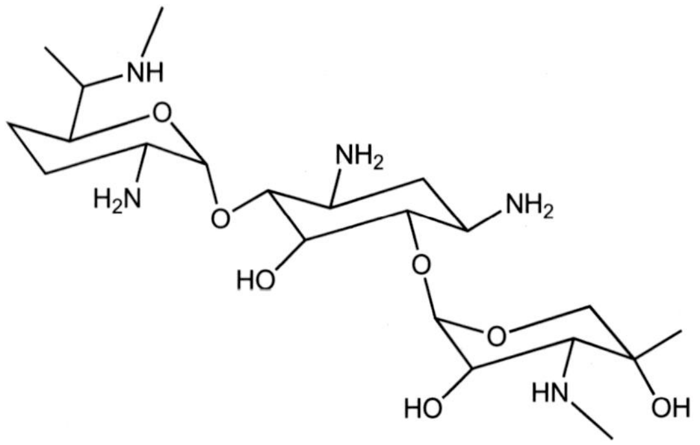

Gentamicin

10.3 Tobramycin

Chemical properties	
Solubility in H_2O	Water soluble
Molecular weight	467.54
pKa	6.7, 8.3, 9.9
Melting point	Not available

Dosing	
Recommended dose, adult	3–5 mg/kg/day
Recommended dose, child	Depends on age

Monitoring	
Sample	Serum, plasma
Effective concentrations	Peak: 5–10 µg/mL Trough: <2 µg/mL
Toxic concentrations	Peak: >12 µg/mL Trough: >2 µg/ml
Methods	Immunoassay, chromatography

Pharmacokinetic properties	
Oral dose absorbed	Administered IV or IM
Time to peak concentration	0.5–1.5 h
Protein bound	< 10%
Volume of distribution	0.2–0.4 L/kg
Half-life, adult	2–3 h
Time to steady state, adult	5–15 h
Half-life, child:	Variable
Time to steady state, child	Variable

Excretion, urine

	% Excreted	Active	Detected in blood
Parent	93	Yes	Yes

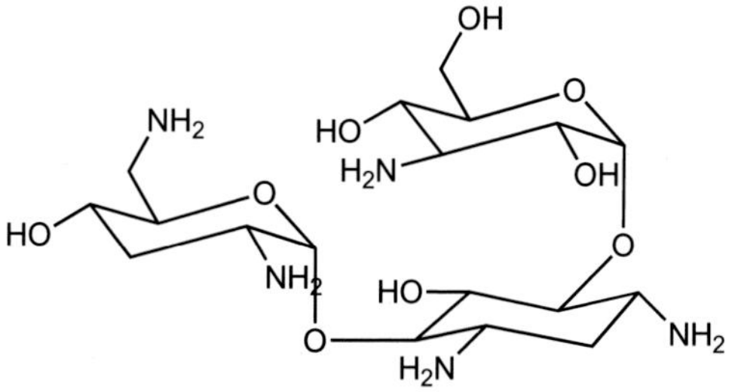

Tobramycin

Currently nine aminoglycosides (amikacin, gentamicin, tobramycin, streptomycin, neomycin, kanamycin, netilmicin, paromomycin and spectinomycin) are approved for use in the United States. Amikacin, gentamicin, and tobramycin are the three most commonly monitored aminoglycosides in clinical laboratories. Aminoglycosides are typically used to treat infections from gram-negative bacteria including *Escherichia coli* and *Pseudomonas aeruginosa*, as well as infections with other susceptible organisms including *Serratia*, *Proteus*, *Acinetobacter*, and *Klebsiella*. These drugs can be used alone or in combination with other antibiotics when treating more serious gram-negative infections and some gram-positive infections (gram-positive bacteria are intrinsically resistant to use of aminoglycosides alone). Once internalized, aminoglycosides bind to the bacterial ribosomal 30S subunit causing misreading of mRNA. This inhibits protein synthesis or leads to the production of defective proteins and subsequently causing death of the microorganism. To elicit its action on a specific organism, gentamicin binds to the aminoacyl-tRNA site (A site) of 16S rRNA on the ribosomal 30S subunit causing inhibition of translocation [1]. Alterations by the microorganism in the ribosomal 30 S subunit contribute to the development of resistant strains. Methylation of the ribosomal 16S target causes resistance to most aminoglycosides [2].

In general, aminoglycosides are poorly absorbed from the gut and for this reason are administered parenterally. These drugs are rapidly absorbed after intramuscular administration with peak concentrations usually achieved within one hour. Following intravenous administration, peak concentrations are usually observed immediately following the completion of the infusion. Two dosing regimens are used. In the first, the dose is divided and administered every 8–12 h. Alternatively, for some patients, the dose is administered in a larger single dose during a longer dosing interval. Nebulized forms of amikacin and tobramycin are also used for patients with cystic fibrosis as a means of treating chronic infections of *Pseudomonas aeruginosa* [3].

Aminoglycosides are hydrophilic with low protein binding (<10%) and freely filtered through the glomerulus in the kidney without being metabolized. These drugs have no stereoisomers, and their disposition is not affected by known genetic polymorphisms in drug transporters or drug-metabolizing enzymes. While most of the drug is excreted into the urine, up to 10% of an intravenously administered dose has been shown to accumulate in the kidney. Patients who have impaired renal function exhibit a prolongation in half-life dependent upon the degree of impairment. Patients who have ascites can have a markedly increased extracellular fluid volume, thus significantly altering the volume of distributions of aminoglycosides and also prolonging half-life. Since dialysis removes aminoglycosides, a rebound phenomenon may occur afterwards. This may alter calculated replacement dosage of aminoglycosides for these patients. Aminoglycosides can be absorbed from peritoneal dialysate [4].

The two major toxicities associated with aminoglycoside therapy are nephrotoxicity and ototoxicity. Nephrotoxicity is encountered in 15–17% of patients treated with conventional divided dose regimens, while ototoxicity is encountered in 20–25% patients. The drugs are taken-up by the epithelial cells of the renal proximal tubules where they bind to acidic phospholipids and megalin in the brush border membrane, and then accumulate. Megalin is a receptor expressed at the apical membrane of renal proximal tubules. Animal studies suggest that by blocking the binding to the megalin receptor, nephrotoxicity can be prevented. [5] The risk of amikacin-induced nephrotoxicity is increased for patients with chronic liver disease and hypoalbuminemia. Serum creatinine is monitored before and after initiating therapy to assess or define nephrotoxicity. The criteria used to define aminoglycoside nephrotoxicity vary between institutions. The following are generally considered significant indicators for nephrotoxicity: an increase in serum creatinine of 5 mg/dL or above when the baseline creatinine is 1.9 mg/dL or lower; an increase of 1 mg/dL or higher when the baseline is between 2.0 and 4.9 mg/dL; and an increase of 1.5 mg/dL or more when the baseline is above 5 mg/dL. Glomerular filtration rate and cystatin C are also used as indicators of nephrotoxicity [6].

In the conventional, multiple dosing protocols, serial trough and peak aminoglycoside levels correlate with nephrotoxicity. With once a day protocols, any drug present at trough is indicative of potential toxicity as no drug should remain. Trough samples should be measured at least once every 3 days for these protocols [7]. Sanchez-Alcaraz et al. studied therapeutic drug monitoring (TDM) of tobramycin in once daily versus twice daily dosage regimens and concluded that once daily regimen of tobramycin is at least as effective as and possibly less toxic than the twice daily regimen. Moreover, using a single daily dose therapy, peak concentration determination is not necessary and only trough concentration should be monitored to ensure levels below 2 μg/mL [8].

Unfortunately, aminoglycoside-induced ototoxicity is often irreversible. Vestibular and cochlear sensory cells are damaged, resulting in both auditory loss and vestibular dysfunction. Initial symptoms include tinnitus, decreased perception of high frequency sound, headache, and vertigo. Total dose, total area under the curve, and duration of therapy seem to correlate better with ototoxicity than peak and trough concentrations. Cellular mechanism of aminoglycoside-induced ototoxicity is complex [9].

The most important drug–drug interactions with aminoglycosides occur with other drugs that are nephrotoxic or ototoxic: for example, cyclosporine, tacrolimus, cisplatin, ethacrynic acid, furosemide, and cephalosporin antibiotics. Neuromuscular blockade has been reported with the co-administration of an aminoglycoside and a calcium channel blocker. This can lead to respiratory depression and neuromuscular blockage and is a particular problem with verapamil. When given with rocuronium, a neuromuscular blocking agent used in intubation, the action of rocuronium has been prolonged. Therefore, neuromuscular blockade should be closely monitored for patients also receiving aminoglycosides [10].

TDM of aminoglycosides are important for patient management, because it may reduce aminoglycoside toxicity. Reduced risk of ototoxicity due to TDM of aminoglycoside in treatment of multi-drug resistant tuberculosis has been reported [11]. Serum is the preferred specimen. With multiple daily treatment, both trough and peak serum concentrations should be measured. Peak serum concentration should be measured 60 min after beginning of the infusion (which usually takes 30 to 45 min). Trough specimens (pre-dose) should be collected within 15–30 min before the next dose is given [12].

The pharmacodynamic advantages of once-daily dosing include enhanced concentration-related bacterial killing, increased aminoglycoside postantibiotic effect, leukocyte enhancement, and decreased adaptive resistance of the bacterial pathogens. The peak concentrations are higher than those resulting from multiple-daily dosing, while trough concentrations are considerably lower after once daily dosing compared to multiple dosing. For once daily dosage, one specimen is taken at a particular time depending on the protocol. For example, using Hartford nomogram, a specimen is collected 6–14h postdose [13]. However, in other protocol for once daily dosing, both peak and trough specimen may be monitored. If aminoglycoside therapy is used for 2 days or less, TDM may not be necessary.

In general, commercially available immunoassays are used for routine TDM of these drugs in clinical laboratories. There are several different platforms of commercially available immunoassays for amikacin, gentamicin and tobramycin. One important issue in determining concentrations of aminoglycosides in serum or plasma is inactivation of aminoglycosides by carbenicillin, ticarcillin and some other beta-lactam antibiotics. Inactivation of tobramycin and gentamicin by these semi-synthetic penicillins were greater than inactivation of amikacin and netilmicin. Because aminoglycosides are often used along with semi-synthetic penicillins, serum collected for TDM of aminoglycosides should be analyzed as soon as possible and if storage of a specimen is needed prior to analysis, the specimen must be frozen [14]. Aminoglycosides can also be analyzed in human serum using liquid chromatography combined with tandem mass spectrometry [15].

References

[1] Yoshizawa S, Fourmy D, Puglisi JD. Structural origins of gentamicin antibiotic action. EMBO J 1998;17:6437–48.

[2] Liou GF, Yoshizawa S, Courvalin P, Gailmand M. Aminoglycosides resistance by Arm A mediated ribosomal 16S methylation in human bacterial pathogens. J Mol Biol 2006;359:358–64.

[3] Sermet-Gaudelus I, Le Cocguic Y, Ferroni A, Clairicia M, et al. Nebulized antibiotics in cystic fibrosis. Paediatr Drugs 2002;4:455–67.

[4] Begg EJ, Barclay ML, Kirkpatrick C. The therapeutic monitoring of antimicrobial agents. Br J Clin Pharamcol 1999;47:23–30.

[5] Nagai J, Takano M. Molecular aspect of renal handling of aminoglycosides and strategies for preventing nephrotoxicity. Drug Metab Pharmacokinet 2004;19:159–70.

[6] Hermida J, Tutor JC. Serum cystatin C for the prediction of glomerular filtration rate with regard to the dose adjustment of amikacin, gentamicin, tobramycin and vancomycin. Ther Drug Monit 2006;28:326–31.

[7] Bartal C, Danon A, Schlaeffer F, Reisenberg K, et al. Pharmacokinetic dosing of aminoglycosides: a controlled study. Am J Med 2003;114:194–8.

[8] Sanchez-Alcaraz A, Vargas A, Quintana MB, Rocher A, et al. Therapeutic drug monitoring of tobramycin: once daily versus twice daily dosage schedule. J Clin Pharam Ther 1998;23:367–73.

[9] Warchol ME. Cellular mechanism of aminoglycoside ototoxicity. Curr Opin Otolaryngol Head Neck Surg 2010;18:454–8.

[10] Gilliard V, Delvaux B, Russell K, Dubois PE. Long lasting potentiation of a single dose of rocuronium by amikacin: a case report. Acta Anaesthesiol Belg 2006;57:157–9.

[11] van Altena R, Dijkstra JA, van der Meer ME, Borjas Howard JF, et al. Reduced chance of hearing loss associated with therapeutic drug monitoring of aminoglycosides in the treatment of multidrug-resistant tuberculosis. Antimicrob Agents Chemother 2017;61:e01400–16.

[12] Balakrishnan I, Shorten RJ. Therapeutic drug monitoring of antimicrobials. Ann Clin Biochem 2016;53:333–46.
[13] Stankowicz MS, Ibrahim J, Brown DL. Once-daily aminoglycoside dosing: an update on current literature. Am J Health Syst Pharm 2015;72(16):1357–13664.
[14] Pickering LK, Rutherford I. Effect of concentration and time upon inactivation of tobramycin, gentamicin, netilmicin, and amikacin by azlocillin, carbenicillin, mecillinam, mezlocillin and piperacillin. J Pharamcol Exp Ther 1981;217:345–9.
[15] Dijkstra JA, Sturkenboom MG, Kv H, Koster RA, et al. Quantification of amikacin and kanamycin in serum using a simple and validated LC-MS/MS method. Bioanalysis 2014;6(16):2125–33.

10.2

Vancomycin

Chemical properties	
Solubility in H_2O	Soluble
Molecular weight	1449.25
pKa	NA[a]
Melting point	NA
Dosing	
Recommended dose, adult	2 g/day IV
Recommended dose, child	20–30 mg/kg/day
Monitoring	
Sample	Serum, plasma
Effective concentrations	Peak: 20–40 μg/mL Trough: 5–10 μg/mL
Toxic concentrations	Peak: >80 μg/mL Trough: >10 μg/mL
Methods	Immunoassay
Pharmacokinetic properties	
Oral dose absorbed	Poorly absorbed from gastrointestinal tract
Time to peak concentration	0.5–1.0 h
Protein bound	55%
Volume of distribution	0.69 L/kg
Half-life, adult	5–11 h
Time to steady state, adult	20–35 h
Half-life, child	3–10 h
Time to steady state, child	15–50 h

Excretion, urine

	% Excreted	Active	Detected in blood
Parent	80–90	Yes	Yes

[a] Not available.

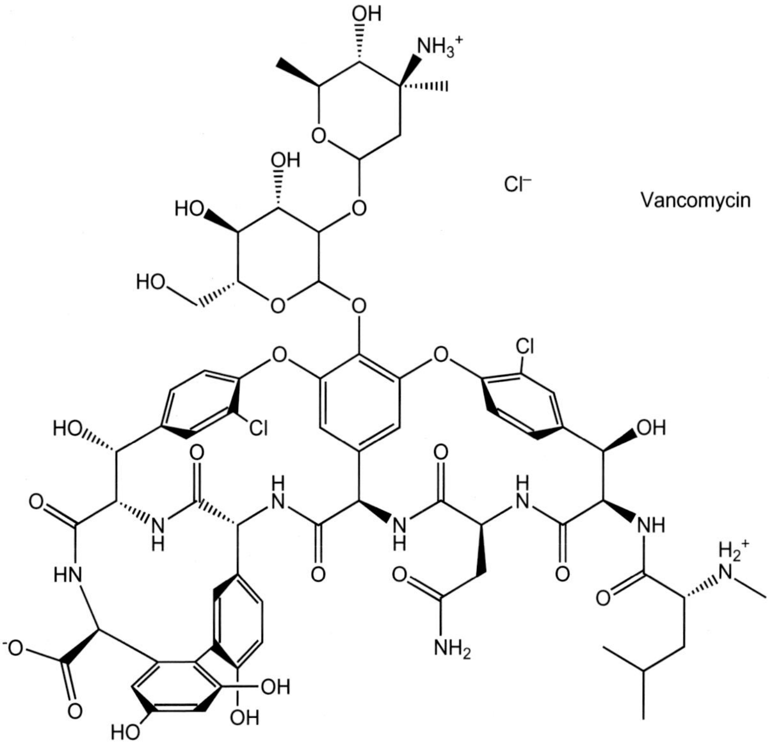

Vancomycin is a branched, glycosylated, non-ribosomal peptide (glycopeptide) first derived in 1956 from the bacterium *Streptomycin orientalis* found in soil of Indonesia and India. The compound was named as vancomycin because of its ability to "vanquish" emerging strains of penicillinase-producing staphylococci that were resistant to penicillin. The drug is effective against gram-positive organisms and has been shown to be active against most strains of the following microorganisms: *Clostridium difficile*, *Listeria monocytogenes*, *Streptococcus pyogenes*, *Streptococcus pneumoniae*, *Streptococcus agalactiae*, *Streptococcus viridans, Streptococcus bovis, Staphylococcus epidermidis*, *Actinomyces* species and *Lactobacillus* species. Vancomycin has been considered a key drug in treating methicillin-resistant *Staphylococcus aureus* (MRSA) [1].

The pharmacological action of vancomycin is due to its ability to inhibit bacterial cell wall biosynthesis. Vancomycin binds non-covalently to undecaprenyl-(muramyl-glucosaminyl)-pentapeptide, a precursor in bacterial cell wall biosynthesis, and inhibits the synthesis of peptidoglycan. This inhibits the transglycosylation and transpeptidation steps the bacteria must perform during cell-wall assembly, and as a result, the cell wall becomes unstable and lyses. Vancomycin increases bacterial cell wall permeability and inhibits RNA [2]. These actions occur at sites distinct from those inhibited by penicillins and cephalosporins. Vancomycin additionally increases bacterial cell wall permeability and inhibits RNA synthesis. Unfortunately, vancomycin resistant MRSA (VRSA) and vancomycin-intermediate resistant MRSA (VISA) bacterial strains are encountered with increasing frequency. Based on cases identified to date, risk factors for developing VRSA may include age, compromised blood flow to the lower limbs, and the presence of chronic ulcers in the patient. Treatment options for such patients are limited [3].

Vancomycin is very poorly absorbed from the gastrointestinal tract after oral administration. Therefore, this route of administration is limited to treatment of *Clostridium difficile* colitis in severally ill patients, where the target is bacteria within the gastrointestinal tract. *Clostridium difficile* associated diarrhea is as serious medical condition with mortality of up to 25% in frail elderly people and can be treated by metronidazole or oral vancomycin administered as a dosage of 125 mg four times a day. Higher dosage may be used for severe episodes [4].

Vancomycin is most often given intravenously for treatment of infections. Intramuscular injection is painful. Vancomycin is available as hydrochloride salt. The serum concentration time curve of vancomycin is best described using a multi-compartmental model that takes into account the distribution of vancomycin. It is distributed in cerebrospinal fluid, pericardial fluid, pleural fluid, ascitic fluid, and synovial fluid. Vancomycin is mostly excreted in urine unchanged by glomerular filtration, but the drug is also found in bile and stool, indicating minor non-renal excretion pathways. Clearance of vancomycin correlates with creatinine clearance. The average volume of distribution of vancomycin in normal adults with ideal body weight is 0.69 (L/kg). The volume of distribution is higher in females, patients over 60 years of age, and obese patients. The half-life of vancomycin varies from 5 to 11 h for adults who have normal renal function. Elderly patients may show longer half-life, even with normal renal function, and need more individualized dose adjustment. In one study, the average half-life of vancomycin in patients >60 years old who had normal creatinine concentration was extended to 17.8 h compared to a mean half-life of 7.5 h in patients 18–59 years [5]. The half-life of vancomycin is also increased in patients who have chronic spinal cord injury, and such patients may require a lower dose.

The volume of distribution and half-life of vancomycin differ in neonates and children compared to adults. Vancomycin pharmacokinetics in critically ill infants shows some interesting changes. In one study, the authors observed that the volume of distribution changed from 0.81 L/kg in day 2 of the treatment in infants to 0.44 L/kg on day 8, and suggested such changes may be related to aggressive fluid resuscitation in the beginning of treatment to these critically ill infants. The half-life also changed from 5.3 h on the second day to 3.4 h on the eighth day [6]. Vancomycin is removed during high-flux hemodialysis using a polysulfone membrane and in one study, the percentage of vancomycin lost ranged from 39.1–55.1% during first session of dialysis. The concentration of vancomycin six hours after hemodialysis ranged from 18.2–45.1 μg/mL and one week after dosing ranged from 8.14–10.1 μg/mL. The authors concluded that 25 mg/kg of vancomycin given during high-flux hemodialysis may provide adequate serum concentration of vancomycin in anuric hemodialysis patients for up to seven days [7].

Vancomycin nephrotoxicity is reported to be <5% when used as monotherapy, but the incidence of nephrotoxicity increased to 8–35% when vancomycin was used in combination with aminoglycosides. Ototoxicity, in the form of tinnitus and loss of hearing of high tone sounds, is usually associated with serum concentrations >80 μg/mL. Additional central nervous system toxicities manifested include lightheadedness, nausea, and vomiting [8]. Rapid infusion of vancomycin may be associated with pruritus, hypotension, and a rash involving the upper torso, head, and neck. Vancomycin "red man" or "red neck" syndrome occurs frequently in normal adults who receive 1000 mg of vancomycin over 1 h, because vancomycin causes an infusion rate-dependent increase in plasma histamine concentration. This phenomenon can be avoided by slower administration of vancomycin over at least 60 min [9].

Heparin and vancomycin may be incompatible if mixed in intravenous solution or infused one after another through a common intravenous line. Luther et al. reported that heparin reduced activity of vancomycin solutions against *S. aureus*, *S. epidermidis*, and *E. faecalis* biofilms [10]. TDM of vancomycin is needed if therapy is continued over 48 h [11]. Both peak and trough concentrations should be monitored. Ranges for peak concentrations of 20–40 μg/mL have been widely quoted, and trough concentrations range of 5–10 μg/mL has reasonable literature support. Unlike for other drugs, the therapeutic targets for vancomycin are variable and even controversial. More recently, emphasis has been placed on trough serum level measurements. In order to improve clinical outcomes of complicated infections, such as bacteremia, endocarditis, osteomyelitis, meningitis, and hospital-acquired pneumonia caused by *S. aureus*, trough serum vancomycin concentrations of 15–20 μg/mL are recommended. Trough serum vancomycin concentrations in that range should achieve an AUC/MIC of >400 for most patients if the MIC is <1 μg/mL [12]. Sometimes during treatment of patients with vancomycin-resistant MRSA, clinicians gradually increase vancomycin target serum vancomycin concentration in the range of 15–20 μg/mL but such levels are associated with a higher risk of nephrotoxicity [13].

Serum is the preferred specimen for vancomycin analysis. Several immunoassays are available for measuring serum vancomycin concentrations. Interferences by paraproteins have been reported to affect vancomycin immunoassays, usually resulting in falsely lower true vancomycin levels. Gunther reported cases of suspected immunoglobulin-mediated interference with the Beckman Coulter particle–enhanced turbidimetric inhibition immunoassay (PETINIA). In a 64 year old patient with elevated IgM of 38.5 g/L (normal: 0.6 to 3.0 g/L), serum vancomycin level was below 4 μg/mL despite increasing dosage of vancomycin. When the specimen was analyzed by another PETINIA assay on Dimension analyzer (Siemens), the vancomycin level was 57.7 μg/mL [14]. Paraprotein caused falsely elevated vancomycin value using iVanco immunoassay on the Architect i2000SR analyzer (Abbott Diagnostics) in an 82 year old man with IgM concentration of 74 g/L and a total protein concentration of 98 g/L. His vancomycin level was >100 μg/mL but on 1:4 dilution the value was 21.8 μg/mL. When a pre-vancomycin infusion specimen was analyzed, a value of 70.3 μg/mL was observed. Suspecting interference, the authors measured free vancomycin in the protein free ultrafiltrate (IgM is absent in the ultrafiltrate due to high molecular weight) and observed free vancomycin value of 14.8 μg/mL. Using previously published equation, the estimated total vancomycin concentration was only 20.8 μg/mL. The authors also demonstrated that this interference can be avoided by using Roche Diagnostics VANC2 assay or a liquid chromatography combined with mass spectrometric assay [15].

References

[1] Holmes NE, Tong SY, Davis JS, van Hal SJ. Treatment of methicillin-resistant Staphylococcus aureus: vancomycin and beyond. Semin Respir Crit Care Med 2015;36:17–30.
[2] Loll PJ, Axelsen PH. The structural biology of molecular recognition by vancomycin. Annu Rev Biophys Biomol Struct 2000;29:265–89.

[3] Appelbaum PC. The emergence of vancomycin-intermediate and vancomycin resistant Staphylococcus aureus. Clin Microbiol Infect 2006;12(Suppl. 1):16–23.
[4] Starr J. Clostridium difficile diarrhoea: diagnosis and treatment. BMJ 2005;331: 498–501.
[5] Guay DR, Vance-Bryan K, Gilliland S, Rodvold K, et al. Comparison of vancomycin pharmacokinetics in hospitalized elderly and young patients using a Bayesian forecaster. J Clin Pharmacol 1993;33:918–22.
[6] Gous AG, Dance MD, Lipman J, Luyt DK, et al. Changes in vancomycin pharmacokinetics in critically ill infants. Anaesth Intensive Care 1995;23:678–82.
[7] Foote EF, Dreitlein WB, Steward CA, Kapoian T, et al. Pharmacokinetics of vancomycin when administered during high flux hemodialysis. Clin Nephrol 1998;50:51–5.
[8] Hadaway L, Chamallas SN. Vancomycin: new perspectives on an old drug. J Infus Nurs 2003;26:278–84.
[9] Polk RE, Healy DP, Schwartz LB, Rock DT, et al. Vancomycin and the red-man syndrome: pharmacodynamics of histamine release. J Infect Dis 1988;157:502–7.
[10] Luther MK, Mermel LA, LaPlante KL. Comparison of linezolid and vancomycin lock solutions with and without heparin against biofilm-producing bacteria. Am J Health Syst Pharm 2017;74:e193–201.
[11] Momattin H, Zogheib M, Homoud A, Al-Tawfiq JA. Safety and outcome of pharmacy-led vancomycin dosing and monitoring. Chemotherapy 2016;61:3–7.
[12] Rybak MJ, Lomaestro BM, Rotschafer JC, Moellering RC, et al. Vancomycin therapeutic guidelines: a summary of consensus recommendations from the infectious diseases Society of America, the American Society of health-system pharmacists, and the society of infectious diseases pharmacists. Clin Infect Dis 2009;49:325–7.
[13] Bruniera FR, Ferreira FM, Saviolli LR, Bacci MR, et al. The use of vancomycin with its therapeutic and adverse effects: a review. Eur Rev Med Pharmacol Sci 2015;19(4): 694–700.
[14] Gunther M, Saxinger L, Gray M, Legatt D, et al. Two suspected cases of immunoglobulin-mediated interference causing falsely low vancomycin concentrations with the Beckman PETINIA method. Ann Pharmacother 2013;47:e19.
[15] Florin L, Vantilborgh A, Pauwels S, Vanwynsberghe T, et al. IgM interference in the Abbott iVanco immunoassay: a case report. Clin Chim Acta 2015;447:32–3.

10.3
Chloramphenicol

Chemical properties	
Solubility in H_2O	2.5 mg/mL
Molecular weight	323.14
pKa:	5.5
Melting point	150.5–151.5 °C
Dosing	
Recommended dose, adult	50–100 mg/kg/day
Recommended dose, child	50 mg/kg/day for >1 month 25 mg/kg/day for <1 month
Monitoring	
Sample	Serum, plasma
Effective concentrations	10–25 μg/mL
Toxic concentrations	>25 μg/mL
Gray Baby Syndrome	40–200 μg/mL
Methods	Immunoassay, HPLC, GC
Pharmacokinetic properties	
Oral dose absorbed	75–90%
Time to peak concentration	2 h
Protein bound	50–60%
Volume of distribution	0.60–1.0 L/kg
Half-life, adult	Approximately 4 h
Time to steady state, adult	7.5–25 h
Half-life, child	<2 wk.: 24 h; 2–4 wk.: 12 h

Excretion, urine

	% Excreted	Active	Detected in blood
Unchanged	5–10	Yes	Yes
As glucuronide	70–80	No	No

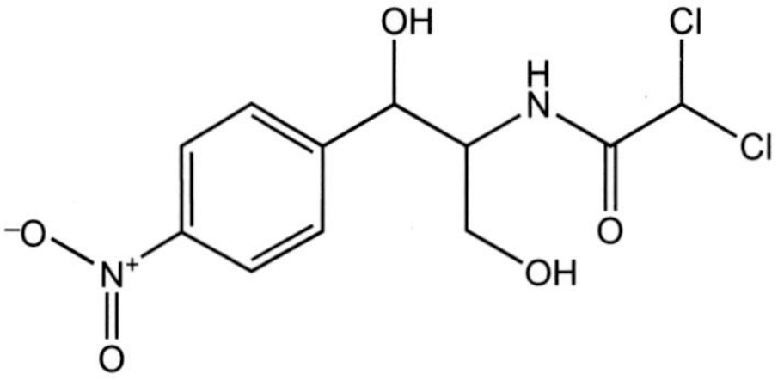

Chloramphenicol

Chloramphenicol is an old antimicrobial agent that is rarely used today. Until 1980, chloramphenicol, ampicillin and trimethoprim/sulfamethoxazole (co-trimoxazole) were the drugs of choice to treat typhoid. Since then, the increasing incidence of strains resistant to chloramphenicol, ampicillin and trimethoprim/sulfamethoxazole has led to the use of fluoroquinolones, particularly ciprofloxacin, for the empirical treatment of enteric fever. Chloramphenicol is potentially very toxic. The most common serious adverse effect is bone marrow suppression. This may occur by two distinct mechanisms: dose-related bone marrow suppression, which usually occurs after more than 7 days of treatment, and fatal irreversible idiosyncratic aplastic anemia. Chloramphenicol-induced aplastic anemia is rare (affecting 1 in approximately 30,000 or more courses of therapy with oral chloramphenicol) but unfortunately cannot be predicted. As a result, chloramphenicol cannot be recommended as a first-line choice for treating respiratory tract infections, meningitis or enteric fever in areas of the world where safer and more effective antibiotics are available. Chloramphenicol is still common in resource-limited countries [1].

More common side-effects of chloramphenicol are skin rashes and gastrointestinal reactions. Occasionally, a few patients on prolonged therapy will develop signs of optic and peripheral neuritis. Most patients receiving chloramphenicol exhibit varying degrees of arrest in red cell maturation; thus, individuals with shortened red cell survival may quickly develop serious anemias. Newborns and young infants are particularly susceptible to a form of cardiovascular collapse known as "gray baby syndrome." The syndrome develops as a consequence of the immaturity of the drug-metabolizing enzymes and is reflected in a decreased ability to form chloramphenicol glucuronides. Unconjugated serum levels of chloramphenicol are elevated into the toxic range. Chloramphenicol should not be administered to newborns or young infants if serum levels cannot be accurately measured, because gray baby syndrome usually occurs when serum chloramphenicol concentration is the range of 40–200 μg/mL. If serum chloramphenicol concentrations are maintained within the therapeutic range 15–25 μg/mL, risk of gray baby syndrome is very low [2].

Chloramphenicol is administered orally, intravenously, or intramuscularly (IM). The palmitate and succinate preparations require in vivo hydrolysis to the biologically active chloramphenicol. IM administration of chloramphenicol has been reported to lead to subtherapeutic concentrations of chloramphenicol during initial therapy in patients with enteric fever. Oral preparations are rapidly absorbed from the gastrointestinal tract, with peak blood levels being reached within 2 h. The bioavailability of oral crystalline chloramphenicol and chloramphenicol palmitate is approximately 80%. Plasma protein binding of chloramphenicol is approximately 60% in healthy adults. The drug is extensively distributed to many tissues and body fluids, including cerebrospinal fluid and breast milk, and it crosses the placenta with an apparent volume of distribution range from 0.6 to 1.0 L/kg. Most of a chloramphenicol dose is metabolized by the liver to inactive products, the chief metabolite being a glucuronide conjugate; only 5 to 15% of chloramphenicol is excreted unchanged in the urine. The elimination half-life is approximately 4 h. Patients with renal insufficiency may exhibit prolonged half-life since chloramphenicol is cleared by renal excretion [3].

Chloramphenicol is a weak inhibitor of cytochrome P450 (CYP) 2C8/9 and 3A4 enzyme systems and may increase serum levels of drugs such as phenytoin, phenobarbital, and warfarin that are metabolized by these systems. Chronic administration of phenytoin or phenobarbital and acute administration of rifampin may shorten the half-life of chloramphenicol, resulting in subtherapeutic concentrations of the antibiotic [4, 5].

Therapeutic drug monitoring (TDM) of chloramphenicol is essential if used in therapy. Schwartz et al. compared a chromatographic method with an immunoassay for chloramphenicol (EMIT) and observed good correlations between two methods based on analysis of 49 serum specimens [6]. In addition, there are many chromatographic methods for TDM of chloramphenicol. Davidson and Fitzpatrick described a rapid column chromatographic protocol for determination of chloramphenicol concentration in serum using mephenesin as the internal standard. The authors used a reverse phase C-18 column and monitored the elution of peak using a UV-detector at 278 nm. Baseline separation was obtained between chloramphenicol and the internal standard and the protocol was free from interferences [7].

References

[1] Eliakim-Raz N, Lador A, Leibovici-Weissman Y, Elbaz M, et al. Efficacy and safety of chloramphenicol: joining the revival of old antibiotics? Systematic review and meta-analysis of randomized controlled trials. J Antimicrob Chemother 2015;70:979–96.

[2] Mulhall A, de Louvois J, Hurley R. Chloramphenicol toxicity in neonates: its incidence and prevention. Br Med J (Clin Res Ed) 1983;287(6403):1424–7.
[3] Ambrose PJ. Clinical pharmacokinetics of chloramphenicol and chloramphenicol succinate. Clin Pharmacokinet 1984;9:222–38.
[4] Balbi HJ. Chloramphenicol: a review. Pediatr Rev 2004;25(8):284–8.
[5] Kelly HW, Couch RC, Davis RL, Cushing AH, Knott R. Interaction of chloramphenicol and rifampin. J Pediatr 1988;112:817–20.
[6] Schwartz JG, Castro DT, Ayo S, Carnahan JJ, Jorgensen JH. A commercial enzyme immunoassay method (EMIT) compared with liquid chromatography and bioassay methods for measurement of chloramphenicol. Clin Chem 1998;34:1872–5.
[7] Davidson DF, Fitzpatrick J. Rapid column chromatographic separation of chloramphenicol in serum. Clin Chem 1986;32:701–2.

10.4

Isoniazid

Chemical properties	
Solubility in H_2O	140 mg/mL
Molecular weight	137.14
pKa	1.82
Melting point	171.4 °C
Dosing	
Recommended dose, adult	5–15 mg/kg (300 mg/day starting dose but it may be increased to 900 mg/day)
Recommended dose, child	5–15 mg/kg/day
Monitoring	
Sample	Serum, plasma (heparin, EDTA)—process and freeze at −70 °C
Effective concentrations	Peak, 2 h post dose 300 mg dose: 3–7 μg/mL 900 mg dose: 9–18 μg/mL
Toxic concentrations	>20 μg/mL
Methods	HPLC, GC, LC–MS/MS
Pharmacokinetic properties	
Oral dose absorbed	90%
Time to peak concentration	0.5–2 h
Protein bound	<10–15%
Volume of distribution	0.57–0.75 L/kg
Half-life, adult	Fast acetylators: 0.5–1.6 h slow acetylators: 2–5 h
Time to steady state, adult	na
Half-life, child	Fast acetylators: 0.5–2.7 h slow acetylators: 2–5.8 h
Time to steady state, child	na

Excretion, urine

	% Excreted	Active	Detected in blood
Unchanged and hydrazone conjugates	6 (fast acetylators); 37 (slow acetylators)	Yes	Yes
Acetylisoniazid	94 (fast acetylators); 63 (slow acetylators)	No	Yes
Isonicotinic acid		No	No

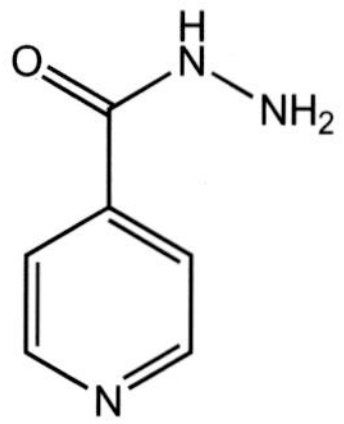

Isoniazid

The first-line drugs used for treatment of drug-sensitive tuberculosis (TB) include isoniazid, rifampin (rifampicin), ethambutol and pyrazinamide. Second-line drugs such as aminosalicylic acid, thioamides, ethionamide, protionamide, rifabutin, rifapentine, cycloserine, capreomycin, and fluoroquinolones are used in treating multi-drug resistant TB. Isoniazid (INH, pyridine-4-carbohydrazide, Nydrazid) is a synthetic drug used alone or with other drugs, in prevention of TB in people exposed to the bacteria or in particularly vulnerable groups (e.g., HIV-infected individuals and children in household of adults with pulmonary TB). Because it eliminates only active bacteria and the bacteria may exist in a resting state for months, long-term INH treatment (6–12 months) is necessary [1]. Typical treatment protocols include an initial 2–6 month treatment with a combination of INH, rifampin and pyrazinamide (Rifater) plus a fourth drug (ethambutol or streptomycin) if exposure to resistant organisms is known or suspected. This regimen is followed by a 9 month (or longer) regimen of INH or a 4 month regimen of rifampin alone or rifampin and INH.

Oral, intravenous and intramuscular are routes of administration for INH, but the most common is the oral route. INH is rapidly absorbed after oral administration with almost 100% bioavailability. Peak serum levels are observed in 0.5–2 h. food significantly reduces bioavailability [1]. In addition, HIV positive patients, especially male patients with low weight may show poor absorption of INH [2]. INH is widely distributed throughout body fluids and shows very low binding with serum proteins. Acetylation is the main metabolic transformation of INH. The elimination half-life of INH and its metabolites is 0.5 to 4 h and their main elimination route is through the kidney, with 75% to 96% of the drug and metabolites excreted in urine within 24 h [1].

Approximately 15–20% patients treated with INH show increases in liver enzymes (alanine and aspartate transaminases), but about 1% patients show significant hepatotoxicity which may be influenced by pharmacogenomics impacting INH metabolism. Polymorphisms of genes encoding *N*-acetyl transferase 2 (NAT2), cytochrome P450 (CYP) 2E1, and glutathione-S-transferase 1 enzymes play important roles in metabolism of INH. NAT2 is responsible for conversion of INH into acetylisoniazide. Genetic polymorphism of NAT2 primarily mediates the variation of individuals into the categories commonly referred to as "slow" vs. "fast" acetylators of INH.

Acetylisoniazid undergoes hydrolysis to isonicotinic acid and acetylhydrazine which can then undergo a second acetylation to diacetylhydrazine or oxidation via CYP2E1 to produce known metabolites that are potentially hepatotoxic. Glutathione in the liver can inactivate such toxins by binding with them. Studies have shown that slow acetylators have a higher risk of INH-induced hepatotoxicity than fast acetylators. Patients with the homozygous CYP2E1 c1/c1 genotype also appear to be at increased risk for drug-induced hepatitis, and periodic liver function testing is appropriate for these patients. The ratio of unchanged drug and hydrazone conjugates to acetylisoniazid and metabolites varies with acetylator phenotype [3]. Children eliminate INH faster than adults, with younger children having a faster elimination rate than older children [4].

INH is an inhibitor of CYP2C9 and may cause increased levels of other drugs metabolized by this enzyme system such as phenobarbital and phenytoin. Administration of INH has also been reported to cause increased levels of carbamazepine, ethosuximide, theophylline, and valproic acid. Drugs known to cause increased INH levels include aminosalicylic acid, procainamide, propranolol, and chlorpromazine. Drugs which decrease serum levels of INH include cyclosporine, ketoconazole, and verapamil. At least one report of severe acetaminophen toxicity has been reported in a patient receiving INH. The mechanism of this interaction appears to be induction of CYP2E1, a mixed-function oxidase that can generate hepatotoxic acetaminophen metabolites [5].

Adverse reactions to INH include rash, hepatitis, anemia, peripheral neuropathy, and mild central nervous system effects. In addition to monitoring INH concentrations, determination of acetylator status may help to establish dosages to ensure achievement of effective concentrations without developing toxicity, although such practice is not currently routine practice. Chromatographic methods are available for TDM of isoniazid [6, 7]. More recently, Prahl et al. described a liquid chromatography combined with tandem mass spectrometric method for simultaneous analysis of isoniazid, rifampin, ethambutol and pyrazinamide. The target peak concentration of isoniazid is 3–7 μg/mL after daily dose of 300 mg [8].

References

[1] Fernandes GFDS, Salgado HRN, Santos JLD. Isoniazid: a review of characteristics, properties and analytical methods. Crit Rev Anal Chem 2017;47:298–308.

[2] McIlleron H, Rustomjee R, Vahedi M, Mthiyane T, et al. Reduced antituberculosis drug concentrations in HIV-infected patients who are men or have low weight: Implications for international dosing guidelines. Antimicrob Agents Chemother 2012;56:3232–8.

[3] Perwitasari DA, Atthobari J, Wilffert B. Pharmacogenetics of isoniazid-induced hepatotoxicity. Drug Metab Rev 2015;47:222–8.

[4] Schaaf HS, Parkin DP, Seifart HI, Werely CJ, et al. Isoniazid pharmacokinetics in children treated for respiratory tuberculosis. Arch Dis Child 2005;90:614–8.

[5] Baciewicz AM, Self TH. Isoniazid interactions. South Med J 1985;78:714–8.

[6] Ray J, Gardiner I, Marriott D. Managing antituberculosis drug therapy by therapeutic drug monitoring of rifampicin and isoniazid. Intern Med J 2003;33:229–34.

[7] Moussa LA, Khassouani CE, Soulaymani R, Jana M, et al. Therapeutic isoniazid monitoring using a simple high-performance liquid chromatographic method with ultraviolet detection. J Chromatogr B 2001;766:181–7.

[8] Prahl JB, Lundqvist M, Bahl JM, Johansen IS, et al. Simultaneous quantification of isoniazid, rifampicin, ethambutol and pyrazinamide by liquid chromatography/tandem mass spectrometry. APMIS 2016;124:1004–15.

10.5

Rifampin

Chemical properties	
Solubility in H_2O	1.4 mg/mL
Molecular weight	822.9
pKa	1.7 (4-OH) and 7.9 (N)
Melting point	183 °C
Dosing	
Recommended dose, adult	10–15 mg/kg/day
Recommended dose, child	10–20 mg/kg/day
Monitoring	
Sample	Serum, plasma (heparin, EDTA)—process and freeze at −70 °C
Effective concentrations	Peak, 2 h postdose; 8–24 μg/mL
Toxic concentrations	>55 μg/mL
Methods	Immunoassay, HPLC, GLC
Pharmacokinetic properties	
Oral dose absorbed	90–95%
Time to peak concentration	1–4 h
Protein bound	86.1–88.9%
Volume of distribution	0.9–1.6 L/kg
Half-life, adult	1.5–5.1 h
Time to steady state, adult	NA
Half-life, child	2.9 h
Time to steady state, child	Variable

Excretion, urine

	% Excreted	Active	Detected in blood
Unchanged	6–30	Yes	Yes
Deacetylrifampin	15	Yes	Yes
3-Formyl derivative	7	No	No

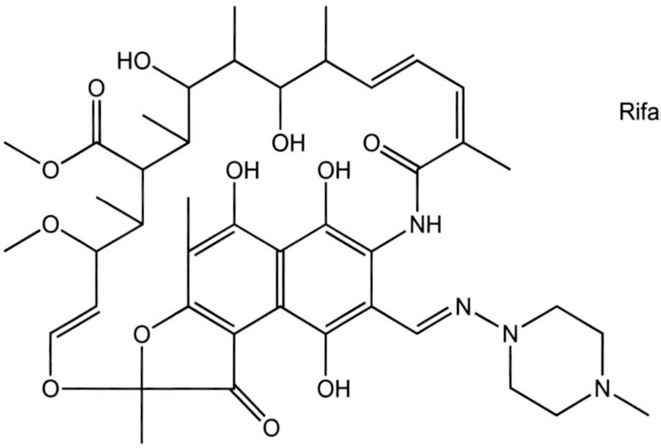

Rifampin

Rifampin (rifampicin) is a semisynthetic antibiotic derivative of rifamycin with bactericidal action. The mechanism is inhibiting DNA-dependent RNA polymerase. Rifampin (Rifadin) is used to treat tuberculosis (TB), as well as leprosy and other infections. It is frequently used in combination with isoniazid (INH) and other drugs. Combination products include rifampin/INH (Rifamate) and rifampin/INH/pyrazinamide (Rifater). Typical treatment protocols include an initial 2–6 month treatment with rifampin/INH/pyrazinamide plus a fourth drug (ethambutol or streptomycin) if exposure to resistant organisms is known or suspected. This regimen is followed by a 9 month (or longer) regimen of INH or a 4 month regimen of rifampin alone or a combination of rifampin and INH. Comparison of the 4 month rifampin to the 9 month INH regimen showed that the rifampin patients had a higher completion percentage and a lower percentage of demonstrated adverse reactions. Meta-analysis of published studies showed that a 3-month regimen of rifampin plus INH and a standard 6–12 month regimen of INH were equivalent in terms of effectiveness and safety [1, 2]. Although rifampin is primarily used for the treatment of TB and other mycobacterial infections, rifampin can also be administered with other antimicrobial agents for the treatment of various infections caused by organisms other than mycobacteria including *Staphylococcus aureus*, *Acinetobacter baumannii*, *Streptococcus pneumoniae*, *Legionella*, *Listeria*, and *Rhodococcus*. Rifampin is also used in treating prosthesis-associated infections caused by surface-adhering microorganisms that form biofilms.

Rifampin is readily absorbed after oral ingestion, although absorption is reduced when the drug is ingested with food or antacids. A standard 600-mg oral produces peak serum concentration of 8–20 μg/mL 1–3 h after administration [3]. The bioavailability of rifampin is approximately 90% in patients with TB but may decrease to 70% or lower after multiple-dose administration. Rifampin is highly bound to plasma proteins (86.1 to 88.9%) and the average volume of distribution is 1 L/kg. It is mainly eliminated by non-renal mechanisms as only approximately 15% of the dose is recovered in urine in unchanged form. The main metabolic pathway is deacetylation by esterases to 25-desacetylrifampin, an active metabolite having 25–50% of the activity of rifampin against clinical isolates. Plasma concentrations of the 25-desacetyl metabolite are relatively small compared to those of rifampin (approximately 10% in healthy subjects). Rifampin is a substrate for transporters such as P-glycoprotein and OATP1B1 (organic anion transporting polypeptide 1B1) and is rapidly eliminated via the bile after its hepatocellular uptake, which is mediated primarily by OATP1B1. Approximately 50% of the dose is excreted via the bile in the feces, mainly as the desacetyl metabolite, and 30% in the urine of which approximately half as unchanged rifampin. The plasma half-life of rifampin usually is around 3 to 4 h but drops to 1–2 h following multiple dosing [4].

Rifampin is a potent inducer of the hepatic and intestinal cytochrome P450 (CYP) enzyme system, which results in many clinically significant drug–drug interactions. It is a potent inducer of CYP3A4 in both the liver and the intestine, thus reducing serum levels of drugs metabolized by CYP3A4. In addition, rifampin also induces enzymes in the CYP2C subfamily, including CYP2C9. Induction of CYP2C9 can reduce plasma concentrations of CYP2C9 substrates such as gliclazide, phenytoin, warfarin, losartan, and celecoxib [5]. Drugs reported to cause increased rifampin levels include probenecid and trimethoprim/sulfamethoxazole. Co-administration with INH increases the potential for hepatotoxicity, and patients receiving INH and rifampin should be monitored for hepatotoxicity. Concurrent use of enalapril and rifampin results in decreased levels of both drugs [6]. Adverse reactions to rifampin include allergic reactions, red-orange coloration of bodily fluids, fever, nausea, immunosuppression, joint pain, headache, anemia, and peripheral neuritis.

Rifapentine (Priftin) is a rifamycin derivative containing a cyclopentyl ring instead of a methyl group at the piperazine moiety. It has a longer half-life (13h) than rifampin and offers the capability of less frequent dosing. Initial trials comparing weekly rifapentine and isoniazid to twice-weekly rifampin-containing regimens demonstrated that the weekly regimen was inferior for persons with cavitary and advanced TB [7].

Therapeutic drug monitoring (TDM) of rifampin is desirable to avoid subtherapeutic blood levels in patients. The expected 2 h post dose concentration is 8–24 μg/mL [8]. Um et al. reported an analytical method for measuring low serum concentrations of anti-tuberculosis drugs (isoniazid, rifampin, ethambutol, pyrazinamide and two metabolites aetylisonizid and 25-desacetylrifampin) in human serum using liquid chromatography coupled with tandem mass spectrometry, because low levels of these drugs have been associated with treatment failures. Interestingly, among 69 patients studied, the prevalence of a low 2h serum concentration of at least one antituberculosis drug was 46.4%. The authors concluded that low levels of anti-tuberculosis drugs among patients suffering from tuberculosis are common and it may be necessary to optimize drug dosages with TDM especially in patients with an inadequate clinical response [9].

References

[1] Ena J, Valls V. Short-course therapy with rifampin plus isoniazid, compared with standard therapy with isoniazid for latent tuberculosis infection: a meta-analysis. CID 2005;40(5):670–6.

[2] Page KR, Sifakis F, Montes de Oca R, Cronin WA, et al. Improved adherence and less toxicity with rifampin vs isoniazid for treatment of latent tuberculosis. Arch Intern Med 2006;166:1863–70.

[3] Lee CY, Huang CH, Lu PL, Ko WC, et al. Role of rifampin for the treatment of bacterial infections other than mycobacteriosis. J Infect 2017;75:395–408.

[4] Verbeeck RK, Günther G, Kibuule D, Hunter C, Rennie TW. Optimizing treatment outcome of first-line anti-tuberculosis drugs: the role of therapeutic drug monitoring. Eur J Clin Pharmacol 2016;72:905–16.

[5] Park JY, Kim KA, Park PW, Park CW, Shin JG. Effect of rifampin on the pharmacokinetics and pharmacodynamics of gliclazide. Clin Pharmacol Ther 2003;74(4):334–40.

[6] Niemi M, Backman JT, Fromm MF, Neuvonen PJ, et al. Pharmacokinetic interactions with rifampicin. Clin Pharmacokinet 2003;42(9):819–50.

[7] Alfarisi O, Alghamdi WA, Al-Shaer MH, Dooley KE, Peloquin CA. Rifampin vs. rifapentine: what is the preferred rifamycin for tuberculosis? Expert Rev Clin Pharmacol 2017;10:1027–36.

[8] Peloquin CA. Therapeutic drug monitoring in the treatment of tuberculosis. Drugs 2002;62:2169–83.

[9] Um SW, Lee SW, Kwon SY, Yoon HI, et al. Low serum concentrations of anti-tuberculosis drugs and determinants of their serum levels. Int J Tuberc Lung Dis 2007;11:972–8.

10.6

Therapeutic drug monitoring of selected antifungal agents

Three main classes of antifungal agents (polyenes, triazoles and echinocandins) are used clinically to treat fungal infections. The polyenes have a broad spectrum of activity that includes yeasts and molds. For the triazoles, susceptibility is more variable and depends on the specific agent. The echinocandins are active against most medically important species of *Aspergillus* and *Candida*, but lack activity against *Cryptococcus*, *Fusarium*, and the mucoraceous molds [1]. Most antifungal agents are not subjected to therapeutic drug monitoring (TDM). However, current evidence indicates that TDM is useful for itraconazole, posaconazole, voriconazole and flucytosine [1].

Fluconazole is a triazole antifungal that is active against most species of *Candida* (with the notable exceptions of *C. krusei* and *C. glabrata*). Fluconazole is also active against *Cryptococcus neoformans* and various dimorphic fungi. Fluconazole is used for the prevention of invasive candidiasis and the treatment of cryptococcal meningitis, coccidioidomycosis, and invasive and superficial candidiasis. Usual dosage is 400–800 mg/day but higher dosages (1200–2000 mg/day) have been used for cryptococcal meningitis. Fluconazole is highly bioavailable and exhibits low protein binding [1]. Routine TDM is not recommended due to wide therapeutic index but may be helpful in patients where absorption is suboptimal such as infants and children [2].

Itraconazole is a triazole antifungal with broad-spectrum antifungal activity. It is active against the most common medically important fungal pathogens, such as *Candida* spp., *C. neoformans* and *Aspergillus* spp. The extent of oral bioavailability of itraconazole is variable and dependent on the specific formulation. The oral bioavailability of itraconazole capsules is increased by food and gastric acidity. The pharmacokinetics of itraconazole is non-linear. Itraconazole accumulates slowly and generally reach concentrations of 0.5–1 mg/L after 7–15 days of dosing. Itraconazole is metabolized via oxidative mechanisms and principally via the cytochrome P450 (CYP) 3A4 enzyme isoform. Itraconazole also inhibits CYP3A4, which leads to a number of clinically relevant drug–drug interactions. Oxidative metabolism generates a multitude of metabolites that are excreted in the urine and feces. One of these metabolites, hydroxy-itraconazole, has antifungal activity that is comparable to itraconazole, and TDM is recommended [1]. The recommended steady state target concentration is 0.5 µg/mL [2].

Voriconazole is a broad-spectrum second-generation triazole antifungal agent that has activity against *Candida* spp. (including fluconazole-resistant species), *C. neoformans*, *Aspergillus* spp., many dimorphic fungi, and several other medically important fungi. Voriconazole is a structural congener of fluconazole. Voriconazole is available as intravenous and oral formulations [1]. The oral bioavailability of voriconazole in a fasting state is approximately 96%, and maximum plasma concentration is observed in <2 h. However, bioavailability is significantly lower if voriconazole is taken with food. Therefore, voriconazole should be taken on an empty stomach. This drug is distributed extensively into tissues with a volume of distribution of 4 L/kg. The estimated protein binding is 58%. Voriconazole is metabolized via oxidative mechanisms. The predominant cytochrome P450 isoenzyme involved in this process is CYP2C19 with CYP3A4 and CYP2C9 playing minor role in metabolism. The estimated half-life is 6–9 h. Genetic variation in gene encoding CYP2C19 enzyme accounts for 30% of the inter-subject variability in voriconazole pharmacokinetics. As many as 20% of non-Indian Asians and 5% of Caucasians or African-Americans are poor metabolizers who may have four times more voriconazole levels compared to normal individuals if given the same dosing. Suggested voriconazole trough concentrations are 0.35–2.2 μg/mL, while toxicity is associated with serum voriconazole level of 5–6 μg/mL [3].

Posaconazole is a broad-spectrum triazole agent that is structurally similar to itraconazole. Posaconazole has activity against a large number of medically important fungal pathogens, including *Candida*, *Aspergillus*, *Cryptococcus* and the mucoraceous molds. Posaconazole is currently only available as an oral suspension (40 mg/mL). Bioavailability is increased if the drug is taken with food. Posaconazole has a long terminal half-life (~34 h) and does not achieve steady-state serum concentrations until the end of the first week of dosing. Posaconazole is primarily metabolized by glucuronidation, with little involvement of oxidative mechanisms. Metabolites are excreted in the feces and urine. Posaconazole inhibits CYP3A4 activity, and dosage adjustment of drugs metabolized via this pathway (most importantly cyclosporine and tacrolimus) is required [1]. Currently routine TDM of posaconazole is not recommended [3].

Flucytosine is a synthetic antimycotic compound, first synthesized in 1957. It has no intrinsic antifungal capacity, but after it has been taken up by susceptible fungal cells, it is converted into 5-fluorouracil which is further converted to metabolites that inhibit fungal RNA and DNA synthesis [4]. Flucytosine is active against the majority of *Candida* spp. and *C. neoformans*, but also has activity against *Aspergillus* spp. and rare dematiaceous fungal pathogens causing chromoblastomycosis. Monotherapy with flucytosine is limited because of the frequent development of resistance. Therefore, this drug always should be used in combination with other antifungal agents to reduce risk of drug resistance. The standard dose is 100–150 mg/kg/day, and is usually administered in three or four divided dosages. A dosage reduction is required in patients with renal impairment.

Flucytosine is a small polar molecule that is cleared via renal mechanisms. Flucytosine has few (if any) direct drug–drug interactions [1]. Currently, TDM of flucytosine is strongly recommended to assure effective drug levels in the individual patient, to avoid resistance, and to prevent serious dose-limiting toxicity. Usually, in patients treated intermittently with flucytosine, target trough concentrations of 25–50 μg/mL and peak concentrations of 50–100 μg/mL, are considered adequate. In patients treated with continuous flucytosine infusion, target serum concentration is 50 μg/mL [4].

The preferred methods for TDM of antifungals are chromatographic methods. Recently Toussaint et al. described a liquid chromatography combined with tandem mass spectrometry method simultaneous measurement of eight antifungals compounds: isavuconazole, voriconazole, posaconazole, fluconazole, caspofungin, flucytosine, itraconazole and its metabolite hydroxy-itraconazole using only 50 μL of plasma [5].

References

[1] Ashbee HR, Barnes RA, Johnson EM, Richardson MD, et al. Therapeutic drug monitoring (TDM) of antifungal agents: guidelines from the British Society for Medical Mycology. J Antimicrob Chemother 2014;69:1162–76.
[2] Hope WW, Billaud EM, Lestner J, Denning DW. Therapeutic drug monitoring for triazoles. Curr Opin Infect Dis 2008;21:580–6.
[3] Hussaini T, Rüping MJ, Farowski F, Vehreschild JJ, Cornely OA. Therapeutic drug monitoring of voriconazole and posaconazole. Pharmacotherapy 2011;31:214–25.
[4] Vermes A, Guchelaar HJ, Dankert J. Flucytosine: a review of its pharmacology, clinical indications, pharmacokinetics, toxicity and drug interactions. J Antimicrob Chemother 2000;46:171–9.
[5] Toussaint B, Lanternier F, Woloch C, Fournier D, et al. An ultra-performance liquid chromatography-tandem mass spectrometry method for the therapeutic drug monitoring of isavuconazole and seven other antifungal compounds in plasma samples. J Chromatogr B Analyt Technol Biomed Life Sci 2017;1046:26–33.

10.7

Therapeutic drug monitoring of antiretroviral drugs

In 1981, the first cases of acquired immunodeficiency syndrome (AIDS) were reported. Soon thereafter, the virus responsible for this disease was isolated and designated as human immunodeficiency virus (HIV-1 and HIV-2). The first antiretroviral agent, zidovudine, received approval in 1987 by the United States Food and Drug Administration (FDA) and subsequently many more antiretroviral agents have been approved by the FDA. Currently, six classes of drugs classified by their mechanism of action are approved by the FDA to treat patients with HIV/AIDS.

The nucleoside reverse transcriptase inhibitors (NRTI) were the first class of antiretroviral drugs available and still comprise key component of most combination regimens. These drugs are phosphorylated intracellularly to their active diphosphate or triphosphate metabolites, which then inhibit the enzymatic action of the HIV reverse transcriptase by incorporating into the nucleotide analogue, thereby causing DNA chain termination or competing with the natural substrate of the virus. These actions halt the conversion of viral RNA into double stranded DNA (Table 10.1). Non-nucleoside reverse transcriptase inhibitors (NNRTIs) are different from NRTIs in that they do not require intracellular phosphorylation to exert their pharmacologic action. NNRTIs are noncompetitive inhibitors of reverse transcriptase.

Protease inhibitors (PIs) exhibit their pharmacologic action late in the HIV replication cycle by binding to HIV proteases, inhibiting the formation of mature, infectious virions. Therapeutic drug monitoring (TDM) is recommended during therapy with PIs. The fusion inhibitor enfuvirtide interferes with this fusion process by binding to the first heptad-repeat (HR1) in the viral envelope glycoprotein gp41, thus preventing conformational changes necessary for the fusion of the viral and cellular membrane. Cross resistance with other antiretroviral drugs has not yet been observed with enfuvirtide, likely due to its unique mechanism of action. Enfuvirtide is added as part of salvage regimens for patients whose HIV has demonstrated resistance to multiple other drug classes. Maraviroc is a CCR5 antagonist approved for use in treatment naïve or treatment-experienced patients infected with CCR5-tropic HIV. Maraviroc selectively binds to human

Table 10.1 Currently used antiretrovirals

Class/Drug	**Therapeutic drug monitoring**
Nucleoside reverse transcriptase (NRT)	
Zidovudine	No
Didanosine	No
Zalcitabine	No
Lamivudine	No
Stavudine	No
Abacavir	No
Tenofovir	No
Emtricitabine	No
Non-nucleoside reverse transcriptase (NNRTI)	
Nevirapine	Limited benefit
Delavirdine	Limited benefit
Efavirenz	Limited benefit
Etravirine	Limited benefit
Rilpivirine	Limited benefit
Protease inhibitors	
Indinavir	Yes
Ritonavir	Yes
Saquinavir	Yes
Nelfinavir	Yes
Amprenavir	Yes
Lopinavir	Yes
Atazanavir	Yes
Darunavir	Yes
Fosamprenavir	Yes
Tipranavir	Yes
Fusion inhibitor	
Enfuvirtide	Not established
CCR5 antagonist	
Maraviroc	Not established
Integrase inhibitors	
Dolutegravir	Not established
Elvitegravir	Not established
Raltegravir	Not established

CCR5 receptor on the cell membrane, thus blocking the interaction of the HIV gp 120 and the CCR5 receptor for CCR5-tropic HIV [1, 2]. The Integrase Strand Transfer Inhibitor (INSTI) class is the newest class of drugs available to treat HIV. INSTIs target the HIV integrase enzyme, which incorporates pro-viral HIV-1 DNA into the host cell genome. The first clinically available INSTI, raltegravir, was approved in 2007, followed by elvitegravir in 2012 and dolutegravir in 2013 [3].

The use of multiple HIV medications from different drug classes (often referred to as highly active antiretroviral therapy or HAART) has led to a significant reduction in the morbidity and mortality of HIV infected individuals by effectively suppressing viral replication. However, treatment failure from HAART occurs due to poor compliance, altered pharmacokinetic parameters, and/or development of viral resistance [4].

A number of reports in the literature indicate that there is a relationship between plasma concentrations of the NNRTIs and PIs and their efficacy and toxicity. For example, small decreases in circulating concentrations of these drugs have resulted in inadequate viral suppression. Studies indicate that a large proportion of patients have trough PI concentrations lower than proposed minimum effective concentrations. In addition, wide inter-individual variability in trough plasma concentrations has been reported, >50% for the NNRTIs and 68.4% for the PIs [2]. The variations observed are likely related, in part, to individual differences in the drugs absorption, distribution, metabolism, and elimination characteristics, as well as drug–drug and drug-food interactions. Most NNRTIs and PIs are metabolized in the liver through the cytochrome P450 (CYP) enzyme system. CYP3A4 is the primary enzyme involved, with CYP2B6, CYP2D6, and CYP2C19 also playing important roles. Numerous other drugs the patients are likely to receive are known to induce or inhibit these enzymes. The resulting interindividual variabilities make these drugs good candidates for TDM. With respect to toxicity, high plasma PI concentrations are associated with renal, urological, and gastrointestinal toxicities, with hyperlipoproteinemia and hyperbilirubinemia noted. Virological failure and toxicity seen in patients receiving the same dose of PIs and NNRTIs indicates that these drugs have narrow therapeutic index, another rationale for TDM. The PIs are also strongly protein-bound to the acute-phase reactant, α_1-acid glycoprotein. That most of the drugs are ~98% bound (the exception being indinavir at 65%) also raises the question if free drug monitoring would be more appropriate for this group [5].

Combined antiretroviral therapy has been shown to significantly reduce HIV transmission from mother to fetus. While the pharmacokinetics of several of the drugs, notably zidovudine, lamivudine; didanosine and stavudine, do not seem to be altered significantly during pregnancy, for others the effects are apparent and variable. For example, the half-life of nevirapine is significantly prolonged during pregnancy, while the C_{max} achieved for indinavir is significantly reduced. HIV-infected women who were pregnant achieved lower serum drug concentrations when given the standard adult doses of nelfinavir and saquinavir compared to non-pregnant women [6]. The pharmacokinetic parameters of antiretrovirals also vary widely in children. Some children require lower dosage than adults, while other children require adult dosage for successful viral control. Both of these populations could benefit from TDM of these drugs.

On the basis of currently published reports, TDM of the NRTIs seems to have little benefit other than to identify non-compliance. These drugs need intracellular activation to triphosphate anabolites and concentrations of intracellular triphosphate correlate poorly with plasma concentrations of parent drugs [5]. Total plasma concentrations of PIs and NNRTIs are monitored when TDM is performed. Colombo et al. studied whether total plasma concentrations of these drugs correlate with cellular concentrations, and found the best correlations for nelfinavir, saquinavir, and lopinavir. There was a moderate correlation between plasma and cellular concentrations for efavirenz, and no correlation was observed for nevirapine. This verifies that plasma concentrations are useful surrogates for cellular concentrations for certain antiretrovirals but not all. For efavirenz and nevirapine, there may be room for improvement if TDM is needed by using a direct measurement of cellular concentrations or another predictive factor such as transporter or metabolic enzyme genotyping [7].

One of the earlier extensive reports of usefulness of TDM of antiretroviral agents was published in 2003. In the ATHENA study, 147 treatment naïve patients (92 patients receiving nelfinavir and 55 patients receiving indinavir) were randomly divided into two groups; either TDM or no TDM. There were fewer episodes of virological suppression failure documented for the 92 nelfinavir recipients who were actively monitored. Similar findings were reported for patients receiving indinavir. The authors concluded that TDM of indinavir and nelfinavir in treatment-naïve patients improves treatment response [8]. A recent study involving 2468 samples collected from 723 patients (68.1% male, median age 43.5 years) taking PIs (atazanavir, lopinavir, ritonavir or darunavir) reported patients in the adherent group had a higher chance of viral control versus the partially non-adherent group and the non-adherent group. The authors concluded that TDM of PIs are very useful in avoiding treatment failure due to non-compliance [9].

Currently there are no commercially available immunoassays for antiretroviral measurements; however, numerous methods are described using high performance liquid chromatography (HPLC) combined with either UV detection or tandem mass spectrometry. The method of Verbesselt et al. permits the simultaneous quantification of eight PIs (amprenavir, indinavir, atazanavir, ritonavir, lopinavir, nelfinavir, nelfinavir M8 metabolite, and saquinavir) in human plasma after liquid–liquid extraction and isocratic chromatographic separation using Allsphere hexyl HPLC column with combined UV and fluorescence detection [10]. Rezk et al. has described a simple liquid–liquid extraction followed by HPLC analysis using a Zorbax-C-18 column (150 × 4.6 cm, particle diameter 3.5 μm) for quantitative determination of 2 NNRTIs (nevirapine and efavirenz) and several PIs (atazanavir, indinavir, amprenavir, saquinavir, nelfinavir, ritonavir and lopinavir). The antiretrovirals were detected at 210 nm using UV detection [11].

The increased availability of HPLC combined with tandem mass spectrometry (LC/MS/MS) has proven to have superior sensitivity for detecting antiretrovirals compared to HPLC combined with UV detection. Koal et al. described an online solid phase extraction LC/MS/MS analysis of 7 PIs and 2 NNRTIs in human plasma [12]. Soldin et al. described an LC/MS/MS for 17 antiretroviral drugs including atazanavir and tipranavir in human plasma or serum [13]. Recently, Tsuchiya et al. published an LC/MS/MS protocol for simultaneous determination of raltegravir, dolutegravir and elvitegravir concentrations in human plasma and cerebrospinal fluid samples [14].

The ideal specimen should be collected as a steady-state trough sample. Steady state is achieved for most of these drugs in <48 h. Exceptions to this rule are nevirapine (an NNRTI) and saquinavir (a PI), because of their longer half-lives. Both of these should not be monitored until 4 days after initiation of the drug therapy [13]. Recommended target concentrations for both protease inhibitors and NNRTI are given in Table 10.2.

Table 10.2 Selected pharmacokinetic parameters and recommended target therapeutic concentrations of protease inhibitors and NNRTI

Drug	Bioavailability (%)	Half-Life[a] (h)	Protein binding (%)	V_d (L/kg)	Target level (trough) (ng/mL)
Nevirapine	50–90	25–30	60	1.4–1.54	3400
Efavirenz	42	40–55	99.5	2–4	1000
Indinavir	30–65	1.4–2.2	60–70	1.2	80–120
Ritonavir	60–80	3–5	97–99	0.4	1000
Lopinavir	Very low oral bioavailability[b]	5–6	98–99	0.74	700
Amprenavir	>70	7–10	90	6.1	150–400
Atazanavir	>60	7–8.6	86	3.6	100
Saquinavir	12	3–13	97	10	100–250
Nelfinavir	>78	3.5–6	98	2–7	700–1000

[a] Following multiple dosing.
[b] Lopinavir has poor oral bioavailability due to limited aqueous solubility but when administered in combination with ritonavir bioavailability and pharmacokinetic parameters are significantly improved.

References

[1] Pau AK, George JM. Antiretroviral therapy: current drugs. Infect Dis Clin North Am 2014;28:371–402.
[2] Molto J, Blanco A, Miranda C, Miranda J, et al. Variability in non-nucleoside reverse transcriptase and protease inhibitors concentrations among HIV-infected adults in routine clinic practices. Br J Clin Pharmacol 2006;62(5):560–6.
[3] Podany AT, Scarsi KK, Fletcher CV. Comparative clinical pharmacokinetics and pharmacodynamics of HIV-1 integrase strand transfer inhibitors. Clin Pharmacokinet 2017;56:25–40.
[4] Andrade HB, Shinotsuka CR, da Silva IRF, Donini CS, et al. Highly active antiretroviral therapy for critically ill HIV patients: a systematic review and meta-analysis. PLoS One 2017;12(10):e0186968.
[5] Back D, Gibbons S, Khoo S. An update on therapeutic drug monitoring for antiretroviral drugs. Ther Drug Monit 2006;28:468–73.
[6] Rakhmanina N, van den Anker, Soldin SJ. Safety and pharmacokinetics of antiretroviral therapy during pregnancy. Ther Drug Monit 2004;26:110–5.
[7] Colombo S, Telenti A, Buclin T, Furrer H, et al. Are plasma levels valid surrogates for cellular concentrations of antiretroviral drugs in HIV-infected patients? Ther Drug Monit 2006;28:332–8.
[8] Burger D, Hugen P, Reiss P, et al. Therapeutic drug monitoring of nelfinavir and indinavir in treatment naïve HIV-1 infected individuals. AIDS 2003;17:1157–65.
[9] Calcagno A, Pagani N, Ariaudo A, Arduino G, et al. Therapeutic drug monitoring of boosted PIs in HIV-positive patients: undetectable plasma concentrations and risk of virological failure. J Antimicrob Chemother 2017;72:1741–4.
[10] Verbesselt R, Van Wijngaerden E, de Hoon J. "Simultaneous determination of 8 HIV protease inhibitors in human plasma by isocratic high performance liquid chromatography with combined UV and fluorescence detection" amprenavir, indinavir, atazanavir, ritonavir, lopinavir, saquinavir, nelfinavir and M8-nelfinavir metabolite. J Chromatogr B Analyt Technol Biomed Life Sci 2007;845:51–60.

[11] Rezk NL, Crutchley RD, Yeh RF, Kashuba AD. Full validation of analytical method for the HIV-protease inhibitor atazanavir in combination with 8 other antiretroviral agents and its application to therapeutic drug monitoring. Ther Drug Monit 2006;28:517–25.
[12] Koal T, Sibum M, Koster E, Resch K, et al. Direct and fast determination of antiretroviral drugs by automated online solid phase extraction-liquid chromatography-tandem mass spectrometry in human plasma. Clin Chem Lab Med 2006;44:299–305.
[13] Gu J, Soldin SJ. Modification of tandem mass spectrometric method to permit simultaneous quantification of 17 anti-HIV drugs which includes atazanavir and tipranavir. Clin Chim Acta 2007;378:222–4.
[14] Tsuchiya K, Ohuchi M, Yamane N, Aikawa H, et al. High-performance liquid chromatography-tandem mass spectrometry for simultaneous determination of raltegravir, dolutegravir and elvitegravir concentrations in human plasma and cerebrospinal fluid samples. Biomed Chromatogr 2018;32(2):https://doi.org/10.1002/bmc.4058.

CHAPTER 11

Therapeutic drug monitoring of antidepressants and psychoactive drugs

Depression is a major problem not only in the United States but throughout the world. It is estimated that 10–20% of adults in the United States experience major depression during their lifetime and ~3% of the population is depressed at any given time. Further evidence of the impact of depression in the United States resides in the fact that 10–15% of all prescriptions are related to the treatment of major depression. Correctly identifying and treating depression is important. Patients with depression are at risk not only of suicide or self-harm attempts, but also at higher risk of developing serious illnesses such as cardiovascular disease and myocardial infarction [1]. The World Health Organization estimates that by 2020, depression will be the second most common cause of premature death or disability. The socioeconomic consequences of depression are very high and there is a clear need for effective antidepressant pharmacotherapy. Until the development of the first generation of antidepressants, the tricyclic antidepressants and monoamine oxidase inhibitors, there were few pharmacologic options for these patients. Since then, newer drugs in multiple classes have been approved for the treatment of depression [2–5]. Drugs available for treating depression can be broadly classified under five categories:

- Monoamine oxidase inhibitors (MAOIs): This group includes phenelzine, selegiline, and tranylcypromine.
- Tricyclic antidepressants (TCAs): This group includes amitriptyline, amoxapine, clomipramine, desipramine, doxepin, imipramine, nortriptyline, protripytline, and trimipramine.
- Selective serotonin reuptake inhibitors (SSRIs): This group includes citalopram, escitalopram, fluoxetine, fluvoxamine, paroxetine, and sertraline.
- Serotonin-norepinephrine reuptake inhibitors (SNRIs): This group includes desvenlafaxine, duloxetine, levomilnacipran, milnacipran, venlafaxine, and vortioxetine.

Therapeutic Drug Monitoring Data
https://doi.org/10.1016/B978-0-12-815849-4.00011-6

- Atypical antidepressants: This group includes bupropion, maprotiline, mirtazapine, nefazodone, and trazodone.

Atypical agents cannot be classified under any category of antidepressants based on mechanism of action. For example, typical antipsychotics (e.g., haloperidol) appear to exert their effects mainly through antagonism of dopamine D2 receptor. In contrast, most atypical psychoactive agents are weak D2 antagonists. Any benefit in mood disorders may relate to their effects on serotonin (5HT) receptors, because all agents block 5HT2A and 5HT2C receptors. Although atypical antidepressants change the levels of one or more neurotransmitters, such as dopamine, serotonin or norepinephrine, these drugs confer a very different pharmacodynamic profile relative to classical antidepressant drugs [6].

In this chapter, major emphasis is placed on TCAs because these drugs require routine therapeutic drug monitoring (TDM). Lithium, which is used for treating bipolar disorder, also requires TDM and is included. Newer antidepressants such as SSRIs and SNRIs in general do not require TDM but under certain clinical situations monitoring may benefit a subset of patients. The newer antidepressants also have better safety profile than the TCAs and MAOIs and have steadily replaced these older medications as first-line therapy for depression. Zahl et al. commented that SSRIs were introduced to the market around 1990. The sales figures of these drugs have subsequently increased year by year, while the sales figures of the TCAs—drugs which can be quite toxic in overdose—have been significantly reduced. Today, SSRIs constitute about two-thirds of the total sales figures of antidepressants in the Nordic countries [7].

TCAs were first introduced in the 1950s and 1960s. As the name indicates, TCAs have a three-ring structure and structurally resemble phenothiazines. It is thought that they restore neurotransmitter levels by blocking reuptake from the synapse. Tertiary amines, including imipramine and amitriptyline, inhibit serotonin uptake to a greater extent than norepinephrine uptake, whereas secondary amines, desipramine and nortriptyline, have a greater effect on the reuptake of norepinephrine. Although TCAs rapidly affect neurotransmitter reuptake, their clinical effect is not seen for several weeks, thus suggesting that their mechanism of action is more complicated than simply inhibiting neurotransmitter reuptake. This fact is further supported by the observation that compounds such as cocaine and amphetamines inhibit neurotransmitter uptake without providing an antidepressant affect. In addition to their use as antidepressants, TCAs are used in the treatment of a number of other disorders such as neuralgic pain, chronic pain, and migraine prophylaxis. In children, the drugs are used to treat obsessive–compulsive disorder, attention-deficit hyperactivity disorder, school phobia, and separation anxiety [8].

Unfortunately, TCAs are associated with high morbidity and mortality. Common side effects include dry mouth, constipation and urinary retention, but more serious side effects such as blurred vision, respiratory depression, hypotension, coma, cardiac arrhythmias, and tachycardia are seen. TCAs are known to lower threshold for seizures [9].

All TCAs are rapidly absorbed after oral administration and bind strongly to plasma albumin (90–95%) at therapeutic plasma concentrations. These drugs also bind to extravascular tissues and as a result show large distribution volumes (10–50 L/kg). TCAs are metabolized largely via cytochrome P450 (CYP) enzymes through demethylation of tertiary TCAs to their secondary amine metabolites which are also pharmacologically active. Subsequent metabolism steps include hydroxylation and glucuronidation, with excretion in the urine serving as a major route for elimination. Plasma concentrations for therapeutic effect are usually stated to be between 50 and 300 ng/mL. Toxic effects and fatalities are expected when plasma concentrations reach approximately 1000 ng/mL but patients with plasma TCA concentrations greater than 450 ng/mL tend to develop cognitive or behavioral toxicity (e.g., agitation, confusion, memory impairment, pacing) [10].

TCA plasma/serum concentrations are affected by a number of other drugs and individual genetic variations. CYP2D6-inducing drugs such as alcohol, barbiturates, carbamazepine, phenobarbital, phenytoin and rifampin increase clearance of TCAs, whereas drugs which inhibit CYP2D6 decrease clearance of TCAs. CYP2D6-inhibiting drugs include amiodarone, beta-adrenergic receptor antagonists, bupropion, celecoxib, chlorpromazine, cimetidine, citalopram, doxorubicin, fluoxetine, haloperidol, quinidine, ranitidine, ritonavir, terbinafine, ticlopidine, and histamine H_1 receptor antagonists [10].

Although there was a trend in the 1980s towards routine TDM of TCAs, the poor correlation between the drug levels and the clinical outcome raised questions about the value of such practice. Nevertheless, the measurement of TCA plasma/serum levels is unequivocally useful in other settings such as assessing patient non-compliance and detecting significant differences in metabolism of TCAs due to age, genetic factors and co-administration of number of other drugs [11–13]. Monitoring is also used to assess overdose and toxicity, but specific electrocardiogram (ECG) changes such as prolonged QRS interval and terminal right axis deviation correlate better with TCA toxicity than plasma levels [14].

Analytical methods for TCAs include immunoassays, high performance liquid chromatography (HPLC), gas chromatography (GC), gas chromatography combined with mass spectrometry, and liquid chromatography coupled to tandem mass spectrometry (LC/MS/MS). HPLC, either alone or coupled to mass spectrometry, is the most commonly used method for TCAs quantification due to significant cross-reactivities of metabolites in immunoassays for the parent drug. As a result, immunoassays are mostly use in case of suspected TCA overdose which typically correlate with total TCA concentration. Therefore, TCA immunoassays should be considered as "total TCA" assay, while liquid chromatographic methods can separately quantify both the parent drug and active metabolite, if applicable. Many drugs are known to interfere in immunoassays and chromatographic methods. The drugs which may cause interference in immunoassays include carbamazepine, quetiapine, phenothiazines, diphenhydramine, cyproheptadine and cyclobenzaprine [15–18].

Starting in the 1990s, new classes of antidepressants have emerged. These include amoxapine and maprotiline which inhibit reuptake of monoamines, trazodone which inhibits serotonin reuptake, bupropion which inhibits reuptake of norepinephrine and dopamine, and venlafaxine and mirtazapine that inhibit uptake of both norepinephrine and serotonin. Also, many antidepressants which selectively inhibit serotonin uptake (SSRIs) are now available, including citalopram, escitalopram, fluoxetine, fluvoxamine, paroxetine, and sertraline. The SSRIs and SNRIs are the most commonly used categories of antidepressants in the United States [4, 5]. Some pharmacokinetic properties of non-TCA antidepressants are given in Table 11.1.

In addition to inhibiting serotonin reuptake, SSRIs interact with serotonin receptors to initiate a pharmacological response. SSRIs have a number of advantages over TCAs including lack of adrenergic, antihistaminic and anticholinergic effects. In addition to their use as antidepressants, SSRIs are also used in the treatment of obsessive–compulsive disorder, panic disorder, bulimia and many other conditions [19]. Although SSRIs have fewer side effects compared to TCAs and monoamine oxidase inhibitors, they are associated with transient nausea, diarrhea, insomnia, somnolence and dizziness. Overdose with SSRIs results in increase of serotonin and can cause serotonin syndrome. The syndrome is associated with mental status changes, agitation, myoclonus, hyperreflexia, diaphoresis, shivering, tremor, diarrhea, incoordination, and high body temperature. Luchini et al. recently reported a case of fetal intoxication by citalopram. The concentration of citalopram

Table 11.1 Pharmacokinetic properties of SSRIs and other non-tricyclic antidepressants

Drug	Active metabolite	Average half-life (h)	V_d (L/kg)	Oral bioavailability	Average protein binding	Therapeutic range (ng/mL)	Toxic level (ng/mL)
Amoxapine	8-hydroxyamoxapine	10	1	90	90	200–400[a]	>600
Bupropion	Hydroxybupropion[b]	1–15	45	90	85	25–100	>400
Citalopram	Norcitalopram[b]	38–48	14	80	50	50–110	>220
Fluoxetine	Norfluoxetine	4–6 days	50	100	94	120–500[a]	>2000
Fluvoxamine	NA	21–43	25	95	77	60–230	>500
Maprotiline	NA	2–58	24	100	88	75–130	>220
Mirtazapine	Normirtazepine	20–40	12	90	85	30–80	>160
Paroxetine	NA	12–44	15	90	95	20–65	>120
Sertraline	Norsertraline[b]	22–36	20	90	98	10–150	>300
Trazodone	NA	4–11	1	80	90	700–1000	>1200
Venlafaxine	*O*-desmethyl venlafaxine	14–18	7	90	27	100–400[a]	>800

NA, no significantly active metabolite.
[a] Total concentration of parent and active metabolite.
[b] Significantly less active than parent drug.

in postmortem blood was 11.60 mg/L [20]. Drug interactions that may lead to serotonin syndrome include MAOIs, tramadol, sibutramine, meperidine, sumatriptan, lithium, St. John's wort, gingko biloba, and atypical antipsychotic agents. A United States Food and Drug Administration warning for SSRIs includes risk of suicidal thinking and behavior in children and adolescents with major depressive disorder (MDD) and other depressive disorders.

Common methods of determination of SSRIs and other non-TCAs include GC and HPLC [21, 22]. The GC methods involve direct assay or drug derivatization before analysis. Drug extractions are either liquid–liquid or solid phase. Single step extractions are generally successful, but methods involving multiple extractions or back-extractions have been described. The methods may involve assay of single drug or multiple drugs. More recently, LC/MS/MS has been applied for simultaneous analysis of 30 antipsychotic drugs in blood including SSRIs [23].

References

[1] Simon GE, Savarino J, Operskalski B, Wang PS. Suicide risk during antidepressant treatment. Am J Psychiatry 2006;163:41–7.

[2] Bateman DN. Tricyclic antidepressant poisoning: central nervous system effects and management. Toxicol Rev 2005;24:181–6.

[3] Watson WA, Litovitz TL, Rodgers Jr. GC, Klein-Schwartz W, Reid N, Youniss J, et al. 2004 annual report of the American association of poison control centers toxic exposure surveillance system. Am J Emerg Med 2005;23:589–666.

[4] Ables AZ, Baughman 3rd OL. Antidepressants: update on new agents and indications. Am Fam Physician 2003;67:547–54.

[5] Pacher P, Kecskemeti V. Trends in the development of new antidepressants. Is there a light at the end of the tunnel? Curr Med Chem 2004;11:925–43.

[6] Shelton RC, Papakostas GI. Augmentation of antidepressants with atypical antipsychotics for treatment-resistant major depressive disorder. Acta Psychiatr Scand 2008;117(4):253–9.

[7] Zahl PH, De Leo D, Ekeberg Ø, Hjelmeland H, Dieserud G. The relationship between sales of SSRI, TCA and suicide rates in the Nordic countries. BMC Psychiatry 2010;10:62.

[8] Kerr GW, McGuffie AC, Wilkie S. Tricyclic antidepressant overdose: a review. Emerg Med J 2001;18(4):236–41.

[9] Feighner JP. Mechanism of action of antidepressant medications. J Clin Psychiatry 1999;60(Suppl. 4):4–11.

[10] Gillman PK. Tricyclic antidepressant pharmacology and therapeutic drug interactions updated. Br J Pharmacol 2007;151:737–48.

[11] Linder MW, Keck Jr. PE. Standards of laboratory practice: antidepressant drug monitoring. National academy of clinical biochemistry. Clin Chem 1998;44:1073–84.

[12] Muller MJ, Dragicevic A, Fric M, Gaertner I, Grasmader K, Hartter S, et al. Therapeutic drug monitoring of tricyclic antidepressants: how does it work under clinical conditions? Pharmacopsychiatry 2003;36:98–104.

[13] Eap CB, Jaquenoud SE, Baumann P. Therapeutic monitoring of antidepressants in the era of pharmacogenetics studies. Ther Drug Monit 2004;26:152–5.

[14] Thanacoody HK, Thomas SH. Tricyclic antidepressant poisoning: cardiovascular toxicity. Toxicol Rev 2005;24:205–14.
[15] Caravati EM, Juenke JM, Crouch BI, Anderson KT. Quetiapine cross-reactivity with plasma tricyclic antidepressant immunoassays. Ann Pharmacother 2005;39:1446–9.
[16] Hendrickson RG, Morocco AP. Quetiapine cross-reactivity among three tricyclic antidepressant immunoassays. J Toxicol Clin Toxicol 2003;41:105–8.
[17] Dasgupta A, McNeese C, Wells A. Interference of carbamazepine and carbamazepine 10,11-epoxide in the fluorescence polarization immunoassay for tricyclic antidepressants: estimation of the true tricyclic antidepressant concentration in the presence of carbamazepine using a mathematical model. Am J Clin Pathol 2004;121:418–25.
[18] Chattergoon DS, Verjee Z, Anderson M, Johnson D, McGuigan MA, Koren G, et al. Carbamazepine interference with an immune assay for tricyclic antidepressants in plasma. J Toxicol Clin Toxicol 1998;36:109–13.
[19] Schatzberg AF. New indications for antidepressants. J Clin Psychiatry 2000;61(Suppl. 11):9–17.
[20] Luchini D, Morabito G, Centini F. Case report of a fatal intoxication by citalopram. Am J Forensic Med Pathol 2005;26:352–4.
[21] Kristoffersen L, Bugge A, Lundanes E, Slordal L. Simultaneous determination of citalopram, fluoxetine, paroxetine and their metabolites in plasma and whole blood by high-performance liquid chromatography with ultraviolet and fluorescence detection. J Chromatogr B Biomed Sci Appl 1999;734:229–46.
[22] Lacassie E, Gaulier JM, Marquet P, Rabatel JF, Lachâtre G. Methods for the determination of seven selective serotonin reuptake inhibitors and three active metabolites in human serum using high-performance liquid chromatography and gas chromatography. J Chromatogr B Biomed Sci Appl 2000;742(2):229–38.
[23] Saar E, Gerostamoulos D, Drummer OH, Beyer J. Identification and quantification of 30 antipsychotics in blood using LC-MS/MS. J Mass Spectrom 2010;45:915–25.

11.1

Amitriptyline

Chemical properties	
Solubility in H_2O	Soluble as hydrochloride salt
Molecular weight	277.39
pKa	9.4
Melting point	196–197 °C
Dosing	
Recommended dose, adult	75–200 mg/day May gradually increase to 300 mg/day
Recommended dose, child	1–2 mg/kg/day
Monitoring	
Sample	Serum, plasma
Effective concentrations	80–200[a] ng/mL
Toxic concentrations	>500[a] ng/mL
Methods	HPLC, immunoassay, GC
Pharmacokinetic properties	
Bioavailability	33–62%
Time to peak concentration	4–8 h
Protein bound	90–98%
Volume of distribution	6–36 L/kg
Half-life, adult	10–28 h
Time to steady state, adult	48–144 h

Excretion, urine

	% Excreted	Window of excretion	Active	Detected in blood
Parent	<1	2–10 days	Yes	Yes
Nortriptyline	3	2	Yes	Yes
10-OH-amitryptiline and 10-hydroxynortriptyline	15–30	2	Yes	Yes
Glucuronides	40–60	2	No	No

[a] Combined amitriptyline and nortriptyline.

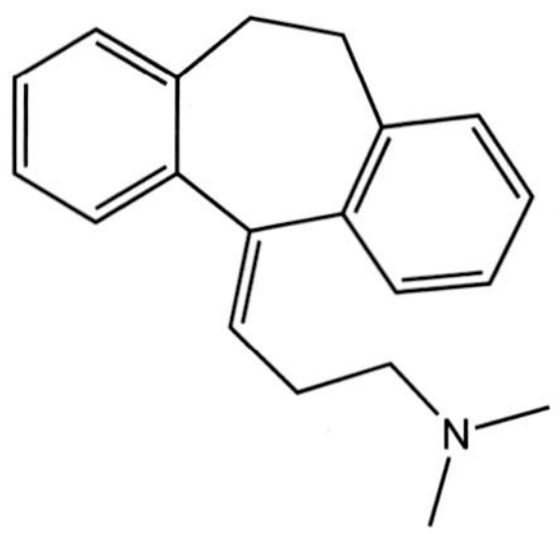

Amitriptyline

Amitriptyline, a tricyclic antidepressant (TCA), is still used clinically for treating depression. However, another major clinical use of amitriptyline is as a prophylactic agent for migraine headache. In the United States, around 30 million people (18% female, 6% male) have one or more migraine headaches per year [1]. TCAs (amitriptyline, nortriptyline, and desipramine) are effective in the treatment of neuropathic pain, fibromyalgia, low back pain, and headaches [2].

For the treatment of depression, a starting dose of 75 mg/d is typically initiated for adults, but lower doses may be initiated for adolescents and older individuals. Therapeutic effects may not be apparent for 2–3 weeks after initiation of drug therapy so it is not unusual for the dose to be gradually increased with changes in symptoms. Amitriptyline is not recommended for use in children under 12 years of age. The oral bioavailability of amitriptyline is highly variable, ranging from 33% to 62% in humans which is responsible for wide inter-individual variation in plasma/serum concentration and therapeutic effects. Amitriptyline is a substrate for P-glycoprotein, which is a major determinant for low and variable bioavailability by increasing the exposure of the drug to intestinal cytochrome P450 (CYP) enzymes by allowing repeated cycling of the drug via diffusion and active efflux [3].

The peak plasma concentration of amitriptyline after oral administration is usually reached in 4–8 h. Amitriptyline is a highly lipophilic compound which is widely distributed throughout the body and extensively bound to tissue and plasma proteins. About one-third to one-half of the drug is excreted within 24 h. The plasma half-life ranges from 10 to 28 h for amitriptyline and from 16 to 80 h for its active metabolite nortriptyline. Amitriptyline's most common side-effects, such as blurred vision, dry mouth, and constipation, are due to its anticholinergic effects. The sedative effect is thought to be mediated by antagonism of histamine receptors. High doses of TCAs are potentially cardiotoxic [4]. Within the TCAs, amitriptyline appears to be particularly toxic in overdose. The principal mechanism of toxicity is cardiac sodium channel blockade, which increases the duration of the cardiac action potential and refractory period and delays atrioventricular conduction. Electrocardiographic changes include prolongation of the PR, QRS and QT intervals, nonspecific ST segment and T wave changes, atrioventricular block, right axis deviation of the terminal 40 ms vector of the QRS complex in the frontal plane (T 40 ms axis), and the Brugada pattern (downsloping ST segment elevation in leads V1–V3 in association with right bundle branch block). Maximal changes in the QRS duration and the T 40 ms axis are usually present within 12 h of ingestion but may take up to a week to resolve. Sinus tachycardia is the most common arrhythmia due to anticholinergic activity and inhibition of norepinephrine uptake. However, bradyarrhythmias (due to atrioventricular block) and tachyarrhythmias (supraventricular and ventricular) may also occur. In general, a QRS duration >100 ms and a rightward T 40 ms axis appear to be better predictors of cardiovascular toxicity than the plasma TCA concentration [5]. Sodium bicarbonate is indicated for treatment of TCA overdose if the QRS duration is more than 100 ms or the terminal right-axis deviation is more than 120 degrees [6]. Contraindications for using TCAs include hypersensitivity to amitriptyline, narrow-angle glaucoma, and use of monoamine oxidase inhibitors within 2 weeks [7].

Amitriptyline is extensively metabolized by liver CYP2D6 isoenzymes into the active metabolite nortriptyline as well to 10-hydroxynortryptiline and 10-hydroxyamitryptiline. These metabolites are then conjugated with glucuronic acid by the action of UDP-glucuronosyltransferase enzyme and conjugated metabolites are also observed in urine [8]. A recent study reported detection of 28 metabolites of amitriptyline after oral administration [9].

The correlation of clinical effectiveness with plasma levels has not been well documented for amitriptyline. TDM is primarily useful to verify compliance or to evaluate toxicity. Qualitative, semiquantitative and quantitative methods are available for the analyses of amitriptyline. Although traditionally therapeutic range of combined amitriptyline and nortriptyline is considered as 80–250 ng/mL, more recent publication indicate that some patient may experience toxicity at levels above 200 ng/mL. Therefore, the therapeutic range should be considered as 80–200 ng/mL. However, concurrent treatment with benzodiazepines and phenothiazines for night-time sedation may shift the lower limit of the therapeutic range to 60 or 70 ng/mL [10].

Common methods of analyses include immunoassay, high performance liquid chromatography, and gas chromatography. Immunoassays suffer from high cross-reactivity from other TCAs as well as their metabolites and are not recommended for routine TDM. Immunoassays designed for the determination of total TCA levels in serum should only be used to assess patients suspected of TCA overdose. High performance liquid chromatography combined with tandem mass spectrometry is the preferred method for TDM of TCAs including amitriptyline and nortriptyline [11, 12].

References

[1] Moras K, Nischal H. Impact of amitriptyline on migraine disability assessment score. J Clin Diagn Res 2014;8(9):KC01–2.

[2] Dharmshaktu P, Tayal V, Kalra BS. Efficacy of antidepressants as analgesics: a review. J Clin Pharmacol 2012;52:6–17.

[3] Abaut AY, Chevanne F, Le Corre P. Oral bioavailability and intestinal secretion of amitriptyline: role of P-glycoprotein? Int J Pharm 2007;330:121–8.

[4] Gupta SK, Shah JC, Hwang SS. Pharmacokinetic and pharmacodynamic characterization of OROS® and immediate-release amitriptyline. Br J Clin Pharmacol 1999;48:71–8.

[5] Thanacoody HK, Thomas SH. Tricyclic antidepressant poisoning: cardiovascular toxicity. Toxicol Rev 2005;24:205–14.

[6] Glauser J. Tricyclic antidepressant poisoning. Cleve Clin J Med 2000;67:704–6.

[7] Bateman DN. Tricyclic antidepressant poisoning: central nervous system effects and management. Toxicol Rev 2005;24:181–6.

[8] Coutts RT, Bach MV, Baker GB. Metabolism of amitriptyline with CYP2D6 expressed in a human cell line. Xenobiotica 1997;27:33–47.

[9] Zhou X, Chen C, Zhang F, Zhang Y, et al. Metabolism and bioactivation of the tricyclic antidepressant amitriptyline in human liver microsomes and human urine. Bioanalysis 2016;8:1365–81.

[10] Ulrich S, Läuter J. Comprehensive survey of the relationship between serum concentration and therapeutic effect of amitriptyline in depression. Clin Pharmacokinet 2002;41:853–76.

[11] Kollroser M, Schober C. Simultaneous determination of seven tricyclic antidepressant drugs in human plasma by direct-injection HPLC-APCI-MS-MS with an ion-trap detector. Ther Drug Monit 2002;24:537–44.

[12] Johnson-Davis KL, Juenke JM, Davis R, McMillin GA. Quantification of tricyclic antidepressants using UPLC-MS/MS. Methods Mol Biol 2012;902:175–84.

11.2

Doxepin

Chemical properties	
Solubility in H_2O	Soluble as HCl salt
Molecular weight	316
pK	8.0
Melting point	184–186 °C
Dosing	
Recommended dose, adult	Typical dosage 75 mg/day for antidepressant effect
Recommended dose, child	1–3 mg/kg/day
Monitoring	
Sample	Serum, plasma
Effective concentrations	150–250[a] ng/mL
Toxic concentrations	>500[a] ng/mL
Methods	HPLC, immunoassay, GC, TLC
Pharmacokinetic properties	
Oral dose absorbed	95%[b]
Time to peak concentration	1–4 h
Protein bound	80%
Volume of distribution	12–28 L/kg
Half-life, adult	12–24 h
Time to steady state, adult	60–120 h
Half-life, child	30–50 h
Time to steady state, child	150–300 h

Excretion, urine

	% Excreted	Window of excretion	Active	Detected in blood
Parent	<1	1–4 days	Yes	Yes
Nordoxepin	<1	1–4 days	Yes	Yes
N-glucuronidedoxepin	18–23	1–4 days	No	No

[a] Combined doxepin and nor-doxepin.
[b] Low bioavailability due to first pass effect.

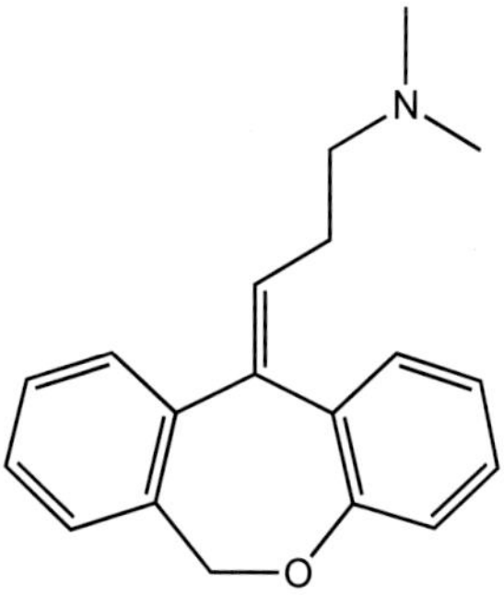

Doxepin

Doxepin, an analog of amitriptyline, is used in the treatment of depression or anxiety. Therapeutic effects may not be apparent for 2–3 weeks following initiation of therapy. Antianxiety effects are evident before antidepressant effects. A topical form of doxepin is used in the treatment of burning mouth syndrome and neuropathic pain. The typical antidepressant dosage of doxepin is 75 mg per day but when used in low dosage (<10 mg/day), it has a high affinity for the histamine H1 receptor but little effect on the serotonergic or adrenergic receptors, thus making doxepin a selective histamine H1 antagonist at these lower doses. It has been hypothesized that blockade of the histamine H1 receptor at a specific time in the circadian cycle, when the release of histamine and wakefulness are both reduced, may promote and maintain sleep. As a result in 2010, the United States Food and Drug Administration approved doxepin (Silenor) in 3 and 6 mg doses for the treatment of insomnia characterized by difficulty with sleep maintenance [1]. The drug is not indicated for use in children less than 12 years of age. Contraindications include hypersensitivity to doxepin, urinary retention, narrow-angle glaucoma, and use of monoamine oxidase inhibitors within 2 weeks. Doxepin has side effects similar to other tricyclic antidepressants (TCAs), with the most common side effects being dry mouth, urinary retention, and constipation. These side effects are due to anticholinergic properties of the drug. Mild overdose may result in drowsiness, stupor, blurred vision, and severe dry mouth. Effects seen in more severe overdose are respiratory depression, hypotension, coma, cardiac arrhythmias, and tachycardia [2].

Doxepin is almost completely absorbed in the intestine, but bioavailability is low due to first-pass metabolism. It binds extensively to proteins and has large volume of distribution. The major route of metabolism is demethylation by cytochrome P450 (CYP) isoenzymes in the liver to a pharmacologically active metabolite, desmethyldoxepin (nordoxepin), which is further metabolized by *N*-oxidation, hydroxylation, and glucuronide conjugation.

Similar to other TCAs, doxepin appears to be substrate for both CYP2D6 and CYP2C19 enzyme isoforms [3]. After a single dose of doxepin, higher bioavailability and lower clearance of doxepin have been found in CYP2D6 poor metabolizers compared to extensive metabolizers. There is a case report of fatal doxepin overdose in a 43 year old man who showed postmortem doxepin and nordoxepin blood concentrations of 2400 and 2900 ng/mL. The authors speculated that the man had defective (CYP2D6*3 or *4) gene that resulted in very low enzymatic activity of CYP2D6 [4].

Effective and toxic ranges are established for the sum of doxepin and nordoxepin, but not for either compound alone. An effective therapeutic range of 150–250 ng/mL (parent plus desmethyl metabolite) has been proposed [5]. Plasma levels do not correlate well with clinical effects so monitoring is most useful in verifying compliance or evaluating toxicity. Immunoassays for total TCAs only estimate a total concentration (doxepin plus metabolite) and are more suitable for diagnosis of overdose. Gas chromatography [6] or high performance liquid chromatography (HPLC) with ultraviolet detection can be used for TDM of doxepin and its metabolites [7]. Liquid chromatography combined with tandem mass spectrometry can also be used for monitoring doxepin and its metabolite concentration in human plasma [8].

References

[1] Yeung WF, Chung KF, Yung KP, Ng TH. Doxepin for insomnia: a systematic review of randomized placebo-controlled trials. Sleep Med Rev 2015;19:75–83.
[2] Bateman DN. Tricyclic antidepressant poisoning: central nervous system effects and management. Toxicol Rev 2005;24:181–6.
[3] Meyer-Barner M, Meineke I, Schreeb KH, Gleiter CH. Pharmacokinetics of doxepin and desmethyldoxepin: an evaluation with the population approach. Eur J Clin Pharmacol 2002;58:253–7.
[4] Koski A, Ojanperä I, Sistonen J, Vuori E, Sajantila A. A fatal doxepin poisoning associated with a defective CYP2D6 genotype. Am J Forensic Med Pathol 2007;28:259–61.
[5] Leucht S, Steimer W, Kreuz S, Abraham D, et al. Doxepin plasma concentrations: is there really a therapeutic range? J Clin Psychopharmacol 2001;21:432–9.
[6] Martinez MA, Sanchez de la Torre C, Almarza E. A comparative solid-phase extraction study for the simultaneous determination of fluvoxamine, mianserin, doxepin, citalopram, paroxetine, and etoperidone in whole blood by capillary gas-liquid chromatography with nitrogen-phosphorus detection. J Anal Toxicol 2004;28:174–80.
[7] Adamczyk M, Fishpaugh JR, Harrington C. Quantitative determination of E- and Z-doxepin and E- and Z-desmethyldoxepin by high-performance liquid chromatography. Ther Drug Monit 1995;17:371–6.
[8] Badenhorst D, Sutherland FC, de Jager AD, Scanes T, et al. Determination of doxepin and desmethyldoxepin in human plasma using liquid chromatography-tandem mass spectrometry. J Chromatogr B Biomed Sci Appl 2000;742:91–8.

11.3

Imipramine

Chemical properties	
Solubility in H_2O (mg/mL)	Soluble as HCl salt
Molecular weight	280.4
pKa	9.5
Melting point (° C)	174–175
Dosing	
Recommended dose, adult	50–100 mg/d
Recommended dose, child	1.5–5.0 mg/kg/d
Monitoring	
Sample	Serum, plasma
Effective concentrations	150–250 ng/mL[a]
Toxic concentrations	>500[a]
Methods	Immunoassays, HPLC, GC, TLC
Pharmacokinetic properties	
Oral dose absorbed	99%[b]
Time to peak concentration	2–6 h
Protein bound	60–92%
Volume of distribution	10–20 L/kg
Half-life, adult	9–24 h
Time to steady state, adult	48–120 h
Half-life, child	6–15 h
Time to steady state, child	48–96 h

Excretion, urine

	% Excreted	Window of excretion	Active	Detected in blood
Parent	<5	3–5 days	Yes	Yes
Desipramine	1–4	3–5 days	Yes	Yes
Hydroxy metabolites	15–35	3–5 days	Yes	Yes
Glucuronides	40–60	3–5 days	No	Yes

[a] Combined imipramine and desipramine.
[b] Low bioavailability due to first pass effect.

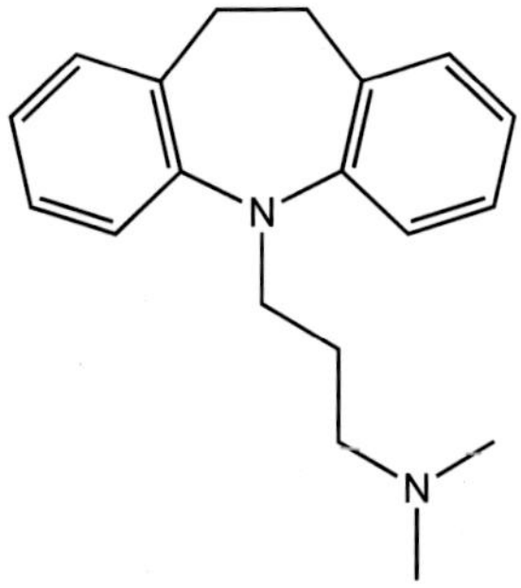

Imipramine

Imipramine was first investigated as a sedative, but today is used for the treatment of major depression. Therapeutic effects of imipramine and its active metabolite desipramine (also sold as a separate drug) may not be evident for 1–3 weeks after drug therapy is initiated. In children over 6 years, imipramine is used as temporary adjunctive therapy in childhood enuresis. Use of the drug is not recommended for children <6 years of age. Imipramine and desipramine are basic drugs which are absorbed from the alkaline environment of small intestine with little or no absorption in the stomach. However, oral bioavailability is low due to first pass metabolism. Peak plasma concentrations are observed 2–6 h after oral administration. Food has no effect on absorption. A high degree of tissue binding is responsible for apparent volumes of distribution of 10–20 L/kg for imipramine and 10–50 L/kg for desipramine. Plasma protein binding of imipramine ranges from 60% to 92%, while for desipramine it ranges from 73% to 92%. Average half-life for imipramine is 20 h and for desipramine is 21 h. Imipramine is extensively metabolized in the liver with less than 5% excreted unchanged. Imipramine is mainly eliminated by demethylation to the active metabolite desipramine and to a lesser extent by aromatic 2-hydroxylation to 2-hydroxyimipramine. Desipramine is metabolized by aromatic 2-hydroxylation to 2-hydroxydesipramine [1].

Similar to other tricyclic antidepressants (TCAs), imipramine has adverse effects including cardiac toxicity at higher plasma levels [2–5]. For therapeutic drug monitoring, concentrations of both imipramine and its active metabolite desipramine should be measured in serum or plasma [5]. The generally accepted therapeutic range of imipramine is 150–250 ng/mL, while toxicity is encountered at levels exceeding 500 ng/mL. Perry et al. commented that therapeutic responses are observed when imipramine and its active metabolite concentrations are between 175 and 350 ng/mL [6]. In children, toxicity may be encountered at a serum concentration of 225 ng/mL. However, children can be safely treated with imipramine when their plasma levels are below 225 ng/mL [7].

TCA immunoassays can be used for measuring combined imipramine and desipramine concentration in serum or plasma. However, gas chromatography or high performance liquid chromatograph is superior because individual concentrations of imipramine and desipramine can be determined. Pommier et al. developed a gas chromatography combined with mass spectrometric method (GC/MS) for the simultaneous determination of imipramine and desipramine in human using deuterated imipramine and deuterated desipramine as internal standards; desipramine and D4-desipramine were converted into their pentafluoropropionyl derivatives prior to GC/MS analysis [8]. Liquid chromatography combined with mass spectrometry can also be used for analysis of TCAs including imipramine and desipramine [9, 10].

References

[1] Sallee FR, Pollock BG. Clinical pharmacokinetics of imipramine and desipramine. Clin Pharmacokinet 1990;18:346–64.

[2] Bateman DN. Tricyclic antidepressant poisoning: central nervous system effects and management. Toxicol Rev 2005;24:181–6.

[3] Liebelt EL, Ulrich A, Francis PD, Woolf A. Serial electrocardiogram changes in acute tricyclic antidepressant overdoses. Crit Care Med 1997;25:1721–6.

[4] Thanacoody HK, Thomas SH. Tricyclic antidepressant poisoning: cardiovascular toxicity. Toxicol Rev 2005;24:205–14.

[5] Muller MJ, Dragicevic A, Fric M, Gaertner I, Grasmader K, Hartter S, et al. Therapeutic drug monitoring of tricyclic antidepressants: how does it work under clinical conditions? Pharmacopsychiatry 2003;36:98–104.

[6] Perry PJ, Zeilmann C, Arndt S. Tricyclic antidepressant concentrations in plasma: an estimate of their sensitivity and specificity as a predictor of response. J Clin Psychopharmacol 1994;14:230–40.

[7] Preskorn SH, Weller EB, Weller RA, Glotzbach E. Plasma levels of imipramine and adverse effects in children. Am J Psychiatry 1983;140:1332–5.

[8] Pommier F, Sioufi A, Godbillon J. Simultaneous determination of imipramine and its metabolite desipramine in human plasma by capillary gas chromatography with mass-selective detection. J Chromatogr B Biomed Sci Appl 1997;703:147–58.

[9] Kollroser M, Schober C. Simultaneous determination of seven tricyclic antidepressant drugs in human plasma by direct-injection HPLC-APCI-MS-MS with an ion-trap detector. Ther Drug Monit 2002;24:537–44.

[10] Crutchfield CA, Breaud AR, Clarke WA. Quantification of tricyclic antidepressants in serum using liquid chromatography electrospray tandem mass spectrometry (HPLC-ESI-MS/MS). Methods Mol Biol 2016;1383:265–70.

11.4

Nortriptyline

Chemical properties	
Solubility in H_2O	Soluble as HCl salt
Molecular weight	263.4
pKa	9.0
Melting point	213–215 °C
Dosing	
Recommended dose, adult	75–150 mg/day
Recommended dose, child	20–100 mg/day
Monitoring	
Sample	Serum, plasma
Effective concentrations	50–100 ng/mL
Toxic concentrations	>500
Methods	HPLC, GC, immunoassay, TLC
Pharmacokinetic properties	
Oral dose absorbed	99%[a]
Time to peak concentration	4–8 h
Protein bound	93–95%
Volume of distribution	20–40 L/kg
Half-life, adult	18–93 h
Time to steady state, adult	100–500 h
Half-life, child	14–22 h
Time to steady state, child	80–120 h

Excretion, urine

	% Excreted	Window of excretion	Active	Detected in blood
Parent	2–5	6 days	Yes	Yes
10-Hydroxynortriptyline	30	6 days	No	Yes
Conjugated 10-hydroxynortriptyline	30	6 days	No	No

[a] Bioavailability is low due to first pass effect.

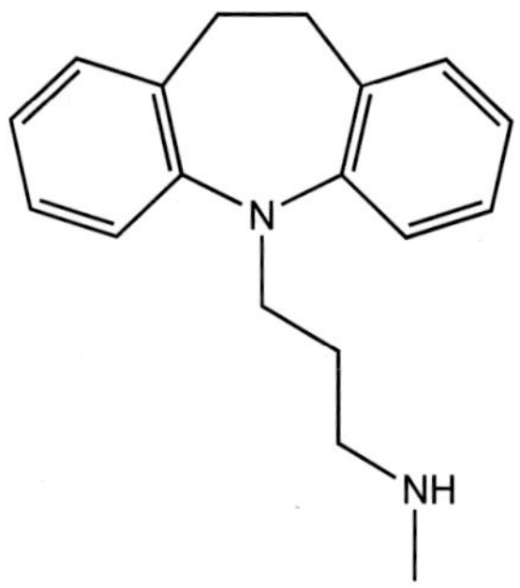

Nortriptyline

Nortriptyline is a tricyclic antidepressant (TCA) that was developed in the 1960s and remains in common use. While nortriptyline is used on its own, it is also produced as an active demethylated metabolite of amitriptyline. Its other uses include the treatment of myofascial pain, neuralgia, burning mouth syndrome, chronic pain, anxiety disorders, enuresis, attention-deficit/hyperactivity disorder (ADHD), and adjunctive therapy for smoking cessation. Nortriptyline's mechanism of action involves inhibition of neurotransmitters reuptake, primarily of norepinephrine and to a lesser extent of serotonin. It is known to inhibit the activity of histamine, serotonin, and acetylcholine and to interfere with the transport, release and storage of catecholamines. The drug shares many of the same pharmacological properties described above for the other TCAs. Adverse effects are a major clinical issue with nortriptyline and other TCAs and occur in as many as 20% of patients. Anticholinergic side effects are common and include dry mouth, constipation, urinary retention and blurred vision. Antagonism of histamine receptors causes sedation. A most common cardiovascular complication of TCAs is postural hypotension caused largely by α1-adrenoceptor blockade, which can be especially problematic in elderly patients who are at risk for injury due to falls. Other cardiac side effects are tachycardia (increased heart rate) and arrhythmias, which are at least partially caused by anticholinergic effects [1, 2].

For treatment of depression, daily oral doses range from 75 to 150 mg in adults and 20 to 100 mg in children. Lower doses are used in children for controlling nocturnal enuresis. The drug is readily absorbed by the intestine. However, oral bioavailability is low due to first pass metabolism.

Nortriptyline is metabolized by the cytochrome P450 (CYP) 2D6 isoenzyme mainly to E-10-hydroxynortriptyline and to a minor extent to the stereoisomer Z-10-hydroxynortriptyline. Both metabolites are active inhibitors of norepinephrine reuptake, with E-10-hydroxynortriptyline having activity equivalent to approximately 50% of nortriptyline. Both metabolites are further metabolized by glucuronidation. Both 10-hydroxynortriptyline and glucuronide forms are excreted in the urine [1]. CYP2D6 polymorphism plays an important role in metabolism of nortriptyline. Asians may need much lower dose to achieve therapeutic plasma concentration because approximately 50% Asians show poor CYP2D6 activity due to genetic mutation (CYP2D6*10 allele) [3].

Although the plasma levels do not correlate well with clinical effects, the recommended levels are 50–100 ng/mL [4]. Therapeutic effects may not be apparent for 2–3 weeks after initiation of therapy and dosing is gradually increased. One study using 18 patients reported best clinical response with nortriptyline plasma concentrations between 50 and 139 ng/mL [5].

A number of chromatography-based methods are described to measure nortriptyline including gas chromatography and high performance liquid chromatography. The most widely used method is high performance liquid chromatography. Immunoassay for estimating total TCAs and specific immunoassays for measuring nortriptyline are available. Most immunoassays for nortriptyline have significant cross reactivity with other TCAs, so the drug history is necessary for accurate interpretation of the results [6, 7]. More recently, liquid chromatography combined with tandem mass spectrometry has been used for simultaneous analysis of TCAs including nortriptyline using dried blood spot [8].

References

[1] Jensen BP, Roberts RL, Vyas R, Bonke G, et al. Influence of ABCB1 (P-glycoprotein) haplotypes on nortriptyline pharmacokinetics and nortriptyline-induced postural hypotension in healthy volunteers. Br J Clin Pharmacol 2012;73:619–28.

[2] Thanacoody HK, Thomas SH. Tricyclic antidepressant poisoning: cardiovascular toxicity. Toxicol Rev 2005;24:205–14.

[3] Yue QY, Zhong ZH, Tybring G, Dalén P, et al. Pharmacokinetics of nortriptyline and its 10-hydroxy metabolite in Chinese subjects of different CYP2D6 genotypes. Clin Pharmacol Ther 1998;64:384–90.

[4] Rubin EH, Biggs JT, Preskorn SH. Nortriptyline pharmacokinetics and plasma levels: implications for clinical practice. J Clin Psychiatry 1985;46:418–24.

[5] Ziegler VE, Clayton PJ, Taylor JR, Tee B, Biggs JT. Nortriptyline plasma levels and therapeutic response. Clin Pharmacol Ther 1976;20:458–63.

[6] Kollroser M, Schober C. Simultaneous determination of seven tricyclic antidepressant drugs in human plasma by direct-injection HPLC-APCI-MS-MS with an ion-trap detector. Ther Drug Monit 2002;24:537–44.

[7] Way BA, Stickle D, Mitchell ME, Koenig JW, Turk J. Isotope dilution gas chromatographic-mass spectrometric measurement of tricyclic antidepressant drugs. Utility of the 4-carbethoxyhexafluorobutyryl derivatives of secondary amines. J Anal Toxicol 1998;22:374–82.

[8] Berm EJJ, Paardekooper J, Brummel-Mulder E, Hak E, Wilffert B, et al. A simple dried blood spot method for therapeutic drug monitoring of the tricyclic antidepressants amitriptyline, nortriptyline, imipramine, clomipramine, and their active metabolites using LC-MS/MS. Talanta 2015;134:165–72.

11.5
Therapeutic drug monitoring of selective serotonin reuptake inhibitors (SSRIs)

The selective serotonin (5-hydroxytryptamine; 5-HT) reuptake inhibitors (SSRIs: citalopram, escitalopram, fluoxetine, fluvoxamine, paroxetine and sertraline) have antidepressant properties. SSRIs are structurally heterogeneous, but share similar actions on 5-HT brain pathways, because they increase the availability of this neurotransmitter at cerebral receptor sites. These drugs do not require routine therapeutic drug monitoring (TDM) like tricyclic antidepressants, but TDM for certain drugs may benefit selected patient population. United States Food and Drug Administration (FDA)-approved immunoassays are not currently available for TDM of SSRIs. Although elderly patients may benefit from TDM of SSRIs, Hermann et al reported a study of 6633 patients that showed clinical follow-up using TDM was under-utilized in patients with advanced age relative to younger patients, despite the importance to monitor closely in the elderly population. The authors further observed that the percentage of samples above the upper end of reference range was twofold higher in patients aged 60 (6.7%) years or older than patients younger than 60 (3.4%) years old [1].

Fluoxetine is a bicyclic derivative of phenylpropylamine and was the first SSRI approved by the FDA [1]. The commercially available fluoxetine is a racemic mixture of two enantiomers (S- and R-fluoxetine), which are approximately equipotent in pharmacological activities. However, the enantiomers of the main demethylated metabolite, norfluoxetine, which are present in the circulation at higher concentrations than those of the parent drug, show marked differences in pharmacological activity. S-norfluoxetine is about 20 times more potent than R-norfluoxetine as a serotonin reuptake inhibitor both in vitro and in vivo [2].

The starting dose of fluoxetine is generally 10–20 mg/day, typically given in the morning to limit interference with sleep in the early course of treatment. The dose may be carefully increased to 20–40 mg/day. Treatment of bulimia and obsessive compulsive disorder may need doses of 60 mg/day or higher. The maximum dosage is 80 mg/day. In children 5–18 years old, the initial dose is 5–10 mg/day with maximum dose of 60 mg/day. After oral administration, fluoxetine is almost completely absorbed, but bioavailability is below 90% due to hepatic first-pass metabolism. The volume of distribution varies from 14 to 100 L/kg. The plasma half-life is 2–4 days, while the half-life of active metabolite is 7–9 days. The drug is mainly metabolized by the liver [3]. TDM may be helpful in selected patient populations such those of advanced age or with liver disease. A high performance liquid chromatography combined with fluorescence detection has been described for simultaneous analysis of fluoxetine and its active metabolite norfluoxetine. The suggested therapeutic range is 120–500 ng/mL for combined concentration of the parent drug and the metabolite [4].

Citalopram is another widely used SSRI. In addition to its use as an antidepressant, its investigational uses include the treatment of dementia, smoking cessation, ethanol abuse, obsessive–compulsive disorder in children, and diabetic neuropathy [5]. Unlike fluoxetine, citalopram is not approved for the use in children. Citalopram is extensively metabolized in the liver by CYP3A4 and 2C19 to *N*-demethylcitalopram and didemethylcitalopram. These metabolites are approximately eight times less potent than citalopram [6]. Although TDM of citalopram is not routinely conducted, it may be useful to estimate fetal exposure of citalopram in pregnant women taking citalopram by measuring citalopram concentrations in sera of pregnant women. A therapeutic range of 50–110 ng/mL has been suggested. Toxicity is encountered at serum levels exceeding 220 ng/mL. A HPLC-ultraviolet method is available for measuring citalopram concentration in serum or plasma [7]. Escitalopram, the active S-enantiomer of the racemic SSRI citalopram, was introduced in 2002. The drug has a long half-life of 27–32 h. The therapeutic range is 15–80 ng/mL and toxicity may occur at serum concentration of 160 ng/mL. TDM may be helpful [8].

Fluvoxamine is FDA-approved for the treatment of obsessive–compulsive disorder in adults and children ≥8 years of age. The medication is also used in the treatment of social anxiety disorder. Its off label uses include the treatment of major depression, panic disorder, and anxiety disorders in children [9]. Initial dose in adults is 50 mg at bedtime and maybe increased to 100–300 mg/day at 4–7-day intervals. In children the initial dose is 25 mg at bedtime and can be increased up to 200 mg/day. Steady-state plasma concentrations are 2–3 times higher in children than those in adolescents, and female children show significantly higher area under the curve for drug metabolism than males [6]. The half-life is 21–43 h and therapeutic range is 60–230 ng/mL. Toxic range is 500 ng/mL and above. TDM may be helpful [8].

Paroxetine was first approved by the FDA for the treatment of social phobia and generalized anxiety disorder. Later on, paroxetine was approved for the treatment of post-traumatic stress disorder, panic disorder with or without agoraphobia and obsessive–compulsive disorder (OCD) in adults. The other uses include treatment of eating disorders, impulse control disorders, self-injurious behavior and premenstrual disorders [10]. The usual dose of paroxetine varies from 10 to 60 mg/day. Paroxetine is extensively metabolized in the liver by oxidation and methylation with subsequent formation of glucuronide and sulfate conjugates which are eliminated in the urine. At high doses, paroxetine exhibits nonlinear pharmacokinetics and may need careful dosage adjustments. The drug has a long half-life of 12–44 h. The therapeutic range is 20–65 ng/mL while toxicity is encountered at level exceeding 120 ng/mL.

Sertraline is used in the treatment of major depression, obsessive–compulsive disorder, panic disorder, post-traumatic stress disorder, premenstrual dysphoric disorder and social anxiety disorder [6]. When administered orally, it is completely absorbed and is highly protein bound. The drug is metabolized by the liver through CYP2C19 and CYP2D6 enzyme systems. One of the major metabolites, norsertraline, has pharmacologic activity. Norsertraline has a half-life of about 3 days which is approximately three times longer than sertraline. Therefore, a 2 week washout period is recommended before initiation of therapy with monoamine oxidase inhibitors. The usual oral dose of sertraline is 25–100 mg/day. The common side effects of sertraline include nausea, dry mouth, fatigue and decreased libido. It is associated with more cases of diarrhea compared to fluoxetine but has fewer cases of anxiety and insomnia [11]. The half-life of sertraline is 22–36 h and suggested therapeutic range is 10–150 ng/mL. Toxicity may be encountered at level 300 ng/mL or higher. TDM may be useful in some patients [8].

As mentioned earlier, TDM of SSRIs are challenging, because immunoassays are not commercially available. Therefore, chromatographic techniques are used for TDM of SSRIs. However, one advantage of chromatographic technique is that multiple drugs can be monitored simultaneously. More recently, liquid chromatography combined with tandem mass spectrometry (LC–MS/MS) has been used for simultaneous determination of multiple antidepressants including SSRIs. Ansermot et al. described a LC–MS/MS method for simultaneous analysis of SSRIs (citalopram, fluoxetine, fluvoxamine, paroxetine and sertraline) and their active metabolites (desmethyl-citalopram and norfluoxetine) in human plasma [12].

References

[1] Hermann M, Waade RB, Molden E. Therapeutic drug monitoring of selective serotonin reuptake inhibitors in elderly patients. Ther Drug Monit 2015;37:546–9.
[2] Altamura AC, Moro AR, Percudani M. Clinical pharmacokinetics of fluoxetine. Clin Pharmacokinet 1994;26:201–14.
[3] Scordo MG, Spina E, Dahl ML, Gatti G, Perucca E. Influence of CYP2C9, 2C19 and 2D6 genetic polymorphisms on the steady-state plasma concentrations of the enantiomers of fluoxetine and norfluoxetine. Basic Clin Pharmacol Toxicol 2005;97: 296–301.
[4] Hiemke C, Härtter S. Pharmacokinetics of selective serotonin reuptake inhibitors. Pharmacol Ther 2000;85:11–28.
[5] Magalhães T, Alves G, Llerena A, Falcão A. Therapeutic drug monitoring of fluoxetine, norfluoxetine and paroxetine: a new tool based on microextraction by packed sorbent coupled to liquid chromatography. J Anal Toxicol 2017;41:631–8.
[6] Ables AZ, Baughman 3rd OL. Antidepressants: update on new agents and indications. Am Fam Physician 2003;67:547–54.
[7] Paulzen M, Goecke TW, Stingl JC, Janssen G, et al. Pregnancy exposure to citalopram - Therapeutic drug monitoring in maternal blood, amniotic fluid and cord blood. Prog Neuropsychopharmacol Biol Psychiatry 2017;79(Pt B):213–9.
[8] Schoretsanitis G, Paulzen M, Unterecker S, Schwarz M, et al. TDM in psychiatry and neurology: a comprehensive summary of the consensus guidelines for therapeutic drug monitoring in neuropsychopharmacology, update 2017; a tool for clinicians. World J Biol Psychiatry 2018;19:162–74.
[9] Figgitt DP, Fluvoxamine MCKJ. An updated review of its use in the management of adults with anxiety disorders. Drugs 2000;60(4):925–54.
[10] Bourin M, Chue P, Guillon Y. Paroxetine: a review. CNS Drug Rev 2001;7:25–47.
[11] Brady K, Pearlstein T, Asnis GM, Baker D, Rothbaum B, Sikes CR, et al. Efficacy and safety of sertraline treatment of posttraumatic stress disorder: a randomized controlled trial. JAMA 2000;283:1837–44.
[12] Ansermot N, Brawand-Amey M, Eap CB. Simultaneous quantification of selective serotonin reuptake inhibitors and metabolites in human plasma by liquid chromatography-electrospray mass spectrometry for therapeutic drug monitoring. J Chromatogr B Analyt Technol Biomed Life Sci 2012;885-886:117–30.

11.6

Lithium

Chemical properties	
Solubility in H_2O	12.8 mg/mL
Molecular weight	73.89
pKa	6.8 (carbonate)
Melting point	618 °C
Dosing	
Recommended dose, adult	10–20 mg/kg/day
Recommended dose, child	15–20 mg/kg/day
Monitoring	
Sample	Serum, plasma
Effective concentrations	0.6–1.2 mmol/L monitor trough at 12 h post dose
Toxic concentrations	>1.5 mmol/L
Methods	Flame emission, atomic absorption, ion-specific electrode
Pharmacokinetic properties	
Oral dose absorbed	80–100%
Time to peak concentration	0.5–3 h
Protein bound	0%
Volume of distribution	0.79 L/kg
Half-life, adult	Average 24 h (varies with renal function)
Time to steady state, adult	3–4 days

Excretion, urine			
	% Excreted	Active	Detected in blood
Unchanged	90–95 (50% first 24 h)	yes	yes

Lithium (chemical symbol: Li) is available for the treatment of bipolar disorders (also called manic-depressive disorder: lifetime prevalence of 1–2%) and was approved by the United States Food and Drug Administration in 1970. The manic state, an incapacitating psychiatric disorder, is characterized by a bizarre increase in psychomotor activity, grandiosity, and emotional lability. Manic attacks are paroxysmal and occur repetitively with relatively normal periods of behavior. The precise mechanism of action of lithium is not completely understood. However, lithium enhances serotonin (5-HT) activity by several mechanisms, including increased synthesis of the neurotransmitter, inhibition of presynaptic serotonin 5-HT1A autoreceptors, and downregulation of postsynaptic 5-HT2 receptors. In addition, lithium may facilitate norepinephrine release and enhance glutamatergic activity as well. Lithium also acts as inhibitor on second messenger systems such as inositol mono-phosphatase and, ultimately, inositol depletion. Lithium is also known to reduce protein kinase C activity, ultimately altering neurotransmission through effects on genomic expression [1].

Lithium is available as lithium carbonate (150, 300 or 600 mg) both as immediate release and extended release formulations. Lithium citrate is used in liquid formulations. Absorption of lithium is complete and not delayed by food intake. Estimated bioavailability is 80–100%. As a water-soluble monovalent cation, lithium is not bound to plasma proteins and distributes widely throughout the body following oral absorption. The volume of distribution for lithium approximates that of total body water (0.79 L/kg). Peak plasma concentration is achieved 0.5–3 h after administration of immediate release formula but may be extended to 2–6 h after taking extended release formula. Lithium is not metabolized and is eliminated almost entirely via renal means. Elimination of lithium via the kidneys is biphasic [2]. The significant predictors of lithium clearance are renal function and body size. The typical values of lithium clearance ranged from 0.41 to 9.39 L/h. The magnitude of inter-individual variability on lithium clearance ranged from 12.7 to 25.1% [3]. The average half-life of lithium is 24 h in young healthy subjects, and steady-state dynamics are ordinarily achieved within 3–4 days [1].

The recommended therapeutic range of lithium in plasma/serum is 0.6–1.2 mmol/L. In one study involving 4359 requests for therapeutic drug monitoring (TDM) of lithium, the authors observed lithium level above therapeutic (>1.2 mmol/L) in 17% cases, below therapeutic (<0.6 mmol/L) in 23%, and within therapeutic range (0.6–1.2 mmol/L) in 60% cases. The authors concluded that TDM for lithium is necessary in order to avoid treatment failure or lithium toxicity [4]. However, consensus has been established among psychiatrists that lithium levels between 0.6 and 0.8 mmol/L are usually desirable and adequate for management of patients receiving lithium for bipolar disorder. However, TDM of lithium along with the need for regular monitoring of renal and endocrine function is also necessary.

In addition, with more complex cases (e.g., atypical presentations) and in special populations (e.g., youth; pregnancy and post-partum; older adults), guidance varied considerably and clear consensus recommendations were more difficult to achieve. In younger adults, desirable plasma lithium levels of 0.6–0.8 mmol/L could be achieved with comparatively lower doses and in the very elderly it may be prudent to target lower plasma levels in the first instance [5]. Lower target ranges (0.5–0.8 mmol/L) have been recommended in the elderly due to increased sensitivity to adverse effects, particularly neurotoxicity.

Undesirable side-effects of lithium can occur at any plasma level; however, they are much more frequent when plasma levels exceed 1.5 mmol/L. Side effects are more likely to occur in children than adults even after adjusting for weight and plasma levels [6]. Practice Guidelines of the American Psychiatric Association recommend testing renal function every 2–3 months during the first 6 months of treatment. Thereafter, renal and thyroid function tests are recommended every 6–12 months or when clinically indicated. Up to 35% of patients treated with lithium may develop hypothyroidism with a greater prevalence in women and those over 50 years of age [7].

It is generally accepted that lithium toxicity may be encountered at serum lithium levels exceeding 1.2 mmol/L. However, Foulser et al. reported a case of a 62-year-old man with a history of bipolar disorder, previously stable on lithium for over 20 years, who presented with a manic relapse and signs of lithium toxicity in the form of a coarse tremor. Serum lithium levels were in the normal therapeutic range. The patient was admitted and, after discontinuation of lithium, his toxicity resolved [8].

In most cases, mild lithium toxicity presents at plasma levels of 1.5–2.5 mmol/L as drowsiness, slurred speech, muscle tremors or coarse tremors. With plasma levels of 2.5–3.5 mmol/L, side-effects are more severe and may include vomiting, tremor, myoclonic jerks and cogwheel rigidity. Levels greater than 3.5 mmol/L are life-threatening. Death usually occurs as a consequence of neurologic and cardiovascular complications. Prompt treatment of lithium intoxication involves discontinuation of the drug and supportive therapy. Enhanced clearance of lithium can be achieved by hemodialysis [9]. One study reported that median lithium level was 2.69 mmol/L (range: 1.50–2.90 mmol/L) in subjects who died from lithium toxicity [10]. In contrast, Haussmann et al. reported a case of non-fatal lithium toxicity with serum lithium level of 5.5 mmol/L [11].

Plasma lithium levels can increase due to renal disease, dehydration and use of several drugs including nonsteroidal anti-inflammatory drugs

(NSAIDs), angiotensin-converting enzyme inhibitors, cyclooxygenase-2 inhibitors, and thiazide diuretics. Prior to initiation of lithium treatment baseline renal function tests, thyroid function tests and an electrocardiogram (for patients over 40 years old) should be performed [1].

Lithium levels should be monitored several days after initiation of therapy, followed by monitoring before and after each dosage change. Monitoring frequency during chronic lithium therapy depends on several factors including stability of lithium levels over time for that patient and the reliability of the patient to notice and report symptoms. Plasma samples obtained 12h after the last doses are the most effective guide to lithium therapy. Although lithium has a narrow therapeutic range, careful monitoring of lithium plasma levels aids the clinician in establishing and maintaining complete control of manic states. No patient should be administered lithium without routine monitoring of lithium levels [11]. A common pre-analytical error in TDM of lithium is blood collection in a lithium-heparin tubes, which will lead to erroneously elevated lithium concentrations.

References

[1] Finley PR. Drug interactions with lithium: an update. Clin Pharmacokinet 2016;55:925–41.
[2] Ward ME, Musa MN, Bailey L. Clinical pharmacokinetics of lithium. J Clin Pharmacol 1994;34:280–5.
[3] Methaneethorn J. Population pharmacokinetic analyses of lithium: a systematic review. Eur J Drug Metab Pharmacokinet 2018;43:25–34.
[4] Sharma S, Joshi S, Chadda RK. Therapeutic drug monitoring of lithium in patients with bipolar affective disorder: experiences from a tertiary care hospital in India. Am J Ther 2009;16:393–7.
[5] Malhi GS, Gessler D, Outhred T. The use of lithium for the treatment of bipolar disorder: Recommendations from clinical practice guidelines. J Affect Disord 2017;217:266–80.
[6] Freeman MP, Freeman SA. Lithium: clinical considerations in internal medicine. Am J Med 2006;119:478–81.
[7] American Psychiatric Association. Practice guideline for the treatment of patients with bipolar disorder (revision). Am J Psychiatry 2002;159(Suppl 4):1–50.
[8] Foulser P, Abbasi Y, Mathilakath A, Nilforooshan R. Do not treat the numbers: lithium toxicity. BMJ Case Rep 2017;2017:. pii: bcr-2017-220079.
[9] Sproule B. Lithium in bipolar disorder—can drug concentrations predict therapeutic effect? Clin Pharmacokinet 2002;41(9):639–60.
[10] Söderberg C, Wernvik E, Jönsson AK, Druid H. Reference values of lithium in postmortem femoral blood. Forensic Sci Int 2017;277:207–14.
[11] Haussmann R, Bauer M, von Bonin S, Lewitzka U. Non-fatal lithium intoxication with 5.5 mmol/L serum level. Pharmacopsychiatry 2015;48:121–2.

Further reading

[12] Ooba N, Tsutsumi D, Kobayashi N, Hidaka S. Prevalence of therapeutic drug monitoring for lithium and the impact of regulatory warnings: analysis using Japanese claims database. Ther Drug Monit 2018;40:252–6.

CHAPTER 12

Therapeutic drug monitoring of cardioactive drugs

Antiarrhythmic drugs are classified according to their electrophysiological effects observed in isolated cardiac tissues. Class I drugs slow conductance by reducing the upstroke velocity of the action potential by rapidly blocking the sodium channel. Drugs in this class include quinidine, procainamide, disopyramide, lidocaine, mexiletine, flecainide and propafenone. These drugs are further sub-classified into class IA through C according to their effects on the duration of the action potential. Quinidine, procainamide, and disopyramide are classified as Class IA drugs, lidocaine and mexiletine as Class 1B drugs, and flecainide and propafenone as Class IC drugs. Class II drugs include the beta-adrenergic receptor antagonists (beta blockers) which block the effect of catecholamines. Class III agents prolong the action potential duration by blocking cardiac potassium channels and include drugs such as amiodarone and sotalol [1, 2]. Dronedarone, a relatively new synthetic non-iodinated derivative of amiodarone, is also considered as a class III agent. Class IV drugs include calcium channel blockers such as verapamil and diltiazem [3]. Digoxin, although an old cardioactive drug, is still a widely prescribed cardiac glycoside and continues to have an important role in long term outpatient management. Digoxin is used in the treatment of patients who continue to have symptoms of heart failure despite therapy with diuretics, angiotensin-converting enzyme (ACE)-inhibitors, and/or beta blockers [4].

The antiarrhythmic drug classes outlined above are based on the most prominent electrophysiologic action of the various drugs. In clinical practice, drugs may exhibit effects on multiple molecular targets, e.g., amiodarone has class I properties (sodium channel blocking) in addition to class II effects. The ability to impact multiple molecular targets can lead to adverse effects, some of them life-threatening.

For several of the older cardioactive drugs, therapeutic drug monitoring (TDM) is an important aspect of patient management because of the clear correlations between the drug level and pharmacological responses. In fact,

Therapeutic Drug Monitoring Data
https://doi.org/10.1016/B978-0-12-815849-4.00012-8

digoxin was one of the first drugs for which routine TDM significantly reduced the incidence of toxicity. When used, Class I antiarrhythmic drugs should be routinely monitored with the exception of propafenone [5]. Quinidine, disopyramide and procainamide have well established therapeutic ranges, and immunoassays are commercially available, although declining clinical use has resulted in fewer clinical laboratories maintaining these assays in-house. The metabolism of procainamide produces a metabolite, *N*-acetylprocainamide (NAPA) that exhibits significant Class III-type antiarrhythmic action. Therefore monitoring of both procainamide and NAPA is essential for efficacy as well as for avoiding undesirable side effects [6].

Beta blockers differ from most other antiarrhythmic agents by not directly modifying ion channel functions. These drugs are also useful in controlling ventricular rate in patients with atrial fibrillation [7]. Because efficacy and side effects of beta blockers correlate poorly with serum drug concentrations, TDM is not usually practiced for these drugs.

For Class III antiarrhythmic drugs there are some indications for TDM for amiodarone; however, routine monitoring of sotalol has not proven useful. Amiodarone has unusual pharmacokinetic parameters with variable bioavailability (20–80%) and a terminal elimination half-life of 35–40 days. The half-life may exceed 100 days after oral administration. A plasma concentration of 0.5 μg/mL is required for efficacy and toxicity is encountered when concentrations exceed 2.5 μg/mL. The therapeutic range for amiodarone is considered to be 1.0–2.5 μg/mL [5]. The Class IV calcium channel blockers are normally monitored using hemodynamic effects rather than TDM. However, TDM of verapamil may have some clinical relevance. Therapeutic range of verapamil is considered as 50–200 ng/mL.

In conclusion, Class I antiarrhythmic drugs and digoxin require routine TDM. The immunoassays that are available are unfortunately subjected to interferences from both endogenous and exogenous factors including other drugs and drug metabolites. For this reason, more sophisticated chromatographic methods offer better sensitivity and specificity for determination of serum concentrations of these drugs. Chromatographic methods also offer the ability to analyze multiple antiarrhythmic drugs simultaneously, especially with liquid chromatography combined with tandem mass spectrometry [7,8].

References

[1] Scholz H. Classification and mechanism of action of antiarrhythmic drugs. Fundam Clin Pharmacol 1994;8:385–90.

[2] Ng GA. Feasibility of selection of antiarrhythmic drug treatment on the basis of arrhythmogenic mechanism—relevance of electrical restitution, wavebreak and rotors. Pharmacol Ther 2017;176:1–12.

[3] Sablayrolles S, Le Grand B. Drug evaluation: dronedarone, a novel non-iodinated antiarrhythmic agent. Curr Opin Investig Drugs 2006;7:842–9.
[4] Dec GW. Digoxin remains useful in the management of chronic heart failure. Med Clin North Am 2003;87:317–37.
[5] Campbell TJ, Williams KM. Therapeutic drug monitoring: antiarrhythmic drugs. Br J Clin Pharmacol 1998;436:307–19.
[6] Ellenbogen KA, Wood MA, Stambler BS. Procainamide: a perspective on its value and danger. Heart Dis Stroke 1993;2:473–6.
[7] Zicha S, Tsuji Y, Shiroshita-Takeshita A, Nattel S. Beta-blockers as antiarrhythmic agents. Handb Exp Pharmacol 2006;171:235–66.
[8] Li S, Liu G, Jai J, Liu Y, et al. Simultaneous determination of ten antiarrhythmic drugs and a metabolite in human plasma by liquid chromatography-tandem mass spectrometry. J Chromatogr B Anal Technol Biomed Life Sci 2007;847(2):174–81.

12.1

Digoxin

Chemical properties

Solubility in H_2O	Insoluble
Molecular weight	780.92
pKa	n/a
Melting point	n/a

Dosing

Recommended dose, adult	0.125–0.25 mg daily
Recommended dose, child	Age dependent

Monitoring

Sample	Serum, plasma (heparin, EDTA)
Effective concentrations, adult	0.5–1.0 ng/mL
Effective concentrations, child	0.5–1.0 ng/mL
Toxic concentrations	>2.0 ng/mL
Methods:	Immunoassays, HPLC, LC–MS/MS

Pharmacokinetic properties

Oral dose absorbed	70–80%
Time to peak concentration	Oral: 1–3 h IV: 1.5–4.0 h IM: 2–6 h
Protein bound	20–30%
Volume of distribution	5.0–7.5 L/kg
Half-life, adult	26–48 h
Time to steady state, adult	7–10 h
Half-life, child	12–48 h
Time to steady state, child	2–10 h

Excretion, urine

	% Excreted	Active	Detected in blood
Parent	0–90	Yes	Yes

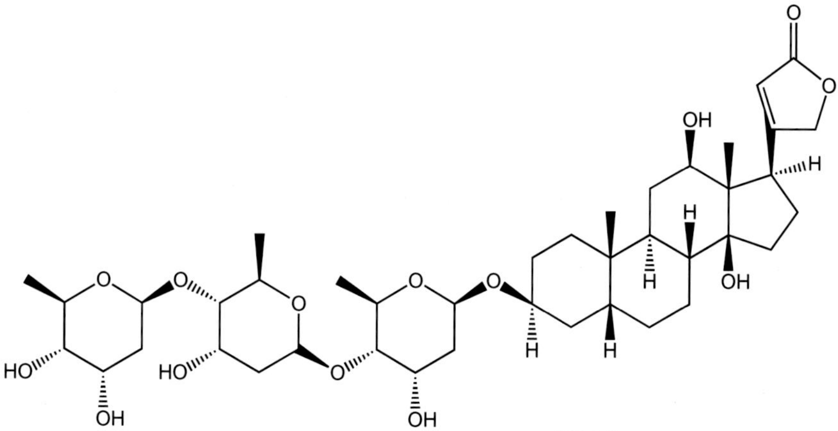

Digoxin

Digoxin (Digitek, Lanoxin, etc.) is a cardiac glycoside used primarily for the long-term management of congestive heart failure. The drug is most effective in low cardiac output heart failure caused by atherosclerotic heart disease, coronary heart disease, hypertension, and congenital heart disease. Digoxin increases the force of cardiac contraction and increases cardiac output. This improves oxygen delivery to the tissues and blood flow to the kidneys, increasing renal filtration and urinary output. The net effect is a reduction in the total circulating blood volume. This improves coronary perfusion by decreasing the heart rate and increasing the diastole filling time. Digoxin is also effective in the treatment of cardiac dysrhythmias such as atrial flutter, atrial fibrillation, and paroxysmal atrial tachycardia. Digoxin is administered to pediatric patients who have severe congenital anomalies which obstruct ventricular outflow or reroute blood flow. Digoxin appears to be of most beneficial in patients with severe heart failure, cardiomegaly, and a third heart sound. Digoxin has a positive inotropic effect on the heart by acting directly on the myocardium to increase the force of cardiac contraction. Two mechanisms are involved in this process: direct inhibition of the membrane-bound Na-K-ATPase and promotion of calcium movement from the extracellular to intracellular cytoplasm [1].

Digoxin has been a key therapeutic for heart failure and atrial tachyarrhythmias for over 200 years following Withering's groundbreaking work depicting the therapeutic benefit of the common botanical foxglove (*Digitalis lanata*) in his 1785 monograph [2]. Currently, digoxin is available in capsule, tablet, and elixir forms. The drug is administered orally or intravenously. The dosage of digoxin should be 0.125–0.25 mg daily in the majority of patients. The lower dose should be used in patients over 70 years of age, those with impaired renal function, or those with a low lean body mass [1]. When digoxin is administered orally, approximately 70–80% of an oral dose of digoxin is absorbed, mainly in the proximal part of the small intestine. The rate of absorption is reduced by the presence of food, delayed gastric emptying, malabsorption syndromes, as well as the presence of certain intestinal microorganisms such as *Eubacterium lenta*. Distribution occurs slowly due to its large apparent volume of distribution (5.0–7.5 L/kg). It is distributed to most tissues in the body with the highest concentration found in the heart and skeletal muscle. At equilibrium, the concentration in cardiac tissue is 15–30 times that in the plasma. The average serum protein binding is 25% (range: 20–30%). The half-life of elimination in healthy persons varies between 26 and 45 h. The main route of elimination is renal excretion of the parent drug, which is closely correlated with the glomerular filtration rate. About 25–28% of digoxin is eliminated by nonrenal routes such as biliary excretion [3].

Metabolism of digoxin produces three types of metabolites. The sequential loss of digitoxose sugars in the liver produces 3-beta-digoxigenin, 3-alpha-digoxigenin, digoxigenin bis and monodigitoxoside. Conjugation with polar groups produces glucuronide and sulfate conjugates of these digitoxose sugars such as glucuronide of 3-alpha-digoxigenin. A third category of digoxin metabolites (e.g., dihydro-digoxin) are produced via reduction of double bonds by gut bacteria [4].

Therapeutic drug monitoring (TDM) of digoxin is essential due to its narrow therapeutic window. Samples for serum digoxin levels should not be collected until after a minimum of 6–8 h after administration of an oral dose so that equilibrium is established; a time period of approximately 12 h is considered optimal. Serum levels of the drug are high and variable during the absorptive and distributive phases. For many years, the therapeutic range of digoxin has been considered as 0.8–2.0 ng/mL despite many studies showing there is a substantial overlap between therapeutic and toxic concentrations. Data from more recent clinical trials support a narrower therapeutic range of digoxin of 0.5–1.0 ng/mL [4].

The principal manifestations of digoxin toxicity include cardiac arrhythmias (ectopic and reentrant cardiac rhythms and heart block), gastrointestinal tract symptoms (anorexia, nausea, vomiting and diarrhea), and neurologic symptoms (visual disturbances, headache, weakness, dizziness and confusion). Most adult patients with clinical toxicity have serum digoxin levels >2 ng/mL. However, patients with conditions such as hypokalemia, hypomagnesemia, or hypothyroidism may show clinical signs of digoxin toxicity at lower serum concentrations [5].

Digoxin toxicity can be treated with Digibind, an antibody fragment preparation that tightly binds digoxin and thereby lowers the amount of bioavailable drug.

Digoxin exhibits numerous interactions with compounds that affect absorption or excretion. The most significant of these are co-administration with diuretics or quinidine. Loop and thiazide diuretics can produce hypokalemia and predispose the heart to digoxin toxicity, whereas quinidine reduces the renal clearance of digoxin by 40–50% [6, 7]. Concurrent use of digoxin in the herbal antidepressant St. John's wort may significantly reduce serum digoxin concentration due to increased clearance of digoxin. The magnitude of interaction is dependent on the concentration of hyperforin in St. John's wort [8].

Immunoassays are widely used for routine TDM of digoxin. Older polyclonal antibody-based digoxin immunoassays were limited by interferences. For example in 1980s–90s, significant interferences of endogenous digoxin-like immunoreactive substances (DLIS) in serum digoxin measurement have been reported. DLIS levels are relatively low in healthy people causing no assay interference but DLIS levels are significantly elevated in volume expanded patients such as patients with uremia, liver disease, hypoalbuminemia, pregnant women and premature babies. However, the development and use of specific monoclonal antibody assays (iDigoxin, Abbott Laboratories) for measurement of digoxin has reduced the issue of interference [9]. Similarly, various diuretics such as spironolactone and potassium canrenoate significantly interfered with digoxin immunoassays, but newer monoclonal antibody based digoxin immunoassays are free from such interferences [10, 11].

Digibind and DigiFab are Fab fragments of antidigoxin antibodies used in treating life-threatening acute digoxin overdoses. These antidotes interfere with digoxin immunoassays, complicating measurement of digoxin levels in patients being treated. McMillin et al. studied the effect of Digibind and DigiFab on 13 different digoxin immunoassays. The magnitude of interference varied significantly with each method. The authors commented that monitoring free digoxin (in the protein free ultrafiltrate) eliminated this interference, because both Digibind and DigiFab due higher molecular weight are absent in protein free ultrafiltrate [12]. However, free digoxin levels are not routinely available in many clinical laboratories.

The Chinese medicine Chan Su contains bufalin which interferes with digoxin immunoassays. Similarly oleander plant poisoning may be rapidly detected using digoxin immunoassays, because the active ingredient oleandrin interferes with digoxin immunoassays [13]. Convallatoxin, the active ingredient of lily of the valley, also interferes with digoxin immunoassays due to structural similarity with digoxin. Fink et al. compared five digoxin assays for rapid detection of convallatoxin. Any digoxin immunoassay capable of rapid detection of convallatoxin is unsuitable for TDM of digoxin in patients who are taking lily of the valley herbal supplements [14]. Although chromatographic methods are rarely used in TDM of digoxin, such methods are free from the interferences described above. Bylda et al. described a protocol for simultaneous quantification of digoxin, digitoxin, and their metabolites in serum using liquid chromatography coupled to tandem mass spectrometry [15].

References

[1] Campbell TJ, MacDonald PS. Digoxin in heart failure and cardiac arrhythmias. Med J Aust 2003;179:98–102.

[2] Stucky MA, Goldberger ZD. Digoxin: its role in contemporary medicine. Postgrad Med J 2015;91(1079):514–8.

[3] Iisalo E. Clinical pharmacokinetics of digoxin. Clin Pharmacokinet 1977;2:1–16.

[4] Stone JA, Soldin SJ. An update on digoxin. Clin Chem 1989;35:1326–31.

[5] Grześk G, Stolarek W, Kasprzak M, Krzyżanowski M, et al. Therapeutic drug monitoring of digoxin-20 years of experience. Pharmacol Rep 2018;70:184–9.

[6] Haji SA, Movahed A. Update on digoxin therapy in congestive heart failure. Am Fam Physician 2000;62:409–16.

[7] Magnani B, Malini PL. Cardiac glycosides: drug interactions of clinical significance. Drug Saf 1995;12:97–109.

[8] Mueller SC, Uehleke B, Woehling H, Petzsch M, et al. Effect of St John's wort dose and preparations on the pharmacokinetics of digoxin. Clin Pharmacol Ther 2004;75:546–57.

[9] Lampon N, Pampin E, Tutor JC. Investigation of possible interference by digoxin-like immunoreactive substances on the architect iDigoxin CMIA in serum samples from pregnant women, and patients with liver disease, renal insufficiency, critical illness and kidney and liver transplant. Clin Lab 2012;58:1301–4.

[10] DeFrance A, Armbruster D, Petty D, Cooper KC. Abbott ARCHITECT clinical chemistry and immunoassay systems: digoxin assays are free of interferences from spironolactone, potassium canrenoate, and their common metabolite canrenone. Ther Drug Monit 2011;33:128–31.

[11] Dasgupta A, Johnson MJ, Sengupta TK. Clinically insignificant negative interferences of spironolactone, potassium canrenoate, and their common metabolite canrenone in new dimension vista LOCI digoxin immunoassay. J Clin Lab Anal 2012;26:143–7.

[12] McMillin GA, Qwen W, Lambert TL, De B, et al. Comparable effects of DIGIBIND and DigiFab in thirteen digoxin immunoassays. Clin Chem 2002;48:1580–4.

[13] Dasgupta A. Therapeutic drug monitoring of digoxin: impact of endogenous and exogenous digoxin-like immunoreactive substances. Toxicol Rev 2006;25:273–81.

[14] Fink SL, Robey TE, Tarabar AF, Hodsdon ME. Rapid detection of convallatoxin using five digoxin immunoassays. Clin Toxicol (Phila) 2014;52:659–63.
[15] Bylda C, Thiele R, Kobold U, Volmer DA. Simultaneous quantification of digoxin, digitoxin, and their metabolites in serum using high performance liquid chromatography-tandem mass spectrometry. Drug Test Anal 2015;7:937–46.

12.2

Digitoxin

Chemical properties	
Solubility in H_2O	Insoluble
Molecular weight	764.92
pKa	n/a
Melting point	n/a
Dosing	
Recommended dose, adult[a]	0.05–0.2 mg/day
Recommended dose, child	Individualized
Monitoring	
Sample	Serum or plasma (heparin, fluoride, oxalate; not EDTA not citrate)-collect sample at trough
Effective concentrations, adult	10–30 ng/mL
Effective concentrations, child	10–30 ng/mL
Toxic concentrations	>30 ng/mL
Methods:	Immunoassay, HPLC, LC–MS/MS
Pharmacokinetic properties	
Oral dose absorbed	90–100%
Time to peak concentration	3–6 (oral) h
Protein bound	97%
Volume of distribution	0.4–1.0 L/kg
Half-life, adult	4–6 days
Time to steady state, adult	3–4 weeks
Half-life, child	4–6 days
Time to steady state, child	3–4 weeks

Excretion, urine

	% Excreted	Active	Detected in blood
Parent	15–20	Yes	Yes
Digoxin	?	Yes	Yes

Slow loading dose: 0.2 mg twice daily for a period of 4 days followed by a maintenance dose. Maintenance dose: 0.05–0.3 mg/day.

[a]Rapid loading dose: 0.6 mg followed by 0.4 mg and then 0.2 mg at intervals of 4–6 h.

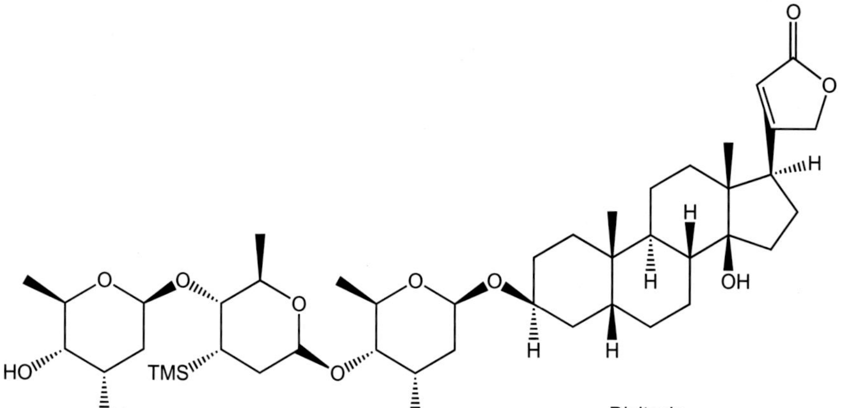

Digitoxin

Digitoxin (Crystodigin, Purodigin) is a cardiac glycoside used primarily for the treatment of congestive heart failure (CHF). Digitoxin is structurally and functionally similar to digoxin but is used less frequently in the United States. It exerts a positive inotropic effect on the heart by acting directly on the myocardium to increase the force of cardiac contraction. Two mechanisms are involved in this process: direct inhibition of membrane-bound sodium- potassium activated adenosine triphosphatase (Na-K-ATPase) and promotion of calcium movement from the extracellular to intracellular cytoplasm [1]. A growing body of evidence indicates that digitoxin cardiac glycoside is a promising anticancer agent when used at therapeutic concentrations [2].

Digitoxin can be administered orally or intravenously. The recommended maintenance dose for an adult is 0.05–0.20 mg per day. The dose for children is dependent on their age. Unlike digoxin, digitoxin is almost completely absorbed (90–100%) from the gastrointestinal tract, due to its highly lipophilic nature. Absorption is retarded by presence of food in the gastrointestinal tract, delayed gastric emptying, and malabsorption syndromes. Following oral administration, the onset of action is within 1–4 h with maximum effect attained within 8–12 h. About 97% of digitoxin is bound to serum protein, mostly albumin. At equilibrium, the concentration of the drug in cardiac tissue is 15–30 times that in the plasma. Digitoxin undergoes extensive hepatic metabolism with one of the metabolites produced being digoxin, itself an active cardiac glycoside. Since digitoxin is not cleared renally, this makes the drug useful in treating CHF patients who also have renal impairment. Both the unchanged parent drug and cardioactive metabolites are excreted into the gut via the bile and are then reabsorbed from the intestine into the enterohepatic circulation. This contributes to the long elimination half-life (4–6 days) of the drug. Steady state is achieved in 3–4 weeks after initiation of digitoxin therapy. Eventually, 80% of the drug is excreted in the urine as inactive metabolites [3]. Concurrent therapy with quinidine increases digitoxin serum levels [4].

The most significant adverse effects associated with digitoxin toxicity are cardiac arrhythmias, including atrioventricular junctional arrhythmias, premature ventricular depolarization, bigeminal rhythms, and atrioventricular blockages. The most common non-cardiac adverse effects are gastrointestinal and include anorexia, nausea, vomiting, and diarrhea. The drug also affects the central nervous system by causing vagal stimulation, disorientation, hallucinations, and visual disturbances. However, hospitalized elderly patients taking digitoxin have a lower rate of toxicity than those taking digoxin.

As with digoxin, life-threatening intoxication with digitoxin can be treated with Digibind, although Digibind has 30–100 times lesser affinity for digitoxin compared to digoxin [5]. Immunoassays or chromatographic methods may be used for serum or plasma digitoxin measurement. The therapeutic range is 10–30 ng/mL [6]. Bremer-Streck et al. described liquid chromatography combined with tandem mass spectrometry method for therapeutic drug monitoring of digitoxin [7].

References

[1] Whayne Jr. TF. Clinical use of digitalis: a state of the art review. Am J Cardiovasc Drugs 2018;18(6):427–40.
[2] Elbaz HA, Stueckle TA, Tse W, Rojanasakul Y, Dinu CZ. Digitoxin and its analogs as novel cancer therapeutics. Exp Hematol Oncol 2012;1(1):4.
[3] Smith TW. Pharmacokinetics, bioavailability and serum levels of cardiac glycosides. J Am Coll Cardiol 1985;5(5 Suppl A):43A–50A.
[4] Kuhlmann J, Dohrmann M, Marcin S. Effects of quinidine on pharmacokinetics and pharmacodynamics of digitoxin achieving steady-state conditions. Clin Pharmacol Ther 1986;39:288–94.
[5] Schmitt K, Tulzer G, Häckel F, Sommer R, Tulzer W. Massive digitoxin intoxication treated with digoxin-specific antibodies in a child. Pediatr Cardiol 1994;15:48–9.
[6] Haustein KO, Winkler U. Therapeutic drug monitoring of digitoxin—results of 3 years' experience. Z Gesamte Inn Med 1989;44:640–3. Article in German.
[7] Bremer-Streck S, Kiehntopf M, Ihle S, Boeer K. Evaluation of a straightforward and rapid method for the therapeutic drug monitoring of digitoxin by LC-MS/MS. Clin Biochem 2013;46(16–17):1728–33.

Further reading

[8] Roever C, Ferrante J, Gonzalez EC, Pal N, Roetzheim RG. Comparing the toxicity of digoxin and digitoxin in a geriatric population: should an old drug be rediscovered? South Med J 2000;93:199–202.

12.3

Disopyramide

Chemical properties	
Solubility in H_2O	1.0 mg/mL
Molecular weight	339.47
pKa	8.34
Melting point	94.5–95.0 °C

Dosing	
Recommended dose, adult	300–600 mg/day; 5–8 mg/kg/day
Recommended dose, child	3.0–6.0 mg/kg/day

Monitoring	
Sample:	Serum, plasma
Effective concentrations	2.0–5.0 μg/mL (trough)
Toxic concentrations	7 μg/mL
Methods	Immunoassay, HPLC, GC

Pharmacokinetic properties	
Oral dose absorbed	>80%
Time to peak concentration	0.5–3.0 h
Protein bound	30–60%
Volume of distribution	0.8–1.9 L/kg
Half-life, adult	Approximately 7 h
Time to steady state, adult	24–48 h
Half-life, child	2.5–3.8 h
Time to steady state, child	Variable

Excretion, urine

	% Excreted	Active	Detected in blood
Parent	50	Yes	Yes
N-des-isopropyl disopyramide	20–30	Yes	Yes

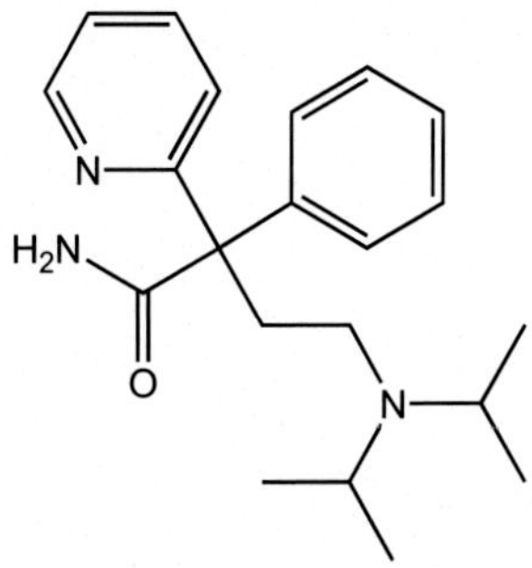

Disopyramide

Disopyramide (4-disiopropylamino-2-phenyl-2-pyridyl butyramide) is a class I antiarrhythmic drug with similar pharmacological effects to those of quinidine and procainamide. Disopyramide is available only in oral form in the United States, with the phosphate salt of disopyramide marketed under the trade name of Norpace (G.D. Searle & Co). A controlled release form is also available containing 100–150 mg of disopyramide. Commercially available disopyramide is a racemic mixture of both dextrorotatory and levorotatory optical isomers, while levorotatory isomer is twice as potent as the dextrorotatory isomer [1, 2].

Disopyramide is well absorbed and undergoes virtually no first-pass metabolism. Oral bioavailability is >80%. Peak concentrations are achieved approximately 0.5–3.0 h after a dose. Absorption is reduced and slightly slowed in patients with acute myocardial infarction. Disopyramide is excreted as unchanged drug (50–65%) or as the metabolite mono-*N*-des-isopropyl disopyramide (mono-*N*-dealkyldisopyramide), an active metabolite with elimination via both renal and biliary routes. Elimination half-life is approximately 7 h in normal subjects, but is prolonged in patients with renal insufficiency (creatinine clearance <60 mL/min). Disopyramide is cleared more rapidly in children than in adults, and therefore children require higher dosages to attain therapeutic concentrations [3]. In patients with renal failure, disopyramide half-life is significantly increased. One study reported a disopyramide average half-life of 12.42 h in patients with renal failure (creatinine clearance <30 mL/min) compared to average half-life of 6.05 h in healthy subjects [4].

Disopyramide exhibits complex protein binding. It is bound to α1-acid glycoprotein (AAG), an acute phase reactant. Protein binding of disopyramide also appears to be concentration dependent when serum drug concentrations are within the therapeutic range. Only unbound disopyramide is pharmacologically active. Thus, free disopyramide concentrations decrease in response to increased concentration of AAG following myocardial infarction, major surgery, and for some uremic patients receiving hemodialysis. Therefore, free disopyramide concentrations are low relative to total concentration in these patients, and therapeutic drug monitoring of free drug could improve dosage adjustment, if needed [3]. Aso et al. observed that the average unbound concentrations of all nine responders in their study were higher than those of the four non-responders and commented that unbound concentrations may better reflect tissue drug concentrations and antiarrhythmic effect than the total disopyramide concentration [5].

The stereochemistry of disopyramide also contributes to its protein binding with greater affinity exhibited towards the S(+)-isomer. Approximately 79% of the S(+)-isomer is bound compared to 66% of the R(−)- isomer [2]. The therapeutic range of disopyramide is considered to be 2.0–5.0 μg/mL. Monitoring free disopyramide concentration is recommended, and levels are expected to be 0.5–2.0 μg/mL. Free disopyramide concentration can be monitored in the protein free ultrafiltrate using commercially available immunoassays [6].

The common adverse effects of disopyramide are mainly due to its anticholinergic effects and include dry mouth, blurred vision and sometimes difficulty in urination. The intensity of the anticholinergic effect of disopyramide depends on both dose and free drug concentrations. For example, dry mouth is associated with free drug concentrations of approximately 0.7 μg/mL, while more severe toxicity such as disorientation and fatigue are not observed until free concentrations exceed 2.0 μg/mL. Between 20% and 40% patients are reported to experience dry mouth and low to moderate urinary retention problems. Negative inotropic effects of disopyramide following therapeutic dosage have been reported. Other complications of disopyramide therapy include the development of congestive heart failure, QRS widening and Q-T prolongation on the electrocardiogram, and worsening of arrhythmia. Disopyramide also has variable effects on blood pressure and heart rate. Disopyramide therapy may also cause hypoglycemia in patients with type 2 diabetes. Such complications usually occur at serum concentrations of 8.0 μg/mL and higher but may also be encountered when serum level is within therapeutic range. Negishi et al. reported a case of a 62-year-old woman with type 2 diabetes taking low-dose glimepiride treatment who suffered from severe hypoglycemia despite serum disopyramide concentration of only 3.7 μg/mL. [7].

Total disopyramide concentrations are reduced in patients taking phenytoin as a result of the induction of hepatic drug metabolizing enzymes by phenytoin [8]. Hemodialysis does not remove disopyramide significantly [3]. Concentrations of disopyramide in serum can be measured by using immunoassays, high performance liquid chromatography, and gas chromatography. Free disopyramide concentration can be measured in the protein free ultrafiltrate because the concentration of free disopyramide, as determined by ultrafiltration, is strongly and directly related to total drug concentration [9].

References

[1] Morady F, Scheinman MM, Desai J. Disopyramide. Ann Intern Med 1982;96:337–43.
[2] Lima JJ, Wenzke SC, Boudoulas H, Schaal SF. Antiarrhythmic activity and unbound concentrations of disopyramide enantiomers in patients. Ther Drug Monit 1990;12:23–8.
[3] Siddoway LA, Woosley RL. Clinical pharmacokinetics of disopyramide. Clin Pharmacokinet 1986;11:214–22.
[4] Nagura Y, Kuno T, Yanai M, Maejima M, et al. Pharmacokinetics and optimum dose of disopyramide in patients with chronic renal failure. Nihon Jinzo Gakkai Shi 1991;33:539–43.
[5] Aso R, Ohashi K, Katoh T, Ogata H. Population pharmacokinetics, protein binding and antiarrhythmic effects of disopyramide enantiomers in arrhythmic patients. Int J Clin Pharmacol Res 2001;21:137–46.
[6] Raghow G, Meyer MC, Straughn AB. Determination of free disopyramide plasma concentrations using ultrafiltration and enzyme multiplied immunoassay. Ther Drug Monit 1985;7:466–71.
[7] Negishi M, Shimomura K, Proks P, Mori M, Shimomura Y. Mechanism of disopyramide-induced hypoglycaemia in a patient with type 2 diabetes. Diabet Med 2009;26:76–8.
[8] Nightingale J, Nappi JM. Effect of phenytoin on serum disopyramide concentrations. Clin Pharm 1987;6:46–50.
[9] Shaw LM, Altman R, Thompson BC, Fields L. Factors affecting the binding of disopyramide to serum proteins. Clin Chem 1985;31:616–9.

12.4
Flecainide

Chemical properties	
Solubility in H_2O	48.4 mg/mL as acetate salt
Molecular weight	414.35
pKa	9.30
Melting point	146 °C
Dosing	
Recommended dose	50–200 mg/day (adult oral dosing)
Monitoring	
Sample	Serum, plasma
Effective concentrations	200–1000 ng/mL
Toxic concentrations	>1500 ng/mL
Methods	HPLC, LC/MS/MS
Pharmacokinetic properties	
Oral dose absorbed	60–80%
Time to peak concentration	2–3 h
Protein bound	40%
Volume of distribution	4.9 L/kg (adults)
Half-life, adult	8–14 h (adults)
Time to steady state, adult	2–3 d
Half-life, child	Not established, lower than adults
Time to steady state, child	Not established

Excretion, urine

	% Excreted	Active	Detected in blood
Unchanged	<5	Yes	Yes
Endoxifen	<5	Yes	Yes
4-Hydroxytamoxifen	<5	Yes	No

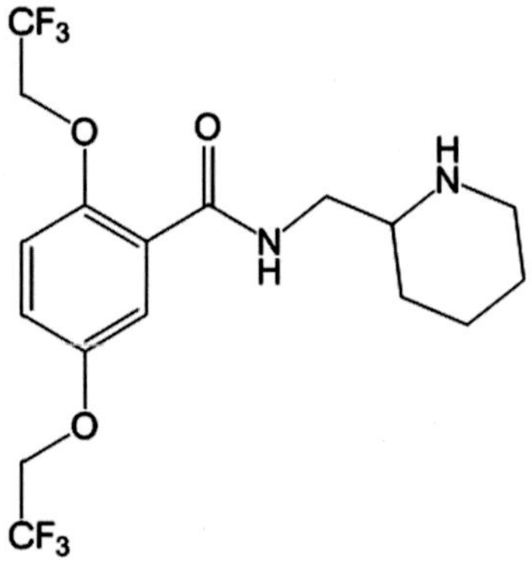

Flecainide

Flecainide (Tambocor) is a Class Ic antiarrhythmic whose clinical use has declined considerably since a landmark trial in the late 1980s showed increased mortality from flecainide therapy in patients recovering from myocardial infarction [1]. Flecainide currently has narrow clinical indications in the prevention of paroxysmal atrial fibrillation and/or atrial flutter and for prophylaxis of paroxysmal supraventricular tachycardia. Other arrhythmias formally treated with flecainide now typically have other therapeutic alternatives.

Flecainide is well absorbed after oral administration and reaches peak concentrations in adults in 2–3 h [2]. Elimination is through a combination of renal clearance (40–45%) and hepatic metabolism (55–60%). Flecainide is mainly metabolized by cytochrome P450 (CYP2D6) , an enzyme that shows pharmacokinetic variation and also potential for drug–drug interactions. Population studies have shown that 5–10% of Caucasians and ~1% of Asians are poor metabolizers of this drug due to low activities related to polymorphisms in CYP2D6. Patients who are poor metabolizers have a significant decrease (42%) in flecainide clearance [3]. Trough concentrations should be 200–1000 ng/mL (0.2–1.0 μg/mL) and severe adverse effect (e.g., ventricular arrhythmias) have occurred occasionally in patients whose serum flecainide concentrations exceeded 1000 ng/mL [4].

The declining clinical use of flecainide means that requests for drug levels are uncommon. A fluorescence polarization immunoassay previously was marketed for flecainide levels but is now unavailable. Flecainide plasma/serum concentrations may be measured with high-performance liquid chromatography or liquid chromatography coupled to tandem mass spectrometry [5].

References

[1] Apostolakis S, Oeff M, Tebbe U, Fabritz L, Breithardt G, Kirchhof P. Flecainide acetate for the treatment of atrial and ventricular arrhythmias. Expert Opin Pharmacother 2013;14:347–57.

[2] Gillis AM, Kates RE. Clinical pharmacokinetics of the newer antiarrhythmic agents. Clin Pharmacokinet 1984;9:375–403.

[3] Doki K, Homma M, Kuga K, Kusano K, Watanabe S, Yamaguchi I, Kohda Y. Effect of CYP2D6 genotype on flecainide pharmacokinetics in Japanese patients with supraventricular tachyarrhythmia. Eur J Clin Pharmacol 2006;62:919–26.

[4] Homma M, Kuga K, Doki K, Katori K, Yamaguchi I, Sugibayashi K, Kohda Y. Assessment of serum flecainide trough levels in patients with tachyarrhythmia. J Pharm Pharmacol 2005;57:47–51.

[5] Slawson MH, Johnson-Davis KL. Quantitation of flecainide, mexiletine, propafenone, and amiodarone in serum or plasma using liquid chromatography-tandem mass spectrometry (LC-MS/MS). Methods Mol Biol 2016;1383:11–9.

12.5

Lidocaine

Chemical properties	
Solubility in H_2O	Soluble
Molecular weight	234.33
pKa	7.86
Melting point	68–69 °C
Dosing	
Recommended dose, adult:	Varies depending of route of administration and clinical condition
Monitoring	
Sample	Serum, plasma, whole blood
Effective concentrations, adult	1.5–5.0 μg/mL
Toxic concentrations	>6 μg/mL
Methods	Immunoassay, HPLC, LC–MS/MS,
Pharmacokinetic properties	
Oral dose absorbed	Not applicable due to poor bioavailability
Time to peak concentration (h):	20–30 min
Protein bound*	60–80% (α1-acid glycoprotein)
Volume of distribution	0.6–4.5 L/kg
Half-life, adult	1.6–1.8 h
Time to steady state, adult	8–10 h

Excretion, urine

	% Excreted	Active	Detected in blood
Parent	2–10	Yes	Yes
Monoethylglycl-xylidine (MEGX)	3–4	Yes	Yes
Glycine xylide	1–3	No	Yes

*Varies with dosage and clinical status.

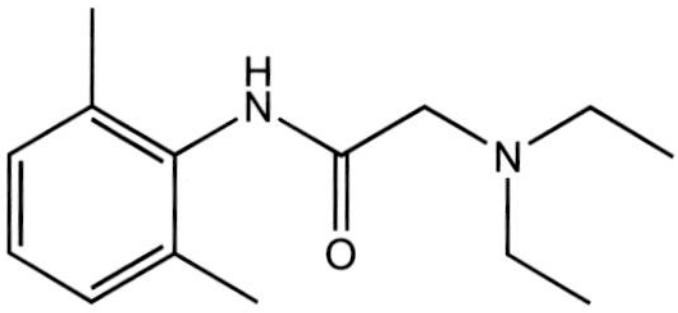

Lidocaine

In the 1940s, Nils Lofgren and Bengt Lundquist developed a xylidine derivative they called lidocaine [1]. Lidocaine (Anestacon, Xylocaine), a sodium channel blocker, is used as a local anesthetic and classified as a type IB antiarrhythmic. Lidocaine undergoes significant first pass metabolism by the liver and is not administered orally. The drug is commonly used for local anesthesia, sometimes in combination with epinephrine (which acts as a vasopressor and extends its duration of action at a site by opposing the local vasodilatory effects of lidocaine).

Intravenous injection of lidocaine is used primarily to suppress ventricular tachycardia and to prevent fibrillation after an acute myocardial infarction. In some coronary care units, lidocaine is administered prophylactically to patients hospitalized with acute chest pains. The drug may be used to control ventricular arrhythmias caused by digitalis toxicity, open-heart surgery, or cardiac catheterization. The minimum lidocaine plasma concentration required to reduce the frequency of premature ventricular contractions significantly (~50%) in most patients has been reported to be between 1.2 and 2.5 μg/mL. Lidocaine is the least cardiotoxic of the currently administered antiarrhythmic drugs. Generally, the drug has minimal adverse effects on the cardiovascular system, where <10% of patients taking lidocaine will develop ventricular arrhythmias. Large doses of lidocaine can depress myocardial contractility, automaticity, and A-V conduction, especially in elderly patients and those with congestive heart failure. Currently, lidocaine is also the most widely used local anesthetic [2].

The pharmacologic effect of lidocaine is mediated through its rapidly acting local anesthetic properties. The drug is a potent suppressor of abnormal cardiac activity. It appears to act exclusively on sodium channels by blocking both activated and inactivated channels. As a result, lidocaine depresses automaticity and reduces the duration of the refractory period in the His-Purkinje system and ventricles. With the exception of ischemic myocardium, therapeutic doses of lidocaine do not depress atrioventricular (A-V) nodal conduction velocity or myocardial contractility [3].

Lidocaine is administered intravenously (IV) or intramuscularly (IM). To produce an effective therapeutic plasma concentration rapidly, it is recommended that an intravenous dose of 1.5 mg/kg within 2–4 min be given intravenously up to three times within 1 h for the initial treatment of ventricular arrhythmias. Infusions of 4 mg/min/70 kg are administered for up to 24 h to maintain normal sinus rhythm. Such dosages produce therapeutic plasma concentrations of 1.5–5.0 μg/mL [4]. The total dosage should not exceed 200–300 mg/h. Weinberg et al. advocated that the ideal lidocaine continuous infusion protocol is a bolus/loading dose of 1 mg/kg, then an infusion at 50 μg/kg per minute infusion for the first hour, then 25 μg/kg per minute for the second hour, then 12 μg/kg per minute for the following 22 h, and finally 10 μg/kg per minute for the remaining 24 h [5]. Lidocaine is quickly absorbed following administration with peak plasma concentration observed 20–30 min after infusion. When lidocaine is given intravenously to normal subjects, the volume of distribution is 0.6–4.5 L/kg. The plasma binding is 60–80% and varies depending on serum level of alpha-1-acid glycoprotein, an acute phase reactant. The elimination half-life is 96–108 min (1.6–1.8 h). The half-life is increased in patients with liver failure (>5 h), heart failure (>2 h), and prolonged infusion; this extended half-life can result in increased plasma levels of the drug. Lidocaine is cleared at a rate of 6.8–11.6 mL/min/kg. Clearance is reduced in patients with congestive heart failure or hepatic disease. Ninety percent of the lidocaine dose is metabolized in the liver. Lidocaine is dealkylated in the liver by mixed-function oxidase to monoethylglycinexylidide and glycine xylidide. These compounds are less bioactive than lidocaine. The dynamic liver function test based on the hepatic conversion of lidocaine to monoethylglycinexylidide (MEGX) can complement established static liver function tests. MEGX test has found widespread application for realtime assessment of hepatic function especially in transplantation and critical care medicine. For this test, blood specimens are collected before and 15 and/or 30 min after injection of an intravenous bolus of a small lidocaine test dose (1 mg/kg) administered over 2 min. Patients with a MEGX 15- or 30-min test value <10 μg/L have a particularly poor 1-year survival rate [6].

The major adverse effects of lidocaine are exerted on the central nervous system. Mild symptoms, such as drowsiness, dizziness, slurred speech, paresthesia, or agitation, occur when plasma concentrations exceed 5 μg/mL. Hearing disturbances, disorientation, muscle twitching, convulsions, respiratory arrest, or coma can occur at higher plasma concentrations. When these adverse effects are observed in patients, the infusion rate should be decreased [5]. Despite the low bioavailability of topical solutions, toxicity can occur following ingestion. Oral ingestion of topical solutions has caused central nervous system toxicity, seizures, and death in children and adults.

In addition, topical preparations of lidocaine are readily absorbed through mucous membranes; this can result in serious toxicity, as has been reported after urethral or rectal administration of topical preparations [7].

Fatal anaphylactic reactions to lidocaine are very rare. Such reactions are characterized by fast onset of symptoms (within seconds up to 30 min after drug exposure) and rapid progression to cardiopulmonary arrest and death within 1 h. Features of cardiovascular, respiratory and neurological system involvements were often seen. Such fatality may occur despite serum lidocaine concentrations within therapeutic range [8].

Immunoassays are available for therapeutic drug monitoring of lidocaine. Free lidocaine concentration in protein free ultrafiltrate can also be measured using immunoassay [9]. Liquid chromatography coupled to tandem mass spectrometry may also be used for monitoring of lidocaine and its metabolites in biological fluids [10].

References

[1] Calatayud J, González A. History of the development and evolution of local anesthesia since the coca leaf. Anesthesiology 2003;98:1503–8.
[2] Pieper JA, Slaughter RL, Anderson GD, Wyman MG, Lalka D. Lidocaine clinical pharmacokinetics. Drug Intell Clin Pharm 1982;16:291–4.
[3] Lauretti GR. Mechanisms of analgesia of intravenous lidocaine. Rev Bras Anestesiol 2008;58:280–6.
[4] Bursi R, Piana C, Grevel J, Huntjens D, Boesl I. Evaluation of the population pharmacokinetic properties of lidocaine and its metabolites after long-term multiple applications of a lidocaine plaster in post-herpetic neuralgia patients. Eur J Drug Metab Pharmacokinet 2017;42:801–14.
[5] Weinberg L, Peake B, Tan C, Nikfarjam M. Pharmacokinetics and pharmacodynamics of lignocaine: a review. World J Anesthesiol 2015;4:17–29.
[6] Oellerich M, Armstrong VW. The MEGX test: a tool for the real-time assessment of hepatic function. Ther Drug Monit 2001;23:81–92.
[7] Fruncillo RJ, et al. CNS toxicity after ingestion of topical lidocaine. N Engl J Med 1982;306:426–7.
[8] Chan TYK. Fatal anaphylactic reactions to lignocaine. Forensic Sci Int 2016;266:449–52.
[9] Beach CL, Ludden TM, Clementi WA, Allerheiligen SR. Measurement of lidocaine free concentration. Ther Drug Monit 1986;8:326–30.
[10] Saluti G, Giusepponi D, Moretti S, Di Salvo A, Galarini R. Flexible method for analysis of lidocaine and its metabolite in biological fluids. J Chromatogr Sci 2016;54:1193–2000.

12.6

Mexiletine

Chemical properties	
Solubility in H_2O	0.54 mg/mL
Molecular weight	179.26
pKa	8.40
Melting point	203–205 °C
Dosing	
Recommended dose	450–600 mg/day (adult)
Monitoring	
Sample	Serum, plasma
Effective concentration	0.5–2.0 μg/mL
Toxic concentrations	>2.0 μg/mL
Methods	GC, HPLC, LC/MS/MS
Pharmacokinetic properties	
Oral dose absorbed	>85%
Time to peak concentration	1–4 h
Protein bound	70%
Volume of distribution	4.5–5.5 L/kg (adults)
Half-life, adult	7–11 h
Time to steady state, adult	2 days
Half-life, child	Not established
Time to steady state, child	Not established

Excretion, urine

	% Excreted	Active	Detected in blood
Unchanged	<15	Yes	Yes
Parahydroxy-mexiletine	<5	No	Yes
Hydroxymethyl-mexiletine	<5	No	Yes

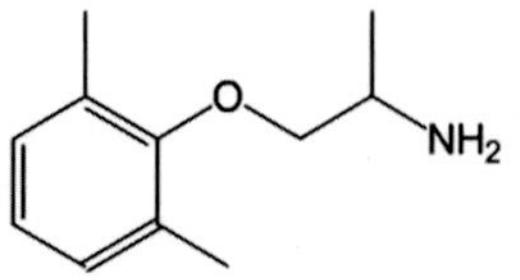

O NH2 Mexiletine

Mexiletine (Mexitil) is an analog of lidocaine, with which it shares similar electrophysiological properties [1]. There are variable and stereoselective differences in the disposition of mexiletine enantiomers, with the *R*-isomer possessing more antiarrhythmic activity. Compared to lidocaine, mexiletine has lower first-pass hepatic metabolism and significantly higher oral bioavailability [2]. This allows for chronic oral administration, something not possible with lidocaine. Mexiletine is approximately 60% bound to plasma proteins and has a volume of distribution of 4.5–5.5 L/kg in adults. The elimination half-life is approximately 7–11 h in adults with normal hepatic and renal function. The half-life is increased in patients with congestive heart failure and following myocardial infarction. Mexiletine is extensively metabolized by the liver to at least 11 non-active metabolites. The major metabolites include parahydroxy-mexiletine, hydroxymethyl-mexiletine, and their corresponding alcohols. Drug–drug interactions involving hepatic metabolism can affect the pharmacokinetics of mexiletine. $<15\%$ of ingested drug is ultimately cleared by the kidneys.

Mexiletine has a therapeutic range of 0.5–2.0 μg/mL, but many patients experience toxicity when the serum concentration just exceeds the upper limit of the range [3]. Tremor and nausea are the most common dose -related adverse effects [1]. Proarrhythmias can occur when plasma/serum concentrations exceed the reference range. Therapeutic drug monitoring has clinical value due to the narrow therapeutic range, potential for toxicity, and presence of factors (e.g., drug–drug interactions or heart failure) that impact mexiletine pharmacokinetics [3]. Currently there is no immunoassay for mexiletine, but chromatographic methods, including high-performance liquid chromatography and the chromatography coupled to tandem mass spectrometry, can be used for determining plasma/serum drug concentrations [4].

References

[1] Monk JP, Brogden RN. Mexiletine. A review of its pharmacodynamic and pharmacokinetic properties, and therapeutic use in the treatment of arrhythmias. Drugs 1990;40:374–411.

[2] Labbe L, Turgeon J. Clinical pharmacokinetics of mexiletine. Clin Pharmacokinet 1999;37:361–84.

[3] Nei SD, Danelich IM, Lose JM, Leung LY, Asirvatham SJ, McLeod CJ. Therapeutic drug monitoring of mexiletine at a large academic medical center. SAGE Open Med 2016;4:2050312116670659.

[4] Slawson MH, Johnson-Davis KL. Quantitation of flecainide, mexiletine, propafenone, and amiodarone in serum or plasma using liquid chromatography-tandem mass spectrometry (LC-MS/MS). Methods Mol Biol 2016;1383:11–9.

12.7

Procainamide

Chemical properties	
Solubility in H_2O	Soluble as HCl
Molecular weight	271.79
pKa	9.23
Melting point	165–169 °C
Dosing	
Recommended dose, adult	Initial infusion dosage: 17 mg/kg
Recommended dose, child	Variable
Monitoring	
Sample	Serum, plasma (heparin, EDTA, and oxalate); collect at trough
Effective concentrations, adult	4–10 μg/mL for procainamide
Effective concentrations, child	Not established
Toxic concentrations	>15 μg/mL for procainamide
Methods	Immunoassays, chromatography
Pharmacokinetic properties	
Oral bioavailability	Almost 100%[a]
Time to peak concentration	Oral: 1–2 h IV: 30 min
Protein bound	15–20%
Volume of distribution	Approximately 2 L/kg
Half-life, adult	2.5–5.0 h
Time to steady state, adult	11–20 h

Excretion, urine

	% Excreted	Active	Detected in blood
Parent	59–75	Yes	Yes
N-acetylprocainamide (NAPA):	7–34	Yes	Yes
p-Aminobenzoic acid:	2–10	No	No
N-desethyl NAPA/ desethyl-procainamide	8–15	?	Yes

[a]Oral formulation no longer available in US but available in Canada.

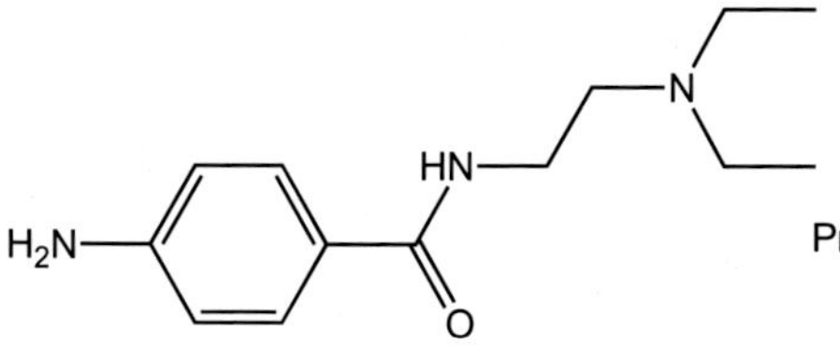

Procainamide

Procainamide (Procanbid) is one of the oldest antiarrhythmic agents, dating back to 1950s but still in clinical use today. Procainamide is classified as a Class IA antiarrhythmic agent and results in prolongation of the QRS interval on the electrocardiogram and subsequent reductions in myocardial excitability and conduction velocity. The electrophysiologic actions of procainamide occur predominately via blockade of sodium channels; however, it also exhibits some potassium channel blockade. Procainamide effectively increases the electrical stimulation threshold of ventricle, causes decreased myocardial excitability and conduction, and prolongs the refractory period of cardiac cells. Additional cardiovascular effects of procainamide include prolongation of the QT interval on the electrocardiogram and direct peripheral vasodilation [1]. Most favorable results have been reported when using procainamide for the treatment of ventricular tachydysrhythmia. Nevertheless, the use of procainamide for the management of arrhythmias has decreased steadily as newer alternatives have become available [1].

When procainamide is selected for the treatment of arrhythmias, a bolus dose of 17 mg/kg should be given via slow intravenous (IV) infusion for initial therapy. Because of its peripheral vasodilatory actions, particularly at high doses or when infused rapidly, procainamide should be infused no faster than 50 mg/min with lower rates (20–30 mg/min) for some patients. After an initial 17 mg/kg dose, a continuous infusion of 1–4 mg/min may be started as clinically necessary. Alternative dosing regimens include sequential doses of 100 mg IV push given every 5 min until one of the following occurs: a total of 17 mg/kg is administered, arrhythmia termination, or development of an adverse effect. Although in the past, procainamide could be administered orally, currently oral procainamide is no longer manufactured in the United States [1].

Procainamide is almost completely absorbed after oral administration, and peak plasma concentrations are generally reached within 1–2 h. Upon intravenous administration, there is a rapid initial distribution phase, which is completed after about 30 min. The apparent volume of distribution is about 2 L/kg body weight. At therapeutic plasma levels, about 15% is bound to plasma proteins. Approximately 50% of administered procainamide is eliminated as unchanged drug via the kidneys. *N*-Acetylprocainamide (NAPA) is the main metabolite and is also pharmacologically active. Approximately 15% (range 7–34% in healthy subjects) of procainamide is recovered as NAPA in urine. [2]. Genetic polymorphism of *N*-acetyltransferase (NAT2) may affect the NAPA/procainamide ratio, with "fast acetylators" (high NAT2 activity) showing higher NAPA concentration. NAT2 metabolizer status can influence the overall antiarrhythmic effect of procainamide. The NAT2* genotype correlates with acetylation of procainamide. [3] The renal clearance of procainamide ranges from 179 to 660 mL/min. Glomerular filtration and active tubular secretion seem to be the most important mechanisms. In patients with low-output cardiac and/or renal impairment, the absorption, distribution and elimination of the drug may be significantly altered [2]. The half-life of procainamide ranges from 2.5 to 5 h in normal renal function but is significantly prolonged in renal dysfunction. The half-life of the active metabolite NAPA is slightly longer (6–8 h in normal renal function) but can be prolonged to >24 h in severe renal impairment [4]. The half-life of NAPA can be prolonged up to 42 h in patients receiving hemodialysis [1].

Therapeutic drug monitoring of both procainamide and NAPA is recommended. Therapeutic range of procainamide is 4–10 μg/mL, while therapeutic concentrations of NAPA range from 6 to 20 μg/mL [5]. Adverse effects are more common when procainamide concentrations exceed 15 μg/mL [1]. However, in patients with renal failure, procainamide concentration may be within therapeutic range, but NAPA concentration may be significantly above the reference range and potentially toxic due to impaired renal clearance. A 65-year-old male on hemodialysis three times a week due to end-stage renal failure experienced significant toxicity despite serum procainamide concentration of 2.6 μg/mL which was below the lower therapeutic range. However, further investigation showed a serum NAPA concentration of 27.7 μg/mL which was significantly above the therapeutic range. His procainamide toxicity was due to NAPA which resolved after discontinuation of procainamide [6].

Adverse effects of therapeutic administration of procainamide include gastrointestinal disturbances, mild CNS effects, and cardiac conduction delay. Serious and life-threatening toxic effects including conduction disturbances, hypotension, and ventricular arrhythmias. These serious adverse effects are usually observed when levels exceed therapeutic range. Chronic administration of procainamide may lead to a systemic lupus erythematosus-like syndrome [7], hypersensitivity (manifesting as fever, agranulocytosis, and hepatic disturbances), and arrhythmias. Clinical toxicity occasionally occurs with the increase in serum levels of quinidine and procainamide when amiodarone is added to the therapy. Therefore, dose reduction of procainamide and quinidine is required [8].

Immunoassays are available for monitoring of procainamide as well as NAPA. The declining therapeutic use of procainamide has resulted in many clinical laboratories no longer performing drug levels in-house. Chromatographic methods can also be used for simultaneous monitoring of both procainamide and NAPA [9].

References

[1] Samarin MJ, Mohrien KM, Oliphant CS. Continuous intravenous antiarrhythmic agents in the intensive care unit: strategies for safe and effective use of amiodarone, lidocaine, and procainamide. Crit Care Nurs Q 2015;38:329–44.
[2] Karlsson E. Clinical pharmacokinetics of procainamide. J Clin Pharmacol 1981;21:20–5.
[3] Okumura K, Kita T, Chikazawa S, Komada F, et al. Genotyping of N-acetylation polymorphism and correlation with procainamide metabolism. Clin Pharmacol Ther 1997;61:509–17.
[4] Mohamed AN, Abdelhady AM, Spencer D, Sowinski KM, et al. Pharmacokinetic modeling and simulation of procainamide and N-acetylprocainamide in a patient receiving continuous renal replacement therapy: a novel approach to guide renal dose adjustments. Am J Kidney Dis 2013;61:1046–8.
[5] Moffett BS, Cannon BC, Friedman RA, Kertesz NJ. Therapeutic levels of intravenous procainamide in neonates: a retrospective assessment. Pharmacotherapy 2006;26:1687–93.
[6] Ashida K, Mine T, Kodani T, Kishima H, Masuyama T. Long QT syndrome caused by N-acetyl procainamide in a patient on hemodialysis. J Cardiol Cases 2015;11:147–9.
[7] Hopkins BE. Procainamide-induced lupus-like syndrome. Med J Aust 1970;2(16):734–5.
[8] Saal AK, Werner JA, Greene HL, Sears GK, Graham EL. Effect of amiodarone on serum quinidine and procainamide levels. Am J Cardiol 1984;53:1264–7.
[9] Ruo TI, Thenot JP, Stec GP, et al. Plasma concentrations of desethyl N-acetylprocainamide in patients treated with procainamide and N-acetylprocainamide. Ther Drug Monit 1981;5:231–7.

12.8

Propranolol

Chemical properties	
Solubility in H_2O	Hydrochloride salt is soluble
Molecular weight	259.34
pKa	9.45
Melting point	54 °C
Dosing	
Recommended dose, adult	10–80 mg/day (1–3 times)
Recommended dose, child	0.5–4.0 mg/kg/day
Monitoring	
Sample	Serum, plasma
Effective concentrations	50–100 ng/mL (trough concentration)
Toxic concentrations	Not established
Methods	HPLC, GC. GC/MS
Pharmacokinetic properties	
Oral dose absorbed	90% (lower systematic availability due to first pass metabolism by liver)
Time to peak concentration	1.0–3.0 h
Protein bound	90–95%
Volume of distribution	3.5–5.0 L/kg
Half-life, adult	2.0–6.0 h
Time to steady state, adult	10–30 h
Half-life, child	3.9–6.4 h

Excretion, urine

	% Excreted	Active	Detected in blood
Parent	1–5	Yes	Yes
4-Hydroxypropanaol	Unknown	Yes	Yes
Naphthoxylactic acid	Unknown	No	Yes

O
OH
N
H
HCl
Propranolol

Propranolol, 1-(isopropylamino)-3-(1-naphthyloxy)-2-propanol, remains one of the most widely used agents that act by blocking beta-adrenergic receptors (beta blockers). The drug is administered orally as hydrochloride salt either in tablet (10–80 mg) or sustained release forms, and is also available as a 1 mg/mL sterile injectable solution for intravenous administration. Propranolol is a nonselective beta-adrenergic receptor antagonist that blocks both beta-1 and beta -2 adrenergic receptors equally. Because of these actions, propranolol has a range of cardiac and non-cardiac indications. It is used for long-term management of patients with angina pectoris. The mechanism of action is a reduction in pulse rate and thus a decrease in myocardial oxygen consumption relative to demand during physical activities in patients. Propranolol is also used in treating cardiac arrhythmias of both atrial and ventricular origin, and other cardiac conditions including atrial fibrillation, atrioventricular (A-V) nodal reentrant tachycardia, and tachyarrhythmia caused by increased sympathetic tone due to stress, exercise, or thyroid storm. Propranolol is also effective in treating tachyarrhythmia due to digitalis toxicity. Propranolol has been shown to reduce cardiovascular mortality in patients following myocardial infarction. It is also used in the treatment of some types of migraine headache. Propranolol is effective in the treatment of mild to moderate hypertension, and is often combined with other agents for this purpose. The antihypertensive mechanism of propranolol is complex and may be a combination of several mechanisms. One key action is thought to relate to the drug's ability to block the release of renin from the juxtaglomerular apparatus [1]. Propranolol may also be a useful adjunct therapy for anxiety disorders [2].

Although the hydrochloride salt of propranolol is water soluble, propranolol itself is lipophilic and has a high oral bioavailability (approximately 90%). Peak concentrations are observed within 1–3 h, and the drug is distributed throughout the body and easily crosses the blood brain barrier [3]. Propranolol has a large apparent volume of distribution, and the drug is strongly bound to serum protein (90–95%), mostly to α_1-acid glycoprotein. In addition, there is stereoselectivity in the protein binding of propranolol such that the (*R*)-isomer is bound to a lesser extent than the (S)-isomer. There is some evidence that protein binding of propranolol varies between different ethnic populations. Chinese are more sensitive to the pharmacological effects of propranolol compared to Caucasians, possibly related to increased free concentrations of both (*R*)-propranolol and (*S*)-propranolol [4]. There is also a gender difference in protein binding of propranolol; females have significantly greater binding of the S-isomer than males [5].

In addition, the intrinsic hepatic clearance of S-propranolol was about 30% lower in the elderly (ages of patients 62–79 years) than in the young, and the elimination half-lives of both enantiomers were two- to threefold prolonged in the elderly compared with the young [6].

After oral administration, propranolol undergoes first pass metabolism so that 25–30% of the dose enters the systematic circulation. Hepatic extraction of propranolol decreases as the dose increases and relatively low bioavailability may be increased by ingestion of food and long term therapy with propranolol. This drug also undergoes a complex pattern of biotransformation with production of several active metabolites. One metabolite, 4-hydroxypropanol, has pharmacological properties comparable to propranolol but a shorter half-life compared to the parent drug. Metabolism of propranolol to the active metabolite 4-hydroxypropanolol is mediated by cytochrome P450 (CYP2D6) and CYP1A2. African-Americans show significant increases in the rate of metabolism compared to Caucasians. Quinidine inhibits CYP2D6 enzyme activity and reduces the conversion of propranolol to 4-hydroxypropanolol [7]. Additional metabolites, including nor-propranolol and α-napthoxy2, 3-propyleneglycol, may also have some pharmacological activity. Most of the unchanged propranolol and propranolol metabolites are excreted in urine in unconjugated and conjugated forms, mostly to glucuronide. Other metabolites identified in urine include naphthoxylactic acid, napthoxycaetic, propranolol glycol, and isopropyl amine. Because propranolol is extensively metabolized by the liver, only a small fraction of dose is excreted as unchanged propranolol. The elimination half-life is 3–6 h but may be prolonged in the presence of hepatic impairment, decreased hepatic blood flow, or hepatic enzyme inhibition. Propranolol metabolism is faster in males than females [8].

Symptoms of propranolol intoxication include weakness, visual disturbances, disorientation, bradycardia, hypotension, bronchospasm, and congestive heart failure. Fatal bronchospasm has been reported in asthmatic patients after a single therapeutic dose. Though widely used, the drug is associated with a higher incidence of serious adverse events including death compared to other beta blockers. Propranolol causes a mild increase in triglycerides and a mild decrease in high density lipoprotein cholesterol [9].

Although therapeutic drug monitoring of propranolol is not widely practiced, serum concentrations may be useful when a patient fails to benefit from a relatively large dosage of propranolol or is suspected of noncompliance. Usually the therapeutic range is considered to be between 50 and 100 ng/mL. Fucci and Offidani reported a case of fatal intoxication with propranolol in a 60 year old man who was found dead in his car with no evidence of trauma or signs of asphyxia. Near the car, a pharmaceutical box of Inderal, a pharmaceutical formulation containing propranolol, was found. He was not taking proprandol but his sister was taking the medication. Postmortem analysis using gas chromatography/mass spectrometry revealed a serum level of 7 μg/mL (7000 ng/mL) and gastric content level of 2000 μg/mL of propranolol, thus establishing the likely cause of death as propranolol poisoning [10].

Propranolol can be measured using gas chromatography, gas chromatography combined with mass spectrometry, and high performance liquid chromatography. A more sophisticated technique such as liquid chromatography coupled to tandem mass spectrometry can also be used for analysis of propranolol along with other beta blockers in various biological matrices [11]. Chromatographic methods are also available for enantioseparation of (*RS*)-propranolol in biological matrix [12].

References

[1] Fuster V, Rydén LE, Cannom DS, Crijns HJ, et al. ACC/AHA/ESC 2006 guidelines for the management of patients with atrial fibrillation: a report of the American College of Cardiology/American Heart Association task force on practice guidelines and the European Society of Cardiology Committee for practice. Circulation 2006;114:e257–354.

[2] Steenen SA, van Wijk AJ, van der Heijden GJ, van Westrhenen R, et al. Propranolol for the treatment of anxiety disorders: systematic review and meta-analysis. J Psychopharmacol 2016;30:128–39.

[3] Cid E, Mella F, Lucchini L, Carcamo M, et al. Plasma concentrations and bioavailability of propranolol by oral rectal and intravenous administration in man. Biopharm Drug Dispos 1986;7:559–66.

[4] Zhou HH, Shay SD, Wood AJ. Contributions of difference in plasma binding of propranolol to ethnic differences in sensitivity. Comparison between Chinese and Caucasians. Chin Med J 1993;106:898–902.

[5] Paxton JW, Briant RH. Alpha-1-acid glycoprotein concentrations and propranolol binding in elderly patients with acute illness. Br J Clin Pharmacol 1984;18:806–10.

[6] Gilmore DA, Gal J, Gerber JG, Nies AS. Age and gender influence the stereoselective pharmacokinetics of propranolol. J Pharmacol Exp Ther 1992;261:1181–6.

[7] Masubuchi Y, Hosokawa S, Horie T, Suzuki T, et al. CYP1A2 and CYP2D6 4-hydroxylate propranolol and both reactions exhibit racial differences. J Pharmacol Exp Ther 2000;294:1099–105.

[8] Yoshimoto K, Echizen H, Chiba K, Tani M, Ishizaki T. Identification of human CYP isoforms involved in the metabolism of propranolol enantiomers—N-desisopropylation is mediated mainly by CYP1A2. Br J Clin Pharmacol 1995;39:421–31.

[9] Love JN, Litovitz TL, Howell JM, Clancy C. Characteristics of fatal beta blockers ingestion: a review of American Association of Poison Control Centers data from 1985 to 1995. J Toxicol Clin Toxicol 1997;35:353–9.

[10] Fucci N, Offidani C. An unusual death by propranolol ingestion. Am J Forensic Med Pathol 2000;23:56–8.
[11] Johnson RD, Lewis RJ. Quantitation of atenolol, metoprolol, and propranolol in postmortem human fluid and tissue specimens via LC/APCI-MS. Forensic Sci Int 2006;156:106–17.
[12] Batra S, Bhushan R. Methods and approaches for determination and enantioseparation of (RS)-propranolol. Biomed Chromatogr 2019;33:e4370.

12.9

Quinidine

Chemical properties	
Solubility in H_2O	0.5 mg/mL
Molecular weight	324.41
pKa	4.3 and 8.4
Melting point	174–175 °C
Dosing	
Recommended dose, adult	10.0–20.0 mg/kg/day
Recommended dose, child	7.7–45.6 mg/kg/day
Monitoring	
Sample	Serum, plasma
Effective concentrations	2.3–5.0 μg/mL (Trough Concentration)
Toxic concentrations	>5.0 μg/mL
Methods	Immunoassay, HPLC
Pharmacokinetic properties	
Oral bioavailability	70%
Time to peak concentration	1.0–1.5 h, quinidine sulfate
Protein bound	70–95%
Volume of distribution	2.0–3.5 L/kg
Half-life, adult	5.0–12 h
Time to steady state, adult	30.0–35.0 h
Half-life, child	2.5–6.7 h
Time to steady state, child	Depends on age

Excretion, urine

	% Excreted	Active	Detected in blood
Parent	20–50	Yes	Yes
3-Hydroxyquinidine	40–50	Yes	Yes
2′-oxo quinidinone	Unknown	No	Yes

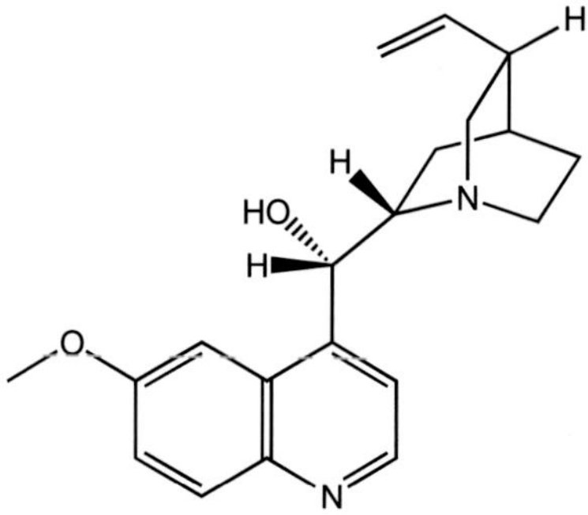

Quinidine

Quinidine, a natural alkaloid, is found along with quinine in the cinchona bark. Quinidine is the dextrorotatory stereoisomer of quinine and has many pharmacological properties of quinine including its antimalarial effect. However, the actions of quinidine on cardiac muscle are more significant than quinine. Historically, quinidine was the first medicine used in the therapy of heart arrhythmias. It was first used in 1749 by a French physician for treatment of atrial fibrillation [1]. Quinidine was used by the 1920s as an antiarrhythmic agent to maintain sinus rhythm after the conversion from atrial flutter or atrial fibrillation and to prevent recurrence of ventricular tachycardia or ventricular fibrillation. Recent studies have demonstrated that quinidine is the only oral medication that has consistently shown efficacy in preventing arrhythmias and terminating storms due to recurrent ventricular fibrillation, in patients with Brugada syndrome, idiopathic ventricular fibrillation, and early repolarization syndrome. Quinidine is also the only antiarrhythmic drug that normalizes the QT interval on the electrocardiogram in patients with the congenital short QT syndrome. However, due to the introduction of newer antiarrhythmic drugs with better safety profiles, the numbers of quinidine prescriptions have decreased significantly [2].

Multiple forms as sulfate or gluconates are available including oral tablets, sustained release formulas, and injectables. Quinidine is usually administered orally as quinidine sulfate in dosage of 300–500 mg up to four times a day. Absolute systemic availability generally is 70% or greater. Plasma peak concentration after oral administration of quinidine sulfate can be observed within 1–2 h; for quinidine gluconate, peak plasma concentrations are observed after 4–6 h. The apparent volume of distribution in healthy persons is 2.0–3.5 L/kg. Quinidine is strongly protein bound (70–95%), mainly to albumin and also alpha-1-acid glycoprotein. Binding to plasma proteins is reduced in patients with cirrhosis, partly because of hypoalbuminemia. Renal insufficiency does not influence protein binding. Quinidine is eliminated by a combination of renal excretion of the intact drug (15–40% of total clearance) and hepatic biotransformation to a variety of metabolites (60–85% of total clearance). Many of the metabolites appear to be pharmacologically active. The elimination half-life is 5–12 h, and the renal clearance is 2.5–5.0 mL/min. Quinidine clearance is reduced in the elderly, in patients with cirrhosis, and in those with congestive heart failure [3, 4]. Conversely, children <12 years often have a shorter half-life largely due to a faster renal excretion.

Both the parent drug and metabolites are excreted in urine. Quinidine is filtered through the glomerulus and secreted by the renal tubules. Since quinidine is a weak base, its reabsorption is reduced and its excretion is increased if the urine is acidic (pH 6–7). When urine pH is 7–8, the renal clearance of quinidine may be reduced by 50%, thus increasing the serum quinidine level. This is due to an increase in the non-ionized fraction of quinidine, resulting in a higher reabsorption of quinidine by renal tubules. Therefore, therapeutic drug monitoring (TDM) is essential to avoid toxicity. Drugs that increase urinary pH, such as acetazolamide and antacids, may cause increased serum levels of quinidine. The quinidine metabolites 3-hydroxyquinidine, 2′-oxoquinidione, and quinidine-*N*-oxide have pharmacological activity [5].

The therapeutic range of quinidine is usually considered between 2.3 μg/mL and 5 μg/mL. Wide variation in serum quinidine levels coupled with a narrow therapeutic range makes this drug a candidate for routine TDM. Quinidine is also strongly bound to α_1-acid glycoprotein (an acute phase reactant), and the variations of pharmacologically free fractions have been reported in altered pathological conditions. The concentrations of α_1-acid glycoprotein are significantly increased in many disease states including myocardial infarction. The significance of this is well demonstrated by a case in which the patient exhibited no symptoms of toxicity despite a total serum quinidine level of 11 μg/mL. This was related to a significant increase in the α_1-acid glycoprotein concentration of 228 mg/dL following an acute myocardial infarction [6].

A serum quinidine level >5 μg/mL is considered potentially toxic; however, toxicity may be encountered at lower concentrations due to elevated concentrations of the active metabolite 3-hydroxyquinidine. This is usually observed in fast metabolizers of quinidine. At steady state, quinidine concentration is significantly higher than 3-hydroxyquinidine, but unbound concentration of 3-hydroxyquinidine may exceed unbound quinidine concentration. Therefore, monitoring both free quinidine and 3-hydroxyquinidine concentrations are clinically useful [7]. Moreover, the free fraction of quinidine is also affected by phenotype of α_1-acid glycoprotein, and TDM of free (unbound) quinidine concentration is therefore potentially useful [8].

Gastrointestinal disturbance is a common adverse effect of quinidine. More serious adverse effects of quinidine toxicity (sometimes referred to as cinchonism) consist of impaired hearing, tinnitus, blurred vision, light handedness, giddiness and tremor. Cardiovascular adverse effects of quinidine are an extension of the pharmacological effects of quinidine and include drug-induced arrhythmias (e.g., torsades de pointes), prolongation of QRS complex due to slowing of intraventricular conduction, and lengthening of QT interval as a result of prolongation of ventricular action potential. Quinidine can produce sinus arrest and sinoatrial block in patients with sick sinus syndrome by depressing pacemaker activity of the sino-atrial node. Toxic levels of quinidine may also decrease blood pressure and depress cardiac contractility. Children are generally more prone to quinidine adverse effects than adults. Quinidine is significantly less bound to serum proteins in children below 2 years of age due to immature protein synthesis. Immaturity of the blood–brain barrier and cardiovascular system may also predispose children to quinidine toxicity. Severe toxicity of quinidine usually occurs in adults at serum levels over 8 μg/mL, although some toxicity may be observed just above therapeutic range of 5 μg/mL. In young children, severe toxicity may be observed after ingestion of only two or more pills, and even ingestion of one pill can cause significant toxicity in a toddler [9].

In the case of serious poisoning with quinidine, older literature indicated that therapy with sodium lactate is beneficial. More recent publications indicate that quinidine-induced proarrhythmia can be treated with sodium bicarbonate which acts as an antagonist to the inhibitory effect of quinidine on sodium conductance [10, 11]. Quinidine is often used along with digoxin, and the drug–drug interaction between quinidine and digoxin is of clinical significance. Concomitant administration may double or triple digoxin serum levels if the quinidine concentration is 2 μg/mL or above and a significant reduction in digoxin dose is recommended if quinidine is introduced in the patient management. It appears that quinidine reduces both renal and non-renal clearance of digoxin, along with altering volume of distribution. [12].

The hepatic metabolism of quinidine is increased by drugs that induce cytochrome P450 (CYP) enzymes including carbamazepine, phenobarbital, phenytoin, and rifampin. These effects reduce the half-life of quinidine. Conversely, inhibitors of CYP enzymes such as cimetidine increase the half-life of quinidine by reducing hepatic metabolism. Quinidine has a pharmacodynamic interaction with warfarin by producing additive hypoprothrombinemic effect. Quinidine may also reduce hepatic synthesis of vitamin-K dependent clotting factors. Quinidine-induced hypoprothrombinemic hemorrhage in patients on chronic warfarin therapy has been described [13]. Since quinidine has anticholinergic properties, it may antagonize the effects of cholinergic drugs (neostigmine) especially in patients with myasthenia gravis. Concurrent administration of quinidine and verapamil may cause increase plasma concentration of quinidine [14].

Usually serum concentration of quinidine is measured using immunoassay or a chromatography-based method. Both quinidine and its active metabolites can be measured together using HPLC [15]. The declining clinical use of quinidine means that many clinical laboratories no longer perform assays in-house.

References

[1] Meyer MC. Generic drug product equivalence: current status. Am J Manag Care 1997;4:1183–9.

[2] Bozic B, Uzelac TV, Kezic A, Bajcetic M. The role of quinidine in the pharmacological therapy of ventricular arrhythmias 'quinidine'. Mini Rev Med Chem 2018;18:468–75.

[3] Ochs HR, Greenblatt DJ, Woo E. Clinical pharmacokinetics of quinidine. Clin Pharmacokinet 1980;5:150–68.

[4] Stanek EJ, Simko RJ, DeNofrio D, Pavri BB. Prolonged quinidine half-life with associated toxicity in a patient with hepatic failure. Pharmacotherapy 1997;17:622–5.

[5] Kavanagh KM, Wyse DG, Mitchell LB, Gilhooly T, et al. Contribution of quinidine metabolites to electrophysiologic responses in human subjects. Clin Pharmacol Ther 1989;46:352–8.

[6] Garfinkel D, Mamelok RD, Blaschke TF. Altered therapeutic range for quinidine after myocardial infarction and cardiac surgery. Ann Intern Med 1987;107:49–50.

[7] Wooding-Scott RA, Visco J, Slaughter RL. Total and unbound concentrations of quinidine and 3-hydroxyquinidine at steady state. Am Heart J 1987;113:302–6.

[8] Li JH, Xu JQ, Cao XM, Ni L, et al. Influence of OEM 1 phenotypes on serum unbound concentration and protein binding of quinidine. Clin Chim Acta 2002;317:85–92.

[9] Huston M, Levinson M. Are one or two dangerous? Quinine and quinidine exposure in toddlers. J Emerg Med 2006;31:395–401.

[10] Kim SY, Benowitz NL. Poisoning due to class IA antiarrhythmic drugs. Quinidine, procainamide and disopyramide. Drug Saf 1990;5:393–420.

[11] Tsai CL. Quinidine cardiotoxicity. J Emerg Med 2005;28:463–5.

[12] Bigger Jr. JT, Leahey Jr. EB. Quinidine and digoxin. An important interaction. Drugs 1982;24:229–339.

[13] Koch-Weser J. Quinidine-induced hypoprothrombinemic hemorrhage in patients on chronic warfarin therapy. Ann Intern Med 1968;68:511–7.
[14] Trohman RG, Estes DM, Castellanos A, Palomo AR, et al. Increased quinidine plasma concentrations during administration of verapamil: a new quinidine-verapamil interaction. Am J Cardiol 1986;57:706–7.
[15] Nielsen F, Nielsen KK, Brøsen K. Determination of quinidine, dihydroquinidine, (3S)-3-hydroxyquinidine and quinidine N-oxide in plasma and urine by high-performance liquid chromatography. J Chromatogr B Biomed Appl 1994;660:103–10.

12.10

Tocainide

Chemical properties	
Solubility in H_2O	10.7 mg/mL
Molecular weight	192.26
pKa	7.75
Melting point	246–266 °C
Dosing	
Recommended dose	1200–1800 mg/day (adult)
Monitoring	
Sample	Serum, plasma
Effective concentration	6–15 μg/mL
Toxic concentrations	Not defined
Methods	GC, HPLC, LC/MS
Pharmacokinetic properties	
Oral dose absorbed	~100%
Time to peak concentration	0.5–2.0 h
Protein bound	5–20%
Volume of distribution	2.5–3.5 L/kg (adults)
Half-life, adult	15 h
Time to steady state, adult	2–3 day
Half-life, child	Not established
Time to steady state, child	Not established

Excretion, urine

	% Excreted	Active	Detected in blood
Unchanged	10%	Yes	Yes
Glucuronidated tocainide carbamic acid	30%	No	No

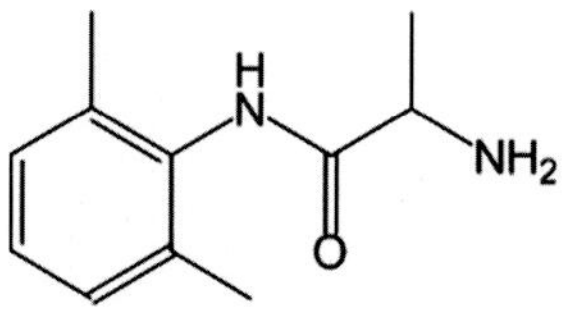

Tocainide

Tocainide is a lidocaine analog and a Class Ib antiarrhythmic agent [1]. Compared to lidocaine, tocainide has less first-pass metabolism and higher oral bioavailability, allowing for chronic oral administration [2]. Tocainide and lidocaine have similar electrophysiological properties. Tocainide has two enantiomers, with the *R*-isomer four-times as potent as the *S*-isomer. Tocainide was historically used for the oral treatment of ventricular arrhythmias. However, numerous reports of serious adverse effects, including agranulocytosis, led to declining therapeutic use [3]. Tocainide is currently not marketed in the United States.

Tocainide has excellent oral bioavailability and is poorly protein bound (5–20%) [2]. Both renal and hepatic failure can impact elimination. The main metabolite is the glucuronidated tocainide carbamic acid. A therapeutic range of 6–15 μg/mL has been proposed. There are no commercially available immunoassays for tocainide determinations in serum but chromatographic methods have been described including gas chromatography and high performance liquid chromatography [4].

References

[1] Holmes B, Brogden RN, Heel RC, Speight TM, Avery GS. Tocainide. A review of its pharmacological properties and therapeutic efficacy. Drugs 1983;26:93–123.

[2] Gillis AM, Kates RE. Clinical pharmacokinetics of the newer antiarrhythmic agents. Clin Pharmacokinet 1984;9:375–403.

[3] Denaro CP, Benowitz NL. Poisoning due to class 1B antiarrhythmic drugs. Lignocaine, mexiletine and tocainide. Med Toxicol Adverse Drug Exp 1989;4:412–28.

[4] Harris SC, Guerra C, Wallace JE. Assay of free and total tocainide by high performance liquid chromatography (HPLC) with ultraviolet (UV) detection. J Forensic Sci 1989;34:912–7.

CHAPTER 13

Immunosuppressants

Immunosuppressive drugs non-specifically diminish immune responses and are widely used in transplantation. They are also used to treat autoimmune disease, allergic disorders, and several other diseases. The discovery in the 1970s that cyclosporine (cyclosporine A; CsA) possessed immunosuppressive activity that specifically targeted T lymphocytes was a major breakthrough in organ transplantation, because it dramatically reduced acute rejection and improved long-term graft and patient survival [1]. The identification of other immunosuppressive drugs that modulate immune responses by additional molecular mechanisms enabled treatment options to evolve and has permitted combination therapies to be individualized based on patient requirements. The number of solid organ transplants in the United States continues to increase each year. In 2018, 36,528 solid organ transplants were performed in the United States but more than 113,000 men, women and children are on the national transplant waiting list as of January 2019. It has been estimated that approximately 20 people die each day waiting for an organ [2].

Immunosuppressive agents used in organ transplantation can be classified according to their mechanism of action. The classes are corticosteroids (prednisone, methylprednisolone and dexamethasone), anti-metabolite/proliferative agents (azathioprine, cyclophosphamide, mycophenolate mofetil, mycophenolate sodium), calcineurin inhibitors (cyclosporine, tacrolimus), and mammalian target of rapamycin (mTOR) inhibitors (sirolimus, everolimus). In addition, polyclonal and monoclonal antibodies against T lymphocyte cell surface antigens are also used in combination with other immunosuppressive agents [3]. A prerequisite for optimizing and individualizing immunosuppressive therapy is reliable and precise methods for monitoring drug concentrations. However, not all immunosuppressive drugs require routine monitoring of blood concentrations. For instance, corticosteroids are dosed based on empirical guidelines and are not routinely monitored. Although methods have been developed to measure blood concentrations of azathioprine, this anti-metabolite is seldom monitored [4]. However, whole blood concentrations of CsA, tacrolimus and sirolimus as well as

Therapeutic Drug Monitoring Data
https://doi.org/10.1016/B978-0-12-815849-4.00013-X

serum or plasma mycophenolic acid (MPA) concentrations are routinely monitored at transplant centers for the following reasons [5]:

- There is a clear relationship between drug concentration and clinical response
- The drugs have a narrow therapeutic index
- The drugs exhibit a high degree of inter- and intra-patient variability
- The pharmacological response can be difficult to distinguish from unwanted side effects
- There is a risk of poor or non-compliance, because the drugs are administered for the lifetime of the graft or patient.
- Sensitive and specific analytical methods are available for routine monitoring of these drugs in clinical laboratories.

The potential for drug interactions among the various classes of immunosuppressive agents provides additional rationale for TDM of immunosuppressive drugs. For instance, CsA inhibits transport of an MPA metabolite from the liver into bile, resulting in lower MPA concentrations when the two drugs are used in combination [6]. Sirolimus aggravates CsA-induced renal dysfunction due to a pharmacokinetic interaction, whereas CsA produces a pharmacodynamic effect that augments sirolimus-induced myelosuppression and hyperlipidemia [7]. Co-administration of tacrolimus and sirolimus, while having no effect on exposure to sirolimus, results in reduced exposure to tacrolimus at sirolimus doses of 2 mg/day and above. Therefore, tacrolimus levels should be monitored when sirolimus is co-administered at doses >2 mg/day and after cessation of corticosteroid treatment [8].

MPA is available as an ester pro-drug and an enteric-coated sodium salt. MPA is a competitive, selective and reversible inhibitor of inosine-5′-monophosphate dehydrogenase (IMPDH), an important rate-limiting enzyme in purine synthesis. As a result, MPA suppresses T- and B- lymphocyte proliferation as well as also decreasing expression of glycoproteins and adhesion molecules responsible for recruiting monocytes and lymphocytes to sites of inflammation and graft rejection. In addition, MPA may destroy activated lymphocytes by induction of a necrotic signal. Improved long-term allograft survival has been demonstrated for therapy with MPA, and it has largely replaced azathioprine in organ transplantation [9]. Therapeutic drug monitoring (TDM) of MPA is essential. However, unlike other immunosuppressants, MPA concentration is measured in serum or plasma. For more details, see monograph on MPA in this chapter.

The calcineurin inhibitors CsA and tacrolimus are available for both oral and intravenous administration. Tacrolimus is about 100-times more potent than CsA and is associated with better long-term graft survival [10]. The calcineurin inhibitors block activation/proliferation of $CD4^+$ and $CD8^+$ T-lymphocytes by inhibiting IL-2 production. They form complexes with specific cytoplasmic binding proteins known as immunophilins [11]. CsA binds cyclophilin, whereas tacrolimus binds the FK506-binding protein-12. The complexes block calcineurin activity and prevent de-phosphorylation of the nuclear factor of activated T cells, and subsequent translocation into the nucleus. This results in down-regulated gene transcription for cytokines such as IL-2 [11]. Nephrotoxicity is a major problem of CsA therapy. Therefore, careful monitoring of whole blood CsA levels is crucial not only to avoid organ rejection but also to avoid nephrotoxicity. Nephrotoxicity with tacrolimus may be less of a problem than with CsA, especially in renal transplantation. Nevertheless, nephrotoxicity may also be encountered with tacrolimus therapy [12].

Sirolimus, a natural product derived from strains of *Streptomycin hygroscopicus*, is available for oral administration. Everolimus is a structural analogue of sirolimus with a 2-hydroxyethyl chain substitution at position 40 of the sirolimus molecule. Everolimus received United States Food and Drug Administration approval in 2010 for therapeutic use as a prophylactic agent in organ transplant and for treatment of advanced kidney cancer. Everolimus has improved pharmacokinetics (much shorter half-life than sirolimus) and greater oral bioavailability than sirolimus [13, 14]. Similar to tacrolimus, sirolimus and everolimus bind to the intracellular immunophilin FK506-binding protein-12. The complexes are highly specific inhibitors of mTOR, a cell cycle serine/threonine kinase involved in the protein kinase B signaling pathway [15]. Both tacrolimus and everolimus thus exert immunosuppressant activity by blocking the proliferative signals of growth factors, thereby preventing cells from entering the S-phase. The effects of these drugs are not limited to IL-2 dependent proliferation of T cells, because everolimus also inhibits growth factor dependent proliferation in hematopoietic as well as nonhematopoietic cells [16, 17]. Therefore, mechanism of immunosuppressants by everolimus is broader than calcineurin inhibitors such as CsA. Because of their complementary mechanisms of action, everolimus can be used along with CsA, allowing for lower doses of CsA and less risk of toxicity [18, 19]. However, like CsA, tacrolimus and sirolimus, whole blood TDM of everolimus is needed [20].

In conclusion, the goal of immunosuppressive drug therapy is to optimize therapeutic effectiveness while minimizing unwanted adverse effects. Therapeutic monitoring of CsA, tacrolimus and sirolimus is now considered an integral part of organ transplant programs, and several arguments have been made for monitoring MPA. Immunoassays are commercially available for TDM of CsA, everolimus, MPA, sirolimus, and tacrolimus. Immunoassays are attractive because they can be semi-automated/automated, have low start-up costs, and do not require highly skilled testing personnel. However, a major drawback is metabolite cross-reactivity, which results in varying degrees of positive bias that is unique to a given immunoassay. Furthermore, cross-reactivity is not always predictable and can vary depending on post-transplant time and organ transplanted. As discussed in Chapter 5, chromatographic methods such as liquid chromatography combined with tandem mass spectrometry (LC–MS/MS) offer high analytical specificity and sensitivity for measurement of immunosuppressant blood levels. Moreover, multiple drugs can be measured simultaneously. For example, CsA, tacrolimus, sirolimus and everolimus are all measured in whole blood, and published methods are available for simultaneous monitoring of all four drugs in a single run [21]. Similarly, LC–MS/MS methods for analysis of MPA are more robust than immunoassays. Recently, Zhang et al. reviewed currently available methods for TDM of immunosuppressants [5].

Drawbacks of LC–MS/MS systems include the high initial capital cost of instrumentation, need for extensive sample cleanup and specialized training. It is critical that each laboratory determine the most appropriate assay system(s) to meet the testing needs of the transplant service and thoroughly validate the analytical performance of each assay.

References

[1] Lindholm A, Albrechtsen D, Tufveson G, Karlberg I, Persson NH, Groth CG. A randomized trial of cyclosporine and prednisolone versus cyclosporine, azathioprine, and prednisolone in primary cadaveric renal transplantation. Transplantation 1992;54:624–31.
[2] U.S. Department of Health and Human Services, n.d. Organ donor statistics, Health Resources & Service Administration https://www.organdonor.gov/statistics-stories/statistics.html; Accessed 31 July 2019
[3] Jasiak NM, Park JM. Immunosuppression in solid-organ transplantation: essentials and practical tips. Crit Care Nurs Q 2016;39:227–40.
[4] Kreuzenkamp-Jansen CW, De Abreu RA, Bokkerink JPM, Trijbels JMF. Determination of extracellular and intracellular thiopurines and methylthiopurines with HPLC. J Chromatogr 1995;672:53–61.
[5] Zhang Y, Zhang R. Recent advances in analytical methods for the therapeutic drug monitoring of immunosuppressive drugs. Drug Test Anal 2018;10:81–94.
[6] Filler G, Lepage N, Delisle B, Mai I. Effect of cyclosporine on mycophenolic acid area under the concentration-curve in pediatric kidney transplant recipients. Ther Drug Monit 2001;23:514–9.
[7] Podder H, Stepkowski SM, Napoli KL, Clark J, et al. Pharmacokinetic interactions augment toxicities of sirolimus/cyclosporine combinations. J Am Soc Nephrol 2001;12:1059–71.

[8] Undre NA. Pharmacokinetics of tacrolimus-based combination therapies. Nephrol Dial Transplant 2003;18(Suppl. 1):i12–5.
[9] Staatz CE, Tett SE. Pharmacology and toxicology of mycophenolate in organ transplant recipients: an update. Arch Toxicol 2014;88:1351–89.
[10] First MR. Tacrolimus based immunosuppression. J Nephrol 2004;17:25–31.
[11] Schreiber SL, Crabtree GR. The mechanism of action of cyclosporin A and FK-506. Immunol Today 1992;13:136–42.
[12] Henry ML. Cyclosporine and tacrolimus (FK506): a comparison of efficacy and safety profiles. Clin Transplant 1999;13:209–20.
[13] Augustine JJ, Hricik DE. Experience with everolimus. Transplant Proc 2004;36(Suppl. 2S):500S–3S.
[14] Kirchner GI, Meier-Wiedenbach I, Manns MP. Clinical pharmacokinetics of everolimus. Clin Pharmacokinet 2004;43:83–95.
[15] Abraham RT, Wiederrecht GJ. Immunopharmacology of rapamycin. Annu Rev Immunol 1996;14:483–510.
[16] Schuler W, Sedrani R, Cottens S, Haberlin B, et al. SDZ RAD a new rapamycin derivative: pharmacological properties in vitro and in vivo. Transplantation 1997;64:36–42.
[17] Kahan BD, Kaplan B, Lorber M, Winkler M, et al. RAD in de novo renal transplantation: comparison of three dosages on the incidence and severity of acute rejection. Transplantation 2001;71:1400–6.
[18] Neumayer HH, Paradis K, Korn A, Jean C, et al. Entry into human study with the novel immunosuppressant SDA RAD in stable renal transplant recipients. Br J Clin Pharmacol 1999;48:694–703.
[19] Deters M, Kirchner G, Resch K, Kaever V. Simultaneous quantification of sirolimus, everolimus, tacrolimus and cyclosporine by liquid chromatography-mass spectrometry. Clin Chem Lab Med 2002;40:285–92.
[20] Mabasa VH, Ensom MH. The role of therapeutic drug monitoring of everolimus in solid organ transplantation. Ther Drug Monit 2005;27:666–76.
[21] Koster RA, Dijkers EC, Uges DR. Robust, high-throughput LC-MS/MS method for therapeutic drug monitoring of cyclosporine, tacrolimus, everolimus, and sirolimus in whole blood. Ther Drug Monit 2009;31:116–25.

13.1

Cyclosporine

Chemical properties	
Solubility in H_2O	Slightly soluble
Molecular weight	1202
pKa	n/a
Melting point	148–151 °C
Dosing	
Recommended dose, adult	4–18 mg/kg/day (oral)
Recommended dose, child	9–28 mg/kg/day (oral)
Monitoring	
Sample	Whole blood
Effective concentrations	Variable
Toxic concentrations	Variable
Methods	Immunoassay, HPLC-UV, LC–MS/MS
Pharmacokinetic properties	
Oral dose absorbed	Variable, ranging from 4% to 89%
Time to peak concentration (h):	1–6 h
Protein bound	90–98% bound to lipoproteins
Volume of distribution	4–6 L/kg
Half-life, adult	6–27 h
Time to steady state, adult	2–6 d
Half-life, child	Age-dependent

Excretion, urine

	% Excreted	Active	Detected in blood
Parent	<1	Yes	Yes
Metabolites (>30)	6	Some	Some

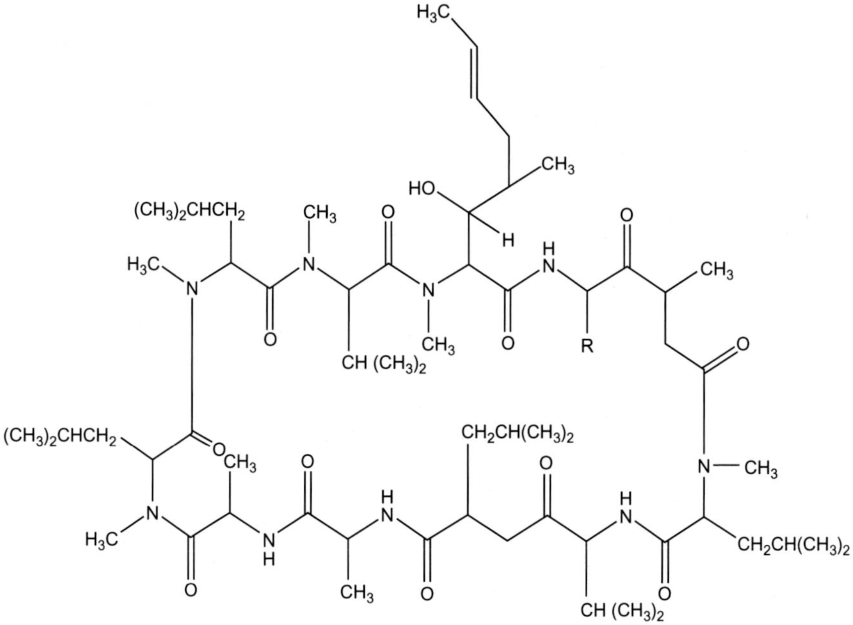

R = D-Ala

Cyclosporine A

Cyclosporine (cyclosporine A; CsA) is a small cyclic polypeptide that was originally isolated from fungal cultures of *Tolypocladium inflatum Gams* in 1970. It is approved in the United States by the Food and Drug Administration (FDA) as an immunosuppressive drug to prolong organ and patient survival in kidney, liver, heart and bone marrow transplants. CsA is also used in treating various autoimmune diseases [1]. CsA is available for both oral and intravenous administration (Sandimmune). A microemulsion oral formulation of CsA, called Neoral, exhibits more reproducible absorption characteristics and has largely replaced Sandimmune. Neoral is available as 25 and 100 mg capsules for oral administration, and as a 100 mg/mL sterile solution for intravenous administration. Several generic microemulsion formulations are also available. CsA freely traverses cell membranes and forms complexes with a cytoplasmic binding protein immunophilin called cyclophilin. CsA-cyclophilin complexes block calcineurin activity and prevent de-phosphorylation of the nuclear factor of activated T cells (NF-AT), and subsequent translocation into the nucleus. The end result is down-regulated cytokine gene transcription (interleukin-2 and others) that leads to a block in the activation/proliferation of $CD4^+$ and $CD8^+$ T lymphocytes [2].

CsA has been used in dermatology since its approval by the FDA in 1997 for treatment of psoriasis. While indicated only for the treatment of moderate to severe psoriasis, CsA has also been used off-label for the treatment of various inflammatory skin conditions, including atopic dermatitis, blistering disorders, and connective tissue diseases [3]. Given its immunomodulatory actions, CsA could be a potential treatment option for Stevens-Johnson syndrome/toxic epidermal necrolysis in addition to best supportive measures [4].

CsA is highly lipophilic, with slow and incomplete oral absorption. Food causes a clinically significant decrease in peak concentration and exposure of CsA. Oral absorption of Sandimmune is low and highly variable, ranging from 4% to 89% in renal and liver transplant patients. Absorption of Neoral is more consistent, averaging approximately 40%. Peak blood concentrations typically occur between 1 and 3 h and 2–6 h following oral administration of Neoral and Sandimmune, respectively [5]. Absorption can be delayed for several hours in a subgroup of patients. Because CsA is lipophilic, it crosses most biologic membranes and has a wide tissue distribution. CsA is extensively distributed in peripheral tissues with an apparent volume of distribution of 3–5 L/kg. CsA distribution in the blood is approximately 41–58% in erythrocytes, 33–47% in plasma, 5–12% in granulocytes, and 4–9% in lymphocytes. In plasma, CsA binds primarily to lipoproteins and secondarily to albumin. Its fraction unbound is approximately 0.1. CsA is extensively metabolized in the liver via the cytochrome P450 (CYP) 3A pathway, and the metabolites are extensively excreted in the bile. The clearance of CsA is 0.3–0.4 L/kg and the half-life ranges from 5 to 27 h [6]. Because the distribution of CsA between plasma and erythrocytes is temperature-dependent and varies with changes in hematocrit, whole blood is the preferred specimen for therapeutic drug monitoring (TDM) of CsA. EDTA-anticoagulated whole blood is the recommended specimen type when measuring CsA concentrations [7].

The specific isoenzyme expression pattern of CYP3A located in the small intestine and liver plays an important role in CsA metabolism. There is also a cellular transporter of immunosuppressive drugs, called P-glycoprotein (also known as multidrug resistance protein 1), which influences CsA metabolism by controlling bioavailability. P-glycoprotein pumps some of the CsA out of enterocytes back into the lumen of the gut. This efflux pump likely contributes significantly to the poor absorption rates observed after oral administration of CsA. CYP3A isoenzymes and P-glycoprotein genetic polymorphisms can also influence the oral bioavailability of CsA and are probably involved in the delayed absorption that has been noted in a subset of patients. CsA is oxidized or *N*-demethylated to >30 metabolites [8]. Zhu et al. observed that the CYP3A5*3 polymorphism is associated with CsA dose-adjusted concentration in renal transplant recipients. Patients carrying the CYP3A5*3/*3 genotype will require a lower dose of CsA to reach target levels compared with the CYP3A5*1/*1 or *1/*3 carriers [9]. CsA clearance is significantly reduced in elderly patients (>65 years of age). Therefore, appropriate dosage adjustment is necessary [10].

Most of the CsA metabolites do not possess immunosuppressive activity and are not clinically significant. Two of the hydroxylated metabolites, AM1 and AM9, exhibit 10–20% of the immunosuppressive activity of the parent compound, and can account for as much as 33% of the whole blood CsA concentration. The major route of CsA elimination is biliary excretion into the feces. As expected, dosage adjustments are necessary in patients with hepatic dysfunction. Only a small fraction (6%) of CsA and metabolites appear in the urine, making dosage adjustments unnecessary in patients with renal insufficiency [8].

Serious side effects related to CsA treatment are concentration-dependent and include nephrotoxicity, neurotoxicity, hepatotoxicity, hirsutism, hypertrichosis, gingival hypertrophy, glucose intolerance, hypertension, hyperlipidemia, hypomagnesemia, hyperuricemia and hypokalemia. In general, over-suppression leads to an increased risk for viral infections and lymphoproliferative disease, especially in children [11].

CsA monitoring is essential to avoid toxicity as well as rejection of the graft. Immunoassays such as ACMIA (affinity chrome-mediated immunoassay), CEDIA (cloned enzyme donor immunoassay), EMIT 2000 (enzyme multiplied immunoassay technique), ADVIA Centaur cycloserine assay, CMIA (chemiluminescent microparticle immunoassay) etc. are commercially available for analysis of CsA in whole blood. All immunoassays require manual pre-treatment of whole blood with an extraction solvent provided by the manufacturer except for the CsA immunoassay (Antibody Conjugate Magnetic Immunoassays: ACMIA) on the Dimension analyzer (Siemens) where pre-treatment relies on on-line mixing and ultrasonic lysing of whole blood. All immunoassays suffer from significant cross-reactivities from CsA metabolites. As a result, significant bias is observed between CsA concentrations observed in whole blood by immunoassay compared to values obtained by chromatographic methods such as high-performance liquid chromatography with ultraviolet detection (HPLC-UV) or liquid chromatography combined with tandem mass spectrometry (LC/MS/MS). For example, Butch et al. evaluated performance of CEDIA cyclosporine PLUS whole blood immunoassay on the Olympus AU400 analyzer and observed metabolite cross-reactivity 8.1% for AM1, 21.7% for AM4n, and 32.5% for AM9. The authors also observed significant negative bias using the immunoassay compared to a specific HPLC-UV method [12]. However, newer CsA immunoassays are more robust. Wallemacq et al. reported findings from multicenter evaluation of Abbott Architect cyclosporine assay (CMIA) using seven clinical laboratories. Values obtained by the immunoassay were compared with corresponding values obtained by LC/MS/MS. The authors observed minimal cross-reactivity of CsA metabolites AM1 (up to 0.2%) and AM9 (up to 2.2%) in the cyclosporine CMIA. Comparison testing with Roche Integra assay (Integra 800; Roche Diagnostics) showed 2.4% cross-reactivity for AM1 metabolite and 10.7% cross-reactivity with the AM9 metabolite. The Architect immunoassay showed an average bias of 31 ng/mL of CsA compared to LC/MS/MS method. The authors concluded that the cross-reactivity of CsA metabolites has been significantly reduced in the Architect cyclosporine immunoassay [13]. Soldin et al. evaluated performance evaluation of a new ADVIA Centaur cyclosporine immunoassay which requires a single step extraction and observed excellent correlation between CsA values obtained by the LC/MS/MS assay and ADVIA Centaur CsA assay with an average bias of only 6% [14].

However, falsely elevated blood CsA levels due to presence of endogenous antibodies (most likely heterophilic antibodies) has been reported to affect the ACMIA cyclosporine assay run on Dimension RXL analyzer (Siemens). De Jonge et al. reported a falsely elevated CsA level of 492 ng/mL in a 77 year old patient. However, using LC/MS/MS, the CsA level was undetectable. In addition, Architect cyclosporine assay also yielded a value lower than the detection limit. Treating the specimen with polyethylene glycol and remeasuring CsA in the supernatant by the same ACMIA assay showed no detectable level of CsA confirming that the interfering substance was a protein, most likely an endogenous antibody [15].

TDM of CsA usually utilizes trough levels. The introduction of Neoral in 1995 has resulted in higher peak concentrations and increased drug exposure, based on area under the concentration time curves (AUC). However, similar trough concentrations are observed for both conventional and microemulsion CsA formulations, demonstrating that trough concentrations are not predictive of total drug exposure. Two hour post-dosing CsA concentrations (called C2 monitoring) have been shown to correlate better with total drug exposure and results in improved clinical outcomes. Another advantage is lower metabolite cross-reactivity with CsA metabolites [16]. Unfortunately, this creates various nursing/phlebotomy challenges because blood samples have to be drawn very close to the 2-h time point after dosing, ideally 10 min on either side of the 2-h mark. Currently, measuring trough CsA levels is more common practice.

Therapeutic ranges for CsA are often organ-specific and can vary widely between transplant centers. In addition, target CsA levels may be lower if CsA is used in addition to another immunosuppressant. Target trough whole blood CsA levels following kidney transplants are typically between 150 and 250 ng/mL shortly after transplant, tapering down below 150 mg/L during maintenance therapy. Recommended levels after liver and heart transplants are 250–350 ng/mL shortly after transplant and <150 mg/L during maintenance therapy. These target ranges were determined by HPLC and will vary when measured by immunoassay, depending on the amount of metabolite cross-reactivity. Min et al. demonstrated that up to 1 month after renal transplantation CsA therapeutic response threshold was 182 ng/mL, while nephrotoxicity threshold was 204 ng/mL. Between 1 month and 3 month after transplantation, the respective therapeutic and toxic threshold for CsA was 175 ng/mL and 189 ng/mL respectively. However, between 3 and 12 months after transplantation, the therapeutic and toxic CsA threshold became 135 ng/mL and 204 ng/mL respectively [17]. Therefore, maintaining CsA level <150 ng/mL in stable transplant patients is justified. However, for C2 monitoring, target concentrations vary between 600 and 1700 ng/mL depending on the type of graft and the time after transplantation [18].

There are clinically significant CsA-drug and CsA-herb interactions. Careful TDM of CsA can identify such interactions. Clinical and experimental data indicate the potential of experiencing greater renal, cardiac, and circulatory risks in individuals treated concurrently with CsA plus NSAIDs (non-steroidal antiinflammatory drugs). The adverse effects are more significant with non-selective COX1/COX 2 inhibitors such as indomethacin or diclofenac compared to COX2 inhibitors such as celecoxib [19]. In addition, interactions between CsA and danazol, diltiazem, erythromycin, fluconazole, itraconazole, ketoconazole, metoclopramide, nicardipine, verapamil, carbamazepine, phenobarbital (phenobarbitone), phenytoin, rifampicin (rifampin) and cotrimoxazole (trimethoprim/sulfamethoxazole) are well documented in a large number of patients [20]. Herbal antidepressant St. John's wort induces CYP enzymes involved in CsA metabolism, thus significantly increasing clearance of the drug. Whole blood levels of CsA may be reduced by 50% due to the CYP-inducing properties of St. John's wort.

Therefore, acute graft rejection may occur in an organ recipient taking and St. John's wort. The magnitude of interaction depends on concentration of hyperforin in the St. John's wort formulation [21].

References

[1] Ponticelli C. Cyclosporine: from renal transplantation to autoimmune diseases. Ann N Y Acad Sci 2005;1051:551–8.
[2] Ho S, Clipstone N, Timmermann L, Northrop J, et al. The mechanism of action of cyclosporin A and FK506. Clin Immunol Immunopathol 1996;80(3 Pt. 2):S40–5.
[3] Dehesa L, Abuchar A, Nuno-Gonzalez A, Vitiello M, Kerdel FA. The use of cyclosporine in dermatology. J Drugs Dermatol 2012;11:979–87.
[4] Ng QX, De Deyn MLZQ, Venkatanarayanan N, Ho CYX, Yeo WS. A meta-analysis of cyclosporine treatment for Stevens-Johnson syndrome/toxic epidermal necrolysis. J Inflamm Res 2018;11:135–42.
[5] Kovarik JM, Mueller EA, van Bree JB, Fluckinger SS, et al. Cyclosporine pharmacokinetics and variability from a microemulsion formulation-a multicenter investigation in kidney transplant patients. Transplantation 1994;58:658–63.
[6] Han K, Pillai VC, Venkataramanan R. Population pharmacokinetics of cyclosporine in transplant recipients. AAPS J 2013;15:901–12.
[7] Wenk M, Follath F, Abisch E. Temperature dependency of apparent cyclosporine A concentrations in plasma. Clin Chem 1983;29:1865.
[8] Chrintians U, Sewing KF. Cyclosporin metabolism in transplant patients. Pharmacol Ther 1993;57:291–345.
[9] Zhu HJ, Yuan SH, Fang Y, Sun XZ, et al. The effect of CYP3A5 polymorphism on dose-adjusted cyclosporine concentration in renal transplant recipients: a meta-analysis. Pharmacogenomics J 2011;11:237–46.
[10] Falck P, Asberg A, Byberg KT, Bremer S, et al. Reduced elimination of cyclosporine A in elderly (>65 years) kidney transplant recipients. Transplantation 2008;86:1379–83.
[11] Graham RM. Cyclosporine: mechanisms of action and toxicity. Cleve Clin J Med 1994;61:308–13.
[12] Butch AW, Fukuchi AM. Analytical performance of the CEDIA cyclosporine PLUS whole blood immunoassay. J Anal Toxicol 2004;28:204–10.
[13] Wallemacq P, Maine GT, Berg K, Rosiere T, et al. Multisite analytical evaluation of Abbott Architect cyclosporine assay. Ther Drug Monit 2010;32:145–51.
[14] Soldin SJ, Hardy RW, Wians FH, Balko JA, et al. Performance evaluation of the new ADVIA Centaur system cyclosporine assay (single-step extraction). Clin Chim Acta 2010;411:806–11.
[15] De Jonge H, Geerts I, Declercq P, de Loor H, et al. Apparent elevation of cyclosporine whole blood concentration in a renal allograft recipient. Ther Drug Monit 2010;32:529–31.
[16] Cantarovick M, Barkun JS, Tchervenkov JI, Besner JG, Aspeslet L, Metrakos P. Comparison of neoral dose monitoring with cyclosporine through levels versus 2-hour postdose levels in stable liver transplant patients. Transplantation 1998;66:1621–7.
[17] Min DI, Perry PJ, Chen HY, Hunsicker LG. Cyclosporine trough concentrations in pediatric allograft rejection and renal toxicity up to 12 months after renal transplantation. Pharmacotherapy 1998;18:282–7.
[18] Sukhavasharin NH, Praditpornsilpa K, Avihingsanon Y, Kuoatawintu P, et al. Study of cyclosporine level at 2 hours after administration in preoperative kidney transplant recipients for prediction of postoperative optimal cyclosporine dose. J Med Assoc Thai 2006;89(Suppl. 2):S15–20.

[19] El-Yazbi AF, Eid AH, El-Mas MM. Cardiovascular and renal interactions between cyclosporine and NSAIDs: underlying mechanisms and clinical relevance. Pharmacol Res 2018;129:251–61.
[20] Campana C, Regazzi MB, Buggia I, Molinaro M. Clinically significant drug interactions with cyclosporin. An update. Clin Pharmacokinet 1996;30:141–79.
[21] Chrubasik-Hausmann S, Vlachojannis J, McLachlan AJ. Understanding drug interactions with St John's wort (*Hypericum perforatum* L.): impact of hyperforin content. J Pharm Pharmacol 2019;71:129–38.

13.2

Everolimus

Chemical properties	
Solubility in H_2O	Practically insoluble
Molecular weight	958
pKa	n/a
Melting point	?
Dosing	
Recommended dose, adult:	1–4 mg/day
Recommended dose, child:	1.6–3.0 mg/m^2/day
Monitoring	
Sample	Whole blood
Effective concentrations	3–8 ng/mL (trough)
Toxic concentrations	Variable
Methods	Immunoassay, HPLC-UV, HPLC–MS
Pharmacokinetic properties	
Oral dose absorbed	Variable
Time to peak concentration	1–3 h
Protein bound	75%
Volume of distribution	4–20 L/kg
Half-life, adult	18–35 h
Time to steady state, adult	4–7 d
Half-life, child:	Similar to adults
Time to steady state, child	3–5 h

Excretion, urine

	% Excreted	Active	Detected in blood
Parent	Unknown	Yes	Yes
Metabolites (>20)	Unknown	Unknown	Some

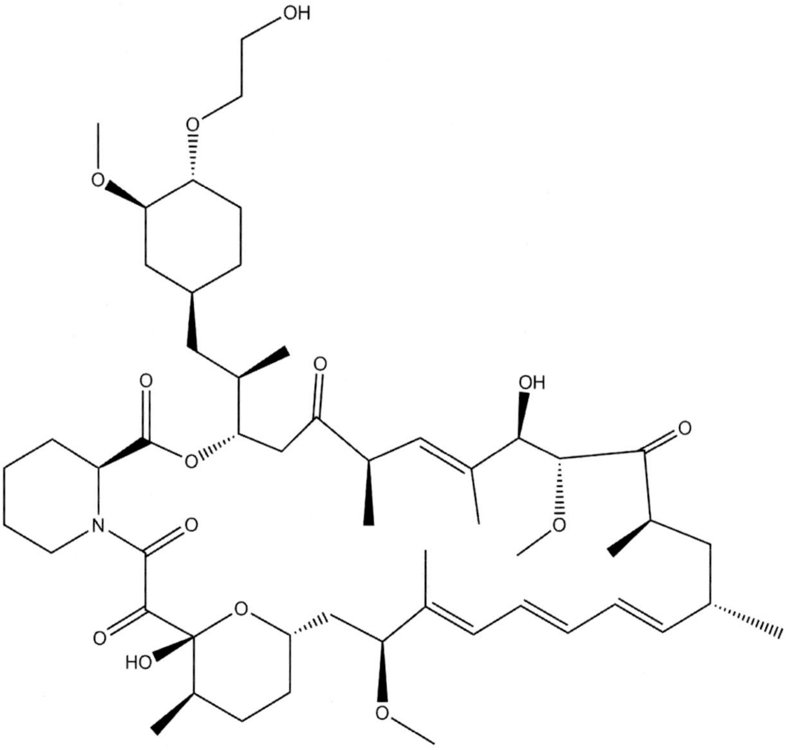

Everolimus

Everolimus (also known as SZD RAD) is a structural analog of sirolimus containing an additional 2-hydroxyethyl group at position 40 of the sirolimus molecule. Everolimus (40-*O*-[2-hydroxy]ethylrampamycin) was formulated to have improved pharmacokinetic characteristics compared with sirolimus. The hydroxyethyl chain increases the polarity of the molecule resulting in greater oral bioavailability. Everolimus is a potent anti-proliferative agent with a mechanism of action similar to sirolimus. As expected, everolimus inhibits the mammalian target of rapamycin (mTOR), a serine–threonine kinase. The mTOR pathway plays a role in regulating cell proliferation and angiogenesis and requires the cellular substrate p70 ribosomal S6 kinase 1 to be phosphorylated for proliferation. Angiogenesis is promoted through the hypoxia-inducible factor-1 and growth factor protein translation. Everolimus blocks cell cycle progression into S-phase. In limited clinical trials, everolimus was shown to be less nephrotoxic than cyclosporine and tacrolimus and may reduce the incidence of allograft vasculopathy in heart transplantation [1–3]. The United States Food and Drug Administration

(FDA) originally approved everolimus in 2020 as an immunosuppressant to treat kidney transplant recipients. Later, the FDA approved everolimus for the treatment of advanced renal cell carcinoma, subependymal giant cell astrocytoma associated with tuberous sclerosis, pancreatic neuroendocrine tumors, and, in combination with exemestane, for treatment of advanced hormone-receptor (HR)-positive, HER2-negative breast cancer [4].

Oral bioavailability of everolimus is around 16%, which is an improvement over the 10% bioavailability of sirolimus [1]. Everolimus can be taken with or without food. After administration of 10 mg everolimus orally, peak blood level can be observed after 2 h. Steady-state levels are reached after 8 days of treatment [3]. Everolimus has a shorter elimination half-life (average 28 h) compared with sirolimus (62 h). This requires twice a day dosing. Intra-patient total drug exposure for everolimus is approximately 40% and can be considerably more variable between patients (>85%) [1]. When combined with cyclosporine, blood concentrations of everolimus are increased due to competition for drug metabolism [5]. At therapeutic concentrations, >75% of everolimus is found within erythrocytes. For this reason whole blood specimens are used for everolimus quantitation. Approximately 75% of the everolimus found within the plasma is bound to protein. Similar to sirolimus, everolimus is metabolized in the intestine and liver by cytochrome P450 (CYP) enzymes. At least 20 metabolites have been identified, with mono-hydroxyl, di-hydroxyl, demethylated and an open ring form being the major metabolites [6]. The immunosuppressive activity and toxicity of everolimus metabolites has not been fully investigated. Metabolites are in relatively low concentrations when monitoring trough blood concentrations.

For TDM of everolimus, a trough concentration of least 3 ng/mL must be reached in the first week of posttransplant for preventing graft rejection [7]. Therapeutic range of everolimus is usually considered as 3–8 ng/mL [8]. Similar to other immunosuppressive agents, the incidence of adverse effects of everolimus is related to dose. Hyperlipidemia is of major concern, with increases in serum triglycerides and cholesterol being observed, especially when everolimus is combined with cyclosporine and corticosteroids. Leukopenia and thrombocytopenia are also observed, and pancytopenia occurs rarely. There is also an increased risk of infectious episodes, such as pneumonia, pharyngitis, sinusitis and multiple herpetic oral lesions.

Immunoassays to measure everolimus are commercially available, although high performance liquid chromatography combined with ultraviolet detection or liquid chromatography combined with tandem mass spectrometry (LC–MS/MS) offer better specificity and sensitivity compared to immunoassays. Seradyn first developed a fluorescence polarization immunoassay (FPIA) (Innofluor Certican Assay System) to measure whole blood everolimus outside the United States on the Abbott TDx analyzer. The FPIA method has a functional sensitivity of 2 ng/mL, which is just below the therapeutic trough blood concentration lower limit of 3 ng/mL [9]. Sallustio et al. observed an average bias of over 30% in everolimus concentration as determined by the Seradyn FPIA and a specific LC–MS/MS for everolimus and concluded that further investigation is needed before this assay can be sued for routine therapeutic monitoring of everolimus [10]. Quantitative Microparticle System (QMS) everolimus assay (Fischer Scientific) received FDA approval for clinical use in 2011 and the assay is linear between 1.5 ng/mL and 20 ng/mL, covering the entire therapeutic range of everolimus. The lower limit of quantitation for this assay is 1.3 ng/mL. The average bias between a LC–MS/MS assay and the QMS assay was 11% indicating that this assay can be used for TDM of everolimus. However, structurally similar sirolimus showed an average 46% cross-reactivity [11]. A new electrochemiluminescence immunoassay (ECLIA) for everolimus (Roche) can also be used for monitoring of everolimus but this method also showed an average 21% positive bias with LC–MS/MS method [12].

References

[1] Kirchner GI, Meier-Wiedenbach I, Manns MP. Clinical pharmacokinetics of everolimus. Clin Pharmacokinet 2004;43:83–95.

[2] Eisen HJ, Tuzcu EM, Dorent R, Kobashigawa J, et alRAD B253 Study Group. Everolimus for the prevention of allograft rejection and vasculopathy in cardiac-transplant recipients. N Engl J Med 2003;349:847–58.

[3] Green MA, Waddell JA, Solimando Jr. DA. Drug monographs: atezolizumab and everolimus. Hosp Pharm 2016;51:810–4.

[4] Hasskarl J. Everolimus recent results. Cancer Res 2014;201:373–92.

[5] Kahan BD, Podbielski J, Napoli KL, et al. Immunosuppressive effects and safety of a sirolimus/cyclosporine combination regimen for renal transplantation. Transplantation 1998;66:1040–6.

[6] Kirchner GI, Winkler M, Mueller L, Vidal C, et al. Pharmacokinetics of SDZ RAD and cyclosporin including their metabolites in seven kidney graft patients after the first dose of SDZ RAD. Br J Clin Pharmacol 2000;50:449–54.

[7] Romagnoli J, Citterio F, Favi E, Salerno MP, et al. Higher incidence of acute rejection in renal transplant recipients with low everolimus exposure. Transplant Proc 2007;39:1823–6.

[8] Mabasa VH, Ensom MH. The role of therapeutic drug monitoring of everolimus in solid organ transplantation. Ther Drug Monit 2005;27:666–76.

[9] Salm P, Warnholtz C, Boyd J, Arabshahi L, et al. Evaluation of a fluorescent polarization immunoassay for whole blood everolimus determination using samples from renal transplant recipients. Clin Biochem 2006;39:732–8.

[10] Sallustio BC, Noll BD, Morris RG. Comparison of blood sirolimus, tacrolimus and everolimus concentrations measured by LC-MS/MS, HPLC-UV and immunoassay methods. Clin Biochem 2011;44:231–6.
[11] Dasgupta A, Davis B, Chow L. Evaluation of QMS everolimus assay using Hitachi 917 analyzer: comparison with liquid chromatography/mass spectrometry. Ther Drug Monit 2011;33:149–54.
[12] Verstraete AG, Rigo-Bonnin R, Wallemacq P, Vogeser M, et al. Multicenter evaluation of a new electrochemiluminescence immunoassay for everolimus concentrations in whole blood. Ther Drug Monit 2018;40:59–68.

13.3

Mycophenolic acid

Chemical properties	
Solubility in H_2O	Slightly soluble
Molecular weight	320
pKa	5.6, 8.5
Melting point	141 °C
Dosing	
Recommended dose, adult	1440–2000 mg/day (oral)
Recommended dose, child	900–1200 mg/m^2/day (oral)
Monitoring	
Sample	Serum, plasma
Effective concentrations	1.0–3.5 μg/mL (trough)
Toxic concentrations	Variable
Methods	Immunoassay, HPLC-UV, HPLC–MS
Pharmacokinetic properties	
Oral dose absorbed	93%
Time to peak concentration	1.0–2.5 h
Protein bound	>98% bound to albumin
Volume of distribution	3.6–4.0 L/kg
Half-life, adult	8–18 h
Time to steady state, adult	?
Half-life, child	Variable
Time to steady state, child	?

Excretion, urine

	% excreted	Active	Detected in blood
Mycophenolic acid	<1	Yes	Yes
Metabolites (≥4)	87	Acyl glucuronide	7-*O*-glucuronide

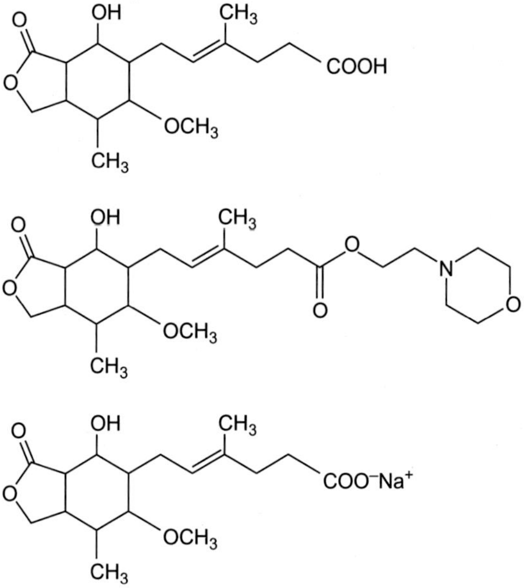

Structures of (from top to bottom) Mycophenolic Acid, Mycophenolic acid mofetil and Mycophenolate Sodium.

Mycophenolic acid (MPA) is a fermentation product of *Penicillium* species that was originally shown to have antibacterial, antifungal and immunosuppressive potential in animal studies. To improve the bioavailability of MPA, mycophenolate mofetil (brand name CellCept), the 2-morpholinoethyl ester of MPA was developed for oral and intravenous administration. Mycophenolate mofetil received United States Food and Drug Administration approval for use as an immunosuppressant with corticosteroids and cyclosporine to prevent organ rejection in 1995. Mycophenolate mofetil is available in 250 mg capsules, 500 mg tablets, as a powder to make an oral suspension at 200 mg/mL, and as a sterile powder (500 mg) for intravenous administration. The sodium salt of MPA, mycophenolate sodium (brand name Myfortic), is available for oral administration in 180 mg and 360 mg delayed-release tablets [1].

Mechanism of immunosuppression of MPA is due to reversible and non-competitive inhibition of inosine monophosphate dehydrogenase enzyme (IMPDH), which is the rate-limiting enzyme in de novo biosynthesis of guanosine nucleotides. Moreover, MPA is fivefold more potent as an inhibitor of the type II isoform of IMPDH compared to the type I isoform of IMPDH. The IMPDH isoform II is expressed mostly on activated lymphocytes, while isoform I is present in most cell types. T- and B-lymphocytes require guanosine nucleotides for new DNA synthesis as well as synthesis of surface antigens and glycosylated membrane proteins. Therefore, MPA has more potent cytostatic effect on lymphocytes than other cells, and this is likely the primary mechanism by which MPA exerts its immunosuppressant effect. In general, during peak serum MPA concentration, approximately 70% of the IMPDH activity is inhibited, while during trough concentration approximately 30% activity is inhibited. A minor metabolite of MPA (mycophenolic acyl glucuronide) can also inhibit IMPDH activity [2].

Mycophenolate mofetil and mycophenolate sodium are rapidly and completely absorbed, and then quickly de-esterified in the blood and tissues to MPA, the active form of the drug. The half-life of mycophenolate mofetil during intravenous administration is <2 min. Following an oral dose of mycophenolate mofetil, MPA reaches a maximum concentration within 1 h [3]. Almost all the drug (>99%) can be found in the plasma compartment. For this reason, serum or plasma is used for MPA quantitation for therapeutic drug monitoring (TDM).

MPA has an elimination half-life of 8–18 h (mycophenolate mofetil) [4]. The main metabolite of MPA is formed via activity of the liver enzyme uridine diphosphate glucuronyl transferase (UGT) where MPA phenolic glucuronide is formed (7-O-MPA β-glucuronide or MPAG). The main isoform responsible for such transformation is UGT1A9, and polymorphism of the gene encoding this enzyme may affect MPAG concentration in serum. Glucuronidation also produces MPA acyl glucuronide which is a minor metabolite. The UGT2B7 isoform is responsible for this transformation. Although MPAG has no pharmacological activity, the minor metabolite MPA acyl glucuronide has pharmacological activity and may also be responsible for adverse gastrointestinal effects. Another minor metabolite, MPA phenyl-glucoside, has been identified in renal transplant recipients. MPAG may reach plasma concentrations 20- to 100-fold higher than MPA [5, 6]. MPAG exhibits significant enterohepatic recirculation with a second MPA plasma peak occurring 4–12 h after drug administration. The kidneys primarily clear MPAG with concentrations rapidly accumulating in patients with severe renal impairment (glomerular filtration rates <25 mL/min). MPA is extensively bound in the circulation to albumin with typical concentrations of free or unbound MPA ranging from 1.25% to 2.5% of the total concentration. Free MPA concentrations are increased in hypoalbuminemia, hyperbilirubinemia and uremia [7]. It has been shown that the immunosuppressive effects of MPA are related to free MPA and not the total drug concentration. In chronic renal failure, the free concentration of MPA can dramatically increase indicating over immunosuppression, while the total MPA concentration is still within the therapeutic range.

Adverse effects from mycophenolate mofetil and mycophenolate sodium are similar. The most common dose-limiting unwanted side effects are diarrhea, nausea, vomiting, and abdominal pain. Marrow suppression and anemia can also occur. An increased risk of cytomegalovirus, *Candida*, and herpes simplex infections has also been reported [3].

Plasma or serum can be used to measure total MPA and free MPA blood concentrations. Plasma from EDTA anticoagulated whole blood is the recommended specimen type, because this specimen can also be used to measure whole blood cyclosporine, tacrolimus, and sirolimus [8]. When monitoring MPA during intravenous infusion of mycophenolate mofetil, whole blood samples should be immediately placed in ice, and the plasma separated within 30 min. This is because mycophenolate mofetil is unstable and rapidly undergoes temperature-dependent degradation to MPA in whole blood samples placed at room temperature.

Trough concentrations of MPA are routinely used for drug monitoring and are generally believed to be a relatively good indicator of total drug exposure [9]. This is somewhat surprising since numerous studies have shown that area under the curve (0–12 h) measurements is more predictive of total drug exposure and acute graft rejection than trough concentrations. In addition, MPA trough concentrations can vary considerably depending upon time after transplantation. Nevertheless, the superiority of area under the curve measurements is probably overshadowed by practical considerations such as additional testing costs and difficulties associated with the collection of multiply timed samples.

When MPA was originally approved for use as mycophenolate mofetil, TDM was considered unnecessary. However, recent studies have found wide variations in total drug exposure (as high as 10-fold) following a fixed dose, suggesting that individualized dosing may be of considerable benefit. A roundtable meeting recently recommended TDM based on the interpatient variability and the significant drug interactions associated with combination immunosuppressive therapy [10]. The generally accepted therapeutic range for trough MPA plasma concentrations is 1.0–3.5 mg/L [8]. Nevertheless, Kiang and Ensom commented that routine monitoring of MPA may not be necessary [11].

A major practical challenge in TDM of MPA is the choice of methodology. MPA has ultraviolet absorption and can be monitored by high performance liquid chromatography combined with ultraviolet detection (HPLC-UV), although liquid chromatography combined with mass spectrometry (LC/MS) or tandem mass spectrometry (LC–MS/MS) are superior techniques compared to HPLC-UV. MPA can also be determined using immunoassays but with a limitation of cross-reactivity of MPA metabolites, typically leading to higher concentrations than obtained with chromatographic methods. MPAG typically shows low cross-reactivity with immunoassays, while the minor acyl glucuronide metabolite may show significant cross-reactivity with certain immunoassays, leading to falsely elevated MPA levels.

Currently, two immunoassays and one enzymatic assay (total MPA assay by Roche) are commercially available for TDM of total MPA. The enzymatic MPA assay is based on the principle of inhibition of an enzymatic reaction by MPA. After sample addition to the reaction mixture, MPA present in the serum inhibits IMPDH II, an enzyme that catalyzes conversion of inosine monophosphate in the presence of nicotinamide adenine dinucleotide (NAD) as a cofactor into xanthine monophosphate. In this reaction NADH, which absorbs at 340 nm, is generated. Therefore, concentration of MPA in the specimen is inversely proportional to the rate of NADH formation. In addition, EMIT (enzyme multiplied immunoassay technique), CEDIA (cloned enzyme donor immunoassay) and PETINIA (particle-enhanced turbidimetric inhibition immunoassay) assays are also available for TDM of mycophenolic acid. Cross-reactivity of the MPA acyl glucuronide metabolite varies across assays. This metabolite accounts for only 10% (range 2.5–37.5%) of AUC 0–12 h of MPA. Shipkova et al. reported concentration dependent cross-reactivity of MPA acyl glucuronide ranging from 135% to 185% with the EMIT assay. The metabolite concentration ranged from 1 to 10 μg/mL [12]. For the CEDIA assay, concentration dependent cross-reactivity of MPA acyl glucuronide (140–215%) was observed where metabolite concentration ranged from 0.5 to 10 μg/mL [13]. In contrast, MPA acyl glucuronide contributes <5% in observed bias for the Roche total MPA enzymatic assay but does not use antibody in the assay design. As a result, Roche total MPA showed lowest bias among all commercially available assays for the acyl glucuronide metabolite [14].

Brown et al. observed a mean positive bias of 14.6% between MPA levels measured by the EMIT assay and a reference LC–MS/MS method [15]. CEDIA assay also showed an average 15.5% positive bias when compared to an HPLC-UV method (UV detection: 254 nm) [16]. PETINIA assay also showed an average 27% positive bias compared to LC–MS/MS method

[17]. However, Roche MPA assay showed only a 5.8% positive bias when compared with results obtained by LC–MS/MS [18]. This is consistent with low cross-reactivity of acyl glucuronide with the enzymatic assay.

Although not routinely monitored, free MPA has clinical utility in uremic patients and patients with hypoalbuminemia. Immunoassays do not have adequate sensitivity to monitor free MPA because free MPA concentration is 2–3% of total concentration. In general patients with albumin concentration below 3.1 g/dL showed elevated percentage of free MPA (>3%), and authors concluded that clinicians should consider monitoring free MPA concentrations in hypoalbuminemic patients with plasma albumin levels below 3.1 g/dL in order to avoid severe MPA toxicity [19]. Both total and free concentration of MPA (in the protein free ultrafiltrate) can be measured by LC–MS/MS [20, 21].

References

[1] Wu JC. Mycophenolate mofetil: molecular mechanisms of action. Perspect Drug Discovery Des 1994;2:185–204.

[2] Allison AC, Eugui EM. Mycophenolate mofetil and its mechanism of action. Immunopharmacology 2000;47:85–118.

[3] Bullingham RE, Nicholls AJ, Kamm BR. Clinical pharmacokinetics of mycophenolate mofetil. Clin Pharmacokinet 1998;34:429–55.

[4] Korecka M, Nikolic D, van Breemen RB, Shaw LM. Inhibition of inosine monophosphate dehydrogenase by mycophenolic acid glucuronide is attributable to the presence of trace quantities of mycophenolic acid. Clin Chem 1999;45:1047–50.

[5] Schutz E, Shipkova M, Armstrong VW, Wieland E, Oellerich M. Identification of a pharmacologically active metabolite of MPA in plasma of transplant recipients treated with mycophenolate mofetil. Clin Chem 1999;45:419–22.

[6] Jeong H, Kaplan B. TDM of mycophenolic mofetil. Clin J Am Soc Nephrol 2007;2:184–91.

[7] Kaplan B, Meier-Kriesche HU, Friedman G, Mulgaonkar S, et al. The effect of renal insufficiency on mycophenolic acid protein binding. J Clin Pharmacol 1999;39:715–20.

[8] Shaw LM, Holt DW, Oellerich M, Meiser B, van Gelder T. Current issues in therapeutic drug monitoring of mycophenolic acid: report of a roundtable discussion. Ther Drug Monit 2001;23:305–15.

[9] Mahalati K, Kahan BD. Pharmacological surrogates of allograft outcome. Ann Transplant 2005;5:14–23.

[10] van Gelder T, Meur YL, Shaw LM, Oellerich M, DeNofrio D, Holt C, Holt DW, Kaplan B, Kuypers D, Meiser B, Toenshoff B, Mamelok RD. Therapeutic drug monitoring of mycophenolate mofetil in transplantation. Ther Drug Monit 2006;28:145–54.

[11] Kiang TK, Ensom MH. Therapeutic drug monitoring of mycophenolate in adult solid organ transplant patients: an update. Expert Opin Drug Metab Toxicol 2016;12:545–53.

[12] Shipkova M, Schutz E, Armstrong VW, Niedmann PD, Oellerich M, Wieland E. Determination of the acyl glucuronide metabolite of MPA in human plasma by HPLC and EMIT. Clin Chem 2000;46:365–72.
[13] Shipkova M, Schutz E, Besenthal I, Fraunberger P, Wieland E. Investigation of the crossreactivity of MPA glucuronide metabolite of mycophenolate mofetil in the CEDIA MPA assay. Ther Drug Monit 2010;32:79–82.
[14] Brandhorst G, Marquet P, Shaw LM, Liebisch G, Schmitz G, Coffing MJ, et al. Multicenter evaluation of a new inosine monophosphate dehydrogenase inhibition assay for quantification of total MPA in plasma. Ther Drug Monit 2008;30:428–33.
[15] Brown NW, Franklin ME, Einarsdottir EN, Gonde CE. An investigation into the bias between liquid chromatography-tandem mass spectrometry and an enzyme multiplied immunoassay technique for the measurement of MPA. Ther Drug Monit 2010;32: 420–6.
[16] Dasgupta A, Johnson M. Positive bias in mycophenolic acid concentrations determined by the CEDIA assay compared to HPLC-UV method: is CEDIA assay suitable for therapeutic drug monitoring of mycophenolic acid? J Clin Lab Anal 2013;27:77–80.
[17] Ham JY, Jung HY, Choi JY, Park SH, et al. Usefulness of mycophenolic acid monitoring with PETINIA for prediction of adverse events in kidney transplant recipients. Scand J Clin Lab Invest 2016;76:296–303.
[18] Parant F, Ranchin B, Gagnieu MC. The Roche Total Mycophenolic Acid® assay: an application protocol for the ABX Pentra 400 analyzer and comparison with LC–MS in children with idiopathic nephrotic syndrome. Pract Lab Med 2017;7:19–26.
[19] Atcheson BA, Taylor PJ, Kirkpatrick CM, Duffull SB, Mudge DW, Pillans PI, et al. Free MPA should be monitored in renal transplant recipients with hypoalbuminemia. Ther Drug Monit 2004;26:284–6.
[20] Streit F, Shipkova M, Armstrong VW, Oellerich M. Validation of a rapid and sensitive liquid chromatography-tandem mass spectrometry method for free and total mycophenolic acid. Clin Chem 2004;50:152–9.
[21] Łuszczyńska P, Pawiński T, Kunicki PK, Sikorska K, Marszałek R. Free mycophenolic acid determination in human plasma ultrafiltrate by a validated liquid chromatography-tandem mass spectrometry method. Biomed Chromatogr 2017;31(10):https://doi.org/10.1002/bmc.3976.

13.4

Sirolimus

Chemical properties	
Solubility in H_2O	Insoluble
Molecular weight	914
pKa	n/a
Melting point	Unknown
Dosing	
Recommended dose, adult	1–6 mg/day
Recommended dose, child	1–3 mg/m^2/day
Monitoring	
Sample	Whole blood
Effective concentrations	5–15 ng/mL (trough) (when used with other drugs) 5–20 ng/mL (trough) when used alone
Toxic concentrations	Variable
Methods	Immunoassay, HPLC-UV, HPLC–MS
Pharmacokinetic properties	
Oral dose absorbed	Variable, ranging from 14% to 41%
Time to peak concentration	1–2 h
Protein bound	92% bound to albumin, alpha-1-acid glycoprotein and lipoproteins
Volume of distribution	4–20 L/kg
Half-life, adult	46–78 h
Time to steady state, adult	5–7 d
Half-life, child	Variable, with a mean of 49 h
Time to steady state, child	?

Excretion, urine

	% Excreted	Active	Detected in blood
Parent	2	Yes	Yes
Metabolites (>7)	<2	Some	Some

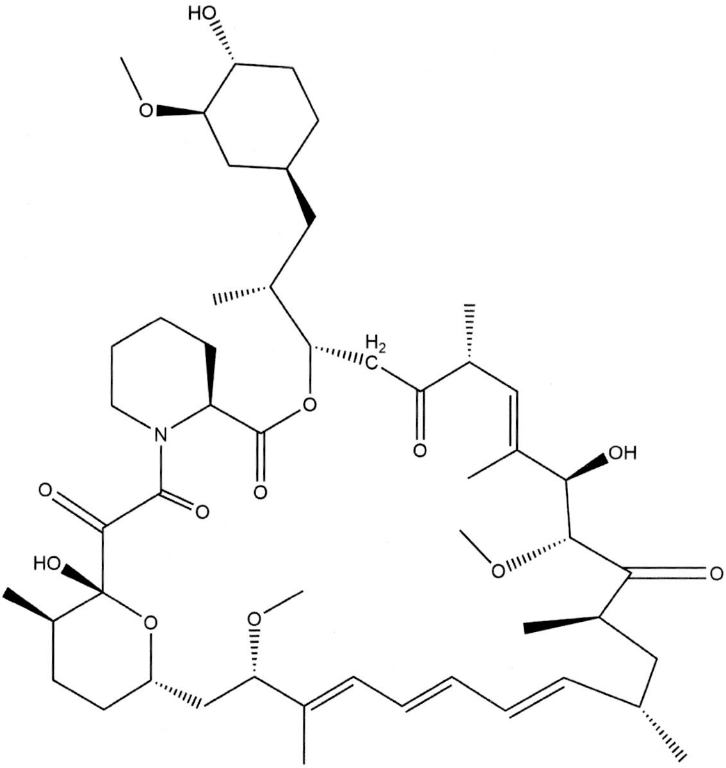

Sirolimus

Sirolimus (also known as rapamycin) is a lipophilic macrocyclic lactone derived from *Streptomyces hygroscopicus*. This actinomycete fermentation product was identified in the early 1970s and was approved by the FDA in 1999 for use with cyclosporine to reduce the incidence of acute rejection in renal transplantation. Sirolimus readily crosses lymphocyte plasma membranes and binds to the intracellular immunophilin, FK506-binding protein-12. In contrast to tacrolimus, sirolimus-immunophilin complexes do not inhibit calcineurin activity. Instead, the complexes are highly specific inhibitors of the mammalian target of rapamycin (mTOR), a cell cycle serine/threonine kinase involved in the protein kinase B signaling pathway. This results in suppressed cytokine-induced T lymphocyte proliferation, with a block in progression from the G1 to S phase of the cell cycle [1]. The mTOR inhibitors work synergistically with the calcineurin inhibitors to produce a profound immunosuppressive effect on T lymphocytes.

Sirolimus (brand name, Rapamune) is available both as tablet and oral solution for oral administration. According to existing evidence, sirolimus is efficacious as a sole immunosuppressive agent in preventing organ rejection, when administered at doses of 1 to 4.2 mg/day. The product label currently recommends doses of 1–6 mg/day, with up to 7 mg/day administered to those with high immunologic risk [2]. Its long half-life of ~60 h allows once a day dosing. Sirolimus is rapidly absorbed from the gastrointestinal tract and peak blood concentrations occur 1–2 h after an oral dose. Oral bioavailability is low, ranging from 5% to 15% and is considerably reduced (~fivefold) when administered within 4 h or concomitantly with cyclosporine [3]. There is a considerable interpatient variability in total drug exposure that can vary by as much as by 50%. Sirolimus is a hydrophobic compound that is extensively distributed to various organs with an estimated volume of distribution of 7–19 L/kg [4]. Sirolimus is primarily bound within erythrocytes (95%), with approximately 3% and 1% partitioning into the plasma and lymphocytes/granulocytes, respectively. Almost all of the plasma sirolimus is bound to proteins, with albumin, alpha-1-glycoprotein and lipoproteins being the major binding proteins [3].

Similar to the calcineurin inhibitors (cyclosporine and tacrolimus), sirolimus is metabolized in the intestine and liver by cytochrome P450 (CYP) enzymes, especially CYP3A isoenzymes. The multidrug efflux pump P-glycoprotein in the gastrointestinal tract also controls metabolism by regulating bioavailability. Sirolimus is hydroxylated and demethylated to more than seven metabolites with the hydroxyl forms being the most abundant. Metabolites represent approximately 55% of whole blood sirolimus levels [5]. The four major metabolites are 16-*O*-demethyl-sirolimus, 39-*O*-demethyl sirolimus, 27, 39-*O*-demethyl-sirolimus and di-hydroxy-sirolimus. The pharmacological activities of these metabolites are <10% of the parent drugs. Sirolimus is eliminated primarily by biliary and fecal pathways, with small quantities appearing in urine. As with the calcineurin inhibitors, dosage adjustments are needed in patients with hepatic dysfunction [4].

Genetic polymorphisms of CYP3A4 and CYP3A5 isoenzymes significantly affect metabolism of sirolimus. One study demonstrated significant individual differences in pharmacokinetic parameters of sirolimus after a single oral dose (5 mg) in 31 healthy subjects which correlated with genetic polymorphisms of CYP3A4 and CYP3A5. Subjects with CYP3A5*1 and CYP3A4*1G allele carriers required higher doses of sirolimus in order to achieve therapeutic concentration. The authors concluded that TDM should be useful in patients taking sirolimus [6].

The incidence of adverse effects is dose-related and includes metabolic, hematological, and dermatological effects. Metabolic side effects include hypercholesterolemia, hyper- and hypokalemia, hypophosphatemia, hyperlipidemia, and abnormal liver function tests. Cytopenias can be problematic, with decreases in leukocyte, erythrocytes and platelet counts being the most common. Skin rashes, acne and mouth ulcers are also observed in patients being switched to sirolimus. For sirolimus, correlation between trough concentration and adverse effects such as thrombocytopenia, leukopenia and hypertriglyceridemia have been reported. As with other immunosuppressive drugs, there is an increased risk of infection and lymphomas. Interstitial pneumonitis is also associated with sirolimus therapy [3].

EDTA anticoagulated whole blood is the recommended specimen matrix [7]. This is because almost all of the sirolimus (~95%) is concentrated in erythrocytes, and plasma levels are too low for most analytical methods. In contrast to the calcineurin inhibitors, there is good correlation between pre-dose sirolimus concentrations and total drug exposure based on area under the curve measurements [8]. This also holds true when sirolimus is used in combination with cyclosporine or tacrolimus [8]. Thus, whole blood trough specimens are recommended when monitoring sirolimus.

Therapeutic monitoring of sirolimus is critical because the administered dose is a poor predictor of total drug exposure due to individual patient variables. Because of the long drug half-life, daily monitoring of sirolimus is typically not necessary. Weekly monitoring of levels may be needed shortly after transplantation followed by monthly monitoring. Target concentrations for sirolimus range between 5 and 15 ng/mL when used in combination with a calcineurin inhibitor. However, recommended target sirolimus concentration is 12–15 ng/mL up to 3 months post-transplant and which can usually be reduced to 7–10 ng/mL after 3 months. Because toxicity may be encountered at whole blood sirolimus concentration exceeding 15 ng/mL, TDM is needed [9]. When used alone, sirolimus has a higher therapeutic range of 12–20 ng/mL [10].

Sirolimus can be measured by immunoassay or high performance liquid chromatography (HPLC) with ultraviolet (UV) or mass spectrometry (MS) detection. A cloned enzyme donor immunoassay (CEDIA) for sirolimus (Microgenics) is also commercially available for use on several Roche automated analyzers. Chemiluminescent microparticle immunoassay (CMIA) for sirolimus is also more recently available which is replacing old MEIA (microparticle enzyme immunoassay) also marketed by Abbott Diagnostics.

Currently available immunoassays have significant cross-reactivity with sirolimus metabolites. The MEIA has 58% and 63% cross-reactivity with 41-*O*-demethyl-sirolimus and 7-*O*-demethyl-sirolimus, respectively The MEIA produces whole-blood sirolimus concentrations that are 9% to 49% higher, respectively, than those obtained by HPLC-UV and HPLC–MS, depending on the study and type of transplant [11, 12]. The CEDIA has 44% cross-reactivity with 11-hydyroxy-sirolimus and 73% cross-reactivity with 41-*O*- and 32-*O*-demethyl-sirolimus. This degree of metabolite cross-reactivity results in significant assay bias of average 20% compared to LC–MS/MS assay. The CMIA assay had a mean percent bias of 21.9% compared to LC–MS/MS assay [13].

As expected, chromatographic methods are superior to immunoassays for TDM of sirolimus because such methods are free from metabolite cross-reactivities. Taking advantage of UV peak absorbance at 278 nm, sirolimus can be measured in whole blood using HPLC-UV. In one study, authors extracted sirolimus from whole blood and analyzed by HPLC using the mobile phase A (68% methanol/2% acetonitrile/30% water) and mobile phase B (30% methanol/42% acetonitrile/28% water). Samples were analyzed using a C-18 column heated at 50 °C [14]. However, LC–MS/MS based method has significant reproducibility and accuracy advantages compared to both immunoassay and conventional HPLC-UV methods for analysis of sirolimus [15]. Published methods are available for LC–MS/MS assay for sirolimus. Moreover, more than one immunosuppressant can be monitored simultaneously using LC–MS/MS. Yuan et al. described a LC–MS/MS method for analysis of sirolimus and everolimus [16]. Recently various analytical methods available for monitoring of immunosuppressants have been reviewed [17].

References

[1] Kimball PM, Derman RK, Van Buren CT, Lewis RM, Katz S, Kahan BD. Cyclosporine and rapamycin affect protein kinase C induction of intracellular activation signal, activator of DNA replication. Transplantation 1993;55:1128–32.

[2] Zimmerman KO, Wu H, Greenberg R, Guptill JT, et al. Therapeutic drug monitoring, electronic health records, and pharmacokinetic modeling to evaluate sirolimus drug exposure-response relationships in renal transplant patients. Ther Drug Monit 2016;38:600–6.

[3] Yatscoff R, LeGatt D, Keenan R, Chackowsky P. Blood distribution of rapamycin. Transplantation 1993;56:1202–6.

[4] Moes DJ, Guchelaar HJ, de Fijter JW. Sirolimus and everolimus in kidney transplantation. Drug Discov Today 2015;20:1243–9.

[5] Gallant-Haidner HL, Trepanier DJ, Freitag DG, Yatscoff RW. Pharmacokinetics and metabolism of sirolimus. Ther Drug Monit 2000;22:31–5.

[6] Zhang J, Dai Y, Liu Z, Zhang M, et al. Effect of CYP3A4 and CYP3A5 genetic polymorphisms on the pharmacokinetics of sirolimus in healthy Chinese volunteers. Ther Drug Monit 2017;39:406–11.
[7] Yatscoff RW, Boeckx R, Holt DW, Kahan BD, LeGatt DF, Sehgal S, Soldin SJ, Napoli K, Stiller C. Consensus guidelines for therapeutic drug monitoring of rapamycin: report of the consensus panel. Ther Drug Monit 1995;17:676–80.
[8] Kahan BD, Napoli KL, Kelly PA, Podbielski J, Hussein I, Urbauer DL, Katz SH, Van Buren CT. Therapeutic drug monitoring of sirolimus: correlations with efficacy and toxicity. Clin Transplant 2000;14:97–109.
[9] Kahan BD, Keown P, Levy GA, Johnston A. Therapeutic drug monitoring of immunosuppressant drugs in clinical practice. Clin Ther 2002;24:330–50.
[10] Wilson D, Johnston F, Holt D, Moreton M, et al. Multi-center evaluation of analytical performance of the microparticle enzyme immunoassay for sirolimus. Clin Biochem 2006;39:378–86.
[11] Vicente FB, Smith FA, Peng Y, Wang S. Evaluation of an immunoassay of whole blood sirolimus in pediatric transplant patients in comparison with high-performance liquid chromatography/tandem mass spectrometry. Clin Chem Lab Med 2006;44:497–9.
[12] Westley IS, Morris RG, Taylor PJ, Salm P, James MJ. CEDIA® sirolimus assay compared with HPLC–MS/MS and HPLC-UV in transplant recipient specimens. Ther Drug Monit 2005;27:309–14.
[13] Holt DW, Mandelbrot DA, Tortorici MA, Korth-Bradley JM, et al. Long-term evaluation of analytical methods used in sirolimus TDM. Clin Transplant 2014;28:243–51.
[14] Andrade MC, Di Marco GS, Felipe CR, Alfieri F, et al. Sirolimus quantification by high-performance liquid chromatography with ultraviolet detection. Transpl Int 2005;18:354–9.
[15] Sallustio BC, Noll BD, Morris RG. Comparison of blood sirolimus, tacrolimus and everolimus concentrations measured by LC–MS/MS, HPLC-UV and immunoassay methods. Clin Biochem 2011;44:231–6.
[16] Yuan C, Payto D, Gabler J, Wang S. A simple and robust LC–MS/MS method for measuring sirolimus and everolimus in whole blood. Bioanalysis 2014;6:1597–604.
[17] Zhang Y, Zhang R. Recent advances in analytical methods for the therapeutic drug monitoring of immunosuppressive drugs. Drug Test Anal 2018;10:81–94.

13.5

Tacrolimus

Chemical properties	
Solubility in H_2O	Practically insoluble
Molecular weight	822
pKa	n/a
Melting point	126 °C
Dosing	
Recommended dose, adult	0.15–0.30 mg/kg/day (oral) 0.05–0.10 mg/kg/day (IV)
Recommended dose, child	0.30 mg/kg/day (oral) 0.05–0.15 mg/kg/day (IV)
Monitoring	
Sample	Whole blood
Effective concentrations	5–15 ng/mL (trough)
Toxic concentrations	Variable
Methods	Immunoassay, HPLC–MS
Pharmacokinetic properties	
Oral dose absorbed	Variable, ranging from 4% to 93%
Time to peak concentration	0.5–6 h
Protein bound:	69–99% bound to albumin, alpha-1-acid glycoprotein and lipoproteins
Volume of distribution	0.3–2.6 L/kg
Half-life, adult	3.9–34.8 h
Time to steady state, adult	2–6 days
Half-life, child	3.8–11.5 h
Time to steady state, child	?

Excretion, urine

	% Excreted	Active	Detected in blood
Parent	<1	Yes	Yes
Metabolites (≥9)	<2	M-II	Some

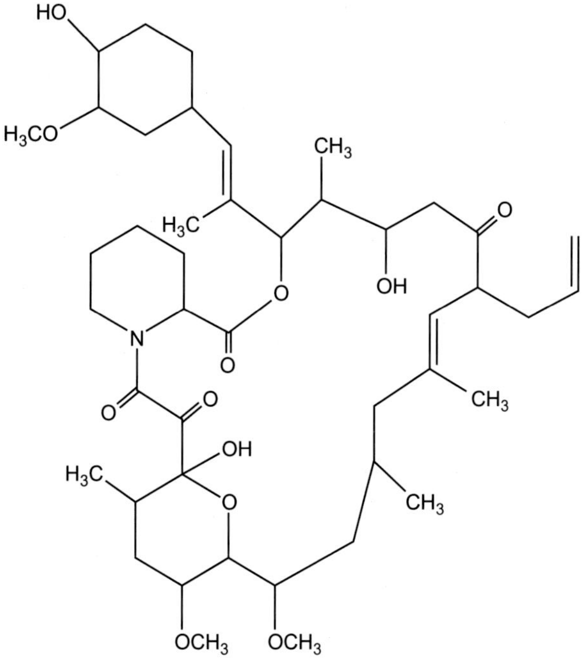

Tacrolimus

Tacrolimus (also known as FK-506) is a macrolide antibiotic that was originally isolated from the fungus *Streptomyces tsukubaensis*. In the United States, tacrolimus (brand name Prograf) was approved for use in liver transplantation in 1994 and in kidney transplantation in 1997. It is approximately 100-times more potent than cyclosporine and is associated with a decrease in acute and chronic rejection, and better long-term graft survival [1]. Tacrolimus is available as 0.5, 1 and 5 mg capsules for oral administration, and as a 5 mg/mL sterile solution for intravenous administration. Tacrolimus forms complexes with the binding protein immunophilin called FK506-binding protein-12 in the cytoplasm of lymphocytes. The complexes block calcineurin activity and prevent de-phosphorylation of the nuclear factor of activated T cells (NF-AT) and subsequent translocation into the nucleus. The end result is down-regulated cytokine gene transcription (interleukin-2 and others) that leads to a block in the activation/proliferation of CD4+ and CD8+ T lymphocytes [2].

Similar to cyclosporine, oral absorption of tacrolimus from the gut is poor and highly variable, averaging 25% [3]. Peak blood concentrations occur within 0.5–6 h. Tacrolimus is primarily bound to albumin, α_1-acid glycoprotein and lipoproteins in the plasma. However, the majority of tacrolimus is found within erythrocytes. The distribution of tacrolimus between plasma and erythrocytes is temperature-dependent and varies with changes in hematocrit. Because of the potential for artifactual redistribution of tacrolimus during specimen processing due to ambient temperature fluctuations, EDTA-anticoagulated whole blood is recommended when measuring tacrolimus concentrations.

Tacrolimus is metabolized by cytochrome P450 (CYP) isoenzymes (predominantly CYP3A) located in the small intestine and liver. Similar to cyclosporine, the bioavailability of tacrolimus is influenced by CYP3A and the multidrug efflux pump (P-glycoprotein) located in intestinal enterocytes. Biotransformation of tacrolimus occurs by demethylation, hydroxylation and oxidative reactions. At least nine metabolites have been identified based on in vitro studies, and all with the exception of 31-*O*-demethyl tacrolimus (M-II) have very little immunosuppressive activity [4]. M-II has been shown in vitro to have the same immunosuppressive activity as the parent compound. Metabolites represent 10–20% of whole blood tacrolimus concentrations. Tacrolimus is eliminated primarily by biliary excretion into the feces. Patients with hepatic dysfunction require dosage adjustments. Very little tacrolimus is found in urine, and blood concentrations are not altered in renal dysfunction.

Polymorphisms of CYP3A5 play a role in pharmacokinetics of tacrolimus. Carriers of CYP3A5 wild type allele (CYP2A5*1) have a higher CYP3A5 expression compared to those patients who are homozygous for non-expression of CYP3A5 activity (CYP3A5*3). Therefore, patients with the CYP3A5*3 allele would exhibit significantly lower tacrolimus clearance than carriers of the wild type gene and would require lower dosage of tacrolimus than patients with normal activity of CYP3A5 [5].

Trough blood tacrolimus concentrations are almost exclusively used for routine monitoring and are believed to be a good indicator of total drug exposure [6]. Initial therapeutic range of tacrolimus was suggested to be 10–25 ng/mL, but levels of approximately 20 ng/mL or higher were found to be frequently associated with toxicity. The therapeutic range of tacrolimus as 5–15 ng/mL has been widely accepted, but lower limits have not been defined clearly [7].

Tacrolimus shares many dose-dependent side effects with cyclosporine. These include nephrotoxicity, neurotoxicity, hepatotoxicity, hypertension, and glucose intolerance. Nephrotoxicity with tacrolimus may be less of a problem than with cyclosporine, especially in renal transplantation.

Diabetogenesis is approximately 3-times more common with tacrolimus than cyclosporine. Hyperkalemia, hyperuricemia, hyperlipidemia, hirsutism, and gingival hypertrophy are also observed following tacrolimus use, but less commonly than with cyclosporine. Alopecia is also associated with tacrolimus use [8].

Monitoring of tacrolimus is an integral part of any organ transplant program because of variable dose-to-blood concentrations and the narrow therapeutic index. Whole blood tacrolimus levels can be measured by chromatography preferably liquid chromatography combined with tandem mass spectrometry (LC–MS/MS). Tacrolimus does not have a suitable ultraviolet (UV) absorption peak and as a result high performance liquid chromatography combined with UV detection cannot be used for TDM of tacrolimus. Semi-automated immunoassays requiring specimen pretreatment are available from Abbott, Siemens and Thermo-Fischer. The Dimension antibody conjugated magnetic immunoassay (ACMIA' marketed by Siemens) is fully-automated and does not require a pretreatment step, allowing whole blood samples to be directly placed on the instrument. Several studies have reported false positive tacrolimus concentrations in patients with low hematocrit values and high imprecision at tacrolimus value <9 ng/mL with the microparticle enzyme immunoassay (MEIA) tacrolimus assay for application on the AxSYM platform (Abbott Laboratories) [9]. Westely et al. evaluated CEDIA tacrolimus assay by measuring values obtained by CEDIA assay, LC–MS/MS and MEIA assay. The authors observed a 33.1% bias with the CEDIA assay and 20.1% bias with the MEIA assay in tacrolimus values compared to the LC–MS/MS method in renal transplant recipients [10]. Bazin et al. evaluated CMIA architect tacrolimus assay and observed an average bias of 20% between values determined by the CMIA assay and LC–MS/MS [11].

Although ACMIA tacrolimus assay does not require pre-treatment of specimen prior to analysis, it is the only assay which is affected by endogenous antibodies including heterophilic antibodies. Altinier et al. described the interference of heterophilic antibody in the ACMIA tacrolimus assay. Sample of a patient showed tacrolimus values in the range of 49–12.5 ng/mL even after interruption of the treatment. The authors confirmed that the elevated tacrolimus levels were due to the presence of heterophilic antibody by treating samples with heterophilic blocking tubes and protein G resin that removed such interference [12]. Another report showed a high tacrolimus value (79.7 ng/mL) in a liver transplant recipient using ACMIA tacrolimus assay despite discontinuation of tacrolimus therapy. The authors identified beta-galactosidase antibodies as the cause of interference because in this assay anti-tacrolimus antibody is conjugated to beta-galactosidase [13]. Rostaing et al. observed a patient with falsely elevated tacrolimus level of 24 ng/mL using the ACMIA tacrolimus assay but not detectable drug using either LC/MS/MS or an EMIT tacrolimus immunoassay. The authors identified positive anti-double stranded DNA autoantibodies as the cause of interference in the ACMIA assay [14].

LC–MS/MS assays are considered as the "gold standard" for TDM of tacrolimus. Recently Mei et al. reported a LC–MS/MS assay for analysis of cyclosporine and tacrolimus in whole blood and commented that average positive bias with CMIA cyclosporine assay was 53.7% and average positive bias was 48.1% for CMIA tacrolimus assay [15]. These positive biases are higher than reported by other authors.

References

[1] First MR. Tacrolimus based immunosuppression. J Nephrol 2004;17:25–31.
[2] Jorgensen KA, Koefoed-Nielsen PB, Karamperis N. Calcineurin phosphatase activity and immunosuppression. A review on the role of calcineurin phosphatase activity and the immunosuppressive effect of cyclosporin A and tacrolimus. Scand J Immunol 2003;57:93–8.
[3] Venkataramanan R, Swaminathan A, Prasad T, Jain A, Zuckerman S, Warty V, McMichael J, Lever J, Burckart G, Starzl T. Clinical pharmacokinetics of tacrolimus. Clin Pharmacokinet 1995;29:404–30.
[4] Kelly P, Kahan BD. Review: metabolism of immunosuppressant drugs. Curr Drug Metab 2002;3:275–87.
[5] Coto E, Tavira B. Pharmacogenetics of calcineurin inhibitors in renal transplantation. Transplantation 2009;88(Suppl):S62–7.
[6] Holt DW. Therapeutic drug monitoring of immunosuppressive drugs in kidney transplantation. Curr Opin Nephrol Hypertens 2002;11:657–63.
[7] McMaster P, Mirza DF, Ismail T, Vennarecci G, et al. Therapeutic drug monitoring of tacrolimus in clinical transplantation. Ther Drug Monit 1995;17:602–5.
[8] Staatz CE, Tett SE. Clinical pharmacokinetics and pharmacodynamics of tacrolimus in solid organ transplantation. Clin Pharmacokinet 2004;43:623–53.
[9] Armedariz Y, Garcia S, Lopez R, Pou L, et al. Hematocrit influences immunoassay performance for the measurement of tacrolimus in whole blood. Ther Drug Monit 2005;27:766–9.
[10] Westley IS, Taylor PJ, Salm P, Morris RG. Cloned enzyme donor immunoassay tacrolimus assay compared with high-performance liquid chromatography-tandem mass spectrometry in liver and renal transplant recipients. Ther Drug Monit 2007;29: 584–91.
[11] Bazin C, Guinedor A, Barau C, Gozalo C, et al. Evaluation of the Architect tacrolimus assay in kidney, liver and heart transplant recipients. J Pharm Biomed Anal 2010;53:997–1002.
[12] Altinier S, Varagnolo M, Zaninotto M, Boccagni P, et al. Heterophilic antibody interference in a non-endogenous molecule assay: an apparent elevation in the tacrolimus concentartion. Clin Chim Acta 2009;402:193–5.

[13] Knorr JP, Grewal KS, Balasubramanian M, Zaki R, et al. Falsely elevated tacrolimus levels caused by immunoassay interference secondary to beta-galactosidase antibodies in an infected liver transplant recipient. Pharmacotherapy 2010;30:954.
[14] Rostaing L, Cointault O, Marquet P, Josse AG, et al. Falsely elevated whole blood tacrolimus concentrations in a kidney transplant patient: potential hazards. Transpl Int 2010;23:227–30.
[15] Mei S, Wang J, Chen D, Zhu L, et al. Simultaneous determination of cyclosporine and tacrolimus in human whole blood by ultra-high performance liquid chromatography tandem mass spectrometry and comparison with a chemiluminescence microparticle immunoassay. J Chromatogr B Analyt Technol Biomed Life Sci 2018;1087–1088: 36–42.

CHAPTER 14
Analgesics

14.1
Acetaminophen

Chemical properties	
Solubility in H_2O	Very slight with heat
Molecular weight	151.16
pKa	9.5
Melting point	109–170 °C
Dosing	
Recommended dose, adult	Single dose of 325 or 500 mg; maximum single dose of 1000 mg. Daily acetaminophen dose should not exceed 4 g
Recommended dose, child	10–15 mg/kg
Monitoring	
Sample	Serum, plasma (heparin, EDTA)
Effective concentrations, adult	10–20 μg/mL
Effective concentrations, child	10–20 μg/mL
Toxic concentrations	>200 μg/mL Use Rumack-Matthew nomogram 4 h post ingestion
Methods	GC, GC/MS, HPLC, immunoassay, LC/MS
Pharmacokinetic properties	
Oral dose absorbed	99%
Time to peak concentration	0.5–2.0 h
Protein bound	20–50%
Volume of distribution	0.9 L/kg
Half-life, adult	1.9–2.5 h
Time to steady state, adult	10–20 h
Half-life, child	1–3 h
Time to steady state, child	10–20 h

Therapeutic Drug Monitoring Data
https://doi.org/10.1016/B978-0-12-815849-4.00014-1

Excretion, urine

	% Excreted	Active	Detected in blood
Parent	4–14	Yes	Yes
glucuronide	30–60	No	Yes
Sulfate	9–23	No	Yes
Cysteinate or mercaptate	14–45	No	Yes

Acetaminophen

Acetaminophen (*N*-acetyl-*p*-aminophenol; Paracetamol, Tylenol) was introduced in the United States in 1960 as a nonprescription analgesic and antipyretic. Acetaminophen is available in several formulations (tablets, suppositories, suspensions). This drug now plays an important role in American health care, with >25 billion doses being used annually as a nonprescription medication. Additionally, over 200 million acetaminophen-containing prescriptions, usually in combination with an opioid (e.g., hydrocodone-acetaminophen; Percocet, Vicodin), are dispensed annually. Although acetaminophen is a relatively safe drug, hepatotoxicity at a higher dose has been reported. The FDA monograph specifies that single-ingredient acetaminophen-containing products that contain 325 mg should be administered in a dose of 325–650 mg every 4 h while symptoms persist, not to exceed 3900 mg in 24 h for not more than 10 days (approved July 8, 1977). Products that contain 500 mg should follow the dosing regimen of adult doses up to 1000 mg, not to exceed 4000 mg in 24 h (approved November 16, 1988). However, more recently, the upper end of acetaminophen dose has been reduced. In 2011, McNeil, the producer of the Tylenol brand of acetaminophen, voluntarily reduced the maximum daily dose of its 500 mg tablet product to 3000 mg/day, and it has pledged to change the labeling of its 325 mg/tablet product to reflect a maximum of 3250 mg/day. Generic manufacturers have not changed their dosing regimens, and they have remained consistent with the established monograph dose of maximum 4 g of acetaminophen per day. Therefore, confusion will be inevitable as both consumers and health care professionals try to determine the proper therapeutic and maximum doses of acetaminophen [1]. Moreover, lack of patient education is another important public health issue because in one study the authors reported that only 25% patients surveyed correctly identified maximum daily dose of acetaminophen as 4g (4000mg) [2].

Acetaminophen is the drug of choice in patients that cannot be treated with non-steroidal anti-inflammatory drugs (NSAID), such as people with bronchial asthma, peptic ulcer disease, hemophilia, salicylate-sensitized people, and children under 12 years of age as well as pregnant or breastfeeding women. It is recommended as a first-line treatment of pain associated with osteoarthritis. The mechanism of action is complex and includes the effects of both peripheral (cyclooxygenase inhibition) and central (cyclooxygenase, serotonergic descending neuronal pathway, L-arginine/nitric oxide pathway, cannabinoid system) antinociception processes and "redox" mechanisms. The antipyretic effect of acetaminophen is mostly due to inhibition of cyclooxygenase-2 effectively in the hypothalamus, thus preventing stimulation of the heat-regulating center. Similarly, a reduction in prostaglandin production in neurons and the vascular endothelium accounts for the drug's analgesic activity. In contrast to salicylates, acetaminophen does not possess anti-inflammatory or anti-thrombotic activity, and is therefore not a NSAID [3, 4].

Acetaminophen is a slightly water-soluble, bitter tasting white crystalline compound that is most often ingested orally. Following oral administration it is rapidly absorbed from the gastrointestinal (GI) tract, its systemic bioavailability being dose-dependent and ranging from 70% to 90%. Its rate of oral absorption is predominantly dependent on the rate of gastric emptying. Therefore, oral absorption of acetaminophen is delayed by food. It distributes rapidly and evenly throughout most tissues and fluids and has a volume of distribution of approximately 0.9 L/kg and approximately 10% to 20% of the drug is bound to red blood cells. The plasma half-life ranges from 1.9 to 2.5 h and the total body clearance from 4.5 to 5.5 mL/kg/min. The plasma half-life is usually normal in patients with mild chronic liver disease, but it is prolonged in those with decompensated liver disease. Acetaminophen is extensively metabolized by the liver, with the major metabolites being the glucuronide (approximately 55% of the dose recovered in urine) and sulfate (30% of the dose). Unchanged drug represents only 4% of the drugs excreted in the urine. A minor fraction of drug is converted to a highly reactive alkylating hepatotoxic metabolite *N*-acetyl-p-benzoquinone imine (NAPQI) which is inactivated with reduced glutathione and excreted in the urine as cysteine (approximately 4%) and mercapturic acid conjugates (approximately 4%). Severe overdose of acetaminophen may cause acute hepatic necrosis due to depletion of glutathione and damage from the reactive NAPQI metabolite. This damage can be prevented by the early administration of sulfhydryl compounds such as methionine and *N*-acetylcysteine [5].

Several studies have demonstrated that up to 5 days of therapeutic doses of oral acetaminophen (4 g per day) are safe even when administered to high risk patients. However, Seifert et al. reported severe acetaminophen toxicity in a 92-year-old, 68 kg woman without known hepatic disease or ethanol abuse. She was treated with 1 g of acetaminophen every 6 h administered intravenously. Her liver enzymes were normal at the initiation of acetaminophen therapy but on day five her aspartate aminotransferase (AST; 4698 U/L: normal range; <35 U/L) and alanine aminotransferase (ALT; 3914 U/L: normal range < 45 U/L) enzymes were significantly elevated along with elevated total bilirubin (1.8 mg/dL: normal range: 0.2–1.2 mg/dL). However, serum acetaminophen concentration was 15.3 μg/mL 26 h after her last dose which was within therapeutic range of 10–20 μg/mL. In contrast, acetaminophen-protein adducts (APAP-CYS), a specific marker of acetaminophen toxicity which is formed due to binding of the reactive metabolite of acetaminophen with hepatic cellular protein was significantly elevated to 4.81 μmol/L (therapeutic <1.1 μmol/L). She was treated with *N*-acetylcysteine and her hepatotoxicity resolved. She was discharged from the hospital 2 days later with APAP-CYS concentration measured by liquid chromatography combined with mass spectrometry (LC/MS/MS) reduced to normal level (0.95 μmol/L). The authors concluded that this was an unusual case of acetaminophen toxicity occurring at the traditional maximum recommended dose for a patient. One of the factors that may contribute to such toxicity is her advanced age [6].

Usually it has been accepted that acute acetaminophen hepatotoxicity is associated with dosage exceeding 150 mg/kg per day or 10 g. However, alcoholics are at higher risk of acetaminophen toxicity due to potential depletion of reduced glutathione and may experience acetaminophen toxicity even after ingestion of therapeutic dosage [7]. The patient who has been overdosed with acetaminophen may exhibit symptoms of gastric irritation, lethargy, and diaphoresis in the first 24 h, and then progress to renal failure, myocardial necrosis, pancreatic inflammation, methemoglobinemia, and hepatotoxicity. Hepatotoxicity is often delayed, corresponding to a dramatic rise in AST and ALT, which may not become evident until 3–4 days following ingestion. Hypoglycemia and thrombocytopenia may also occur [8].

Monitoring of acetaminophen is seldom necessary during therapeutic administration; however, serum concentrations are helpful in predicting the probability of hepatotoxicity following an overdose, especially if the time since ingestion is known. To assess the likelihood of hepatotoxicity, a plasma specimen must be obtained between 4 and 24 h postingestion and the concentration interpreted using a modified Rumack-Matthew Nomogram. Alternatively, multiple samples may be collected and the half-life calculated. The antidote for acetaminophen poisoning, *N*-acetyl-L-cysteine, is most effective if administered within 8–10 h of the ingestion [9]. In general, if serum acetaminophen concentration 4 h after ingestion is >150 μg/mL, administration of antidote is recommended. However, there are no clear guidelines for interpreting acetaminophen levels obtained between 2 and 4 h post ingestion. Yarema et al. commented that an acetaminophen concentration cutpoint of 100 μg/mL (662 μmol/L) at 2–4 h after an acute ingestion as a threshold for repeat testing and/or treatment would occasionally miss potentially toxic exposures [10]. Immunoassays are commercially available for measuring serum or plasma acetaminophen concentrations.

References

[1] Krenzelok EP, Royal MA. Confusion: acetaminophen dosing changes based on NO evidence in adults. Drugs RD 2012;12:45–8.

[2] Herndon CM, Dankenbring DM. Patient perception and knowledge of acetaminophen in a large family medicine service. J Pain Palliat Care Pharmacother 2014;28:109–16.

[3] Jóźwiak-Bebenista M, Nowak JZ. Paracetamol: mechanism of action, applications and safety concern. Acta Pol Pharm 2014;71:11–23.

[4] Graham GG, Scott KF. Mechanism of action of paracetamol. Am J Ther 2005;12:46–55.

[5] Forrest JA, Clements JA, Prescott LF. Clinical pharmacokinetics of paracetamol. Clin Pharmacokinet 1982;7:93–107.

[6] Seifert SA, Kovnat D, Anderson VE, Green JL, et al. Acute hepatotoxicity associated with therapeutic doses of intravenous acetaminophen. Clin Toxicol (Phila) 2016;54:282–5.

[7] Tanaka E, Yamazaki K, Misawa S. Update: the clinical importance of acetaminophen hepatotoxicity in non-alcoholic and alcoholic subjects. J Clin Pharm Ther 2000;25:325–32.

[8] Sheen C, Dillon J, Bateman D, Simpson K, Macdonald T. Paracetamol toxicity: epidemiology, prevention and costs to the health-care system. QJM 2002;95(9):609–19.

[9] Jones AL. Mechanism of action and value of N-acetylcysteine in the treatment of early and late acetaminophen poisoning: a critical review. J Toxicol Clin Toxicol 1998;36:277–85.

[10] Yarema MC, Green JP, Sivilotti M, Johnson DW, et al. Can a serum acetaminophen concentration obtained less than 4 hours post-ingestion determine which patients do not require treatment with acetylcysteine? Clin Toxicol (Phila) 2017;55:102–8.

14.2

Acetylsalicylic acid

Chemical properties	
Solubility in H_2O	0.33 mg/mL
Molecular weight	180.15
pKa	3.5
Melting point	135 °C
Dosing	
Recommended dose, adult	30–70 mg/kg/d
Recommended dose, child	14–25 mg/kg/d
Monitoring	
Sample	Serum, plasma (EDTA)
Effective concentrations, adult	20–100 μg/mL
Effective concentrations, child	20–100 μg/mL
Toxic concentrations	>300 μg/mL
Methods	GC, HPLC, immunoassay
Pharmacokinetic properties	
Oral dose absorbed	80–100%
Time to peak concentration	1–2 h
Protein bound	50–90%
Volume of distribution	0.15–0.20 L/kg
Half-life, adult	2–3 h[a]
Time to steady state, adult	10–22.5 h
Half-life, child	2–3 h[a]
Time to steady state, child	10–15 h

Excretion, urine

	% Excreted	Active	Detected in blood
Parent	10–80	Yes	Yes
Salicylic acid	5–15	Yes	Yes
Salicyluric acid glucuronide	15–40	No	No
Gentisinic acid:	<1	No	Trace

Effective levels as anti-inflammatory: 100–250 μg/mL in adults and children.
Overdose treatment: charcoal, bicarbonate; hemodialysis in severe cases.
[a] Elimination half-life increased with high doses (5–6 h with 1000 mg)

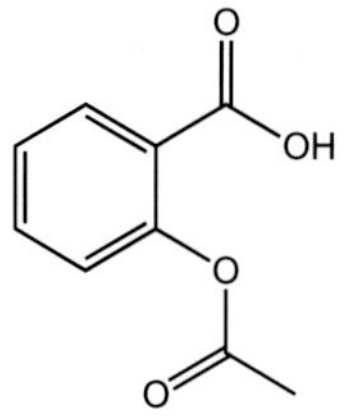

Aspirin

Acetylsalicylic acid (aspirin; 2-acetoxybenzoic acid) was one of the first medications to be widely used as an anti-inflammatory and analgesic. Its origins are reputed to date back centuries and to many cultures. The discovery of aspirin stretches back more than 3500 years to when bark from the willow tree was used as a pain reliever and antipyretic. In 1831, Joanna Pagenstecher, a Swiss pharmacist, prepared a distillate from the flowers of the meadowsweet plant (*Spiraea ulmaria*). In 1835, Karl Lowig took Pagenstecher's aldehyde distillate and prepared an acid which he called spirsaure. Then in 1853, Charles Frederick Gerhard added an acetyl group to Lowig's spirsaure thus forming acetylsalicylic acid. Salicylic acid itself is used only externally, as it is extremely irritating. Various derivatives, primarily salicylic acid esters (e.g., methyl salicylate) and salicylate esters of organic acids (acetylsalicylic acid or aspirin), have been developed for internal use. We now know that Lowig's acid is salicylic acid, but the name "aspirin" remains in common usage. The "a" in aspirin refers to the acetyl group, and the "spir" refers to "spirsaure", which derives its name from the genus of the meadowsweet flower (*Spiraea*). Several plants contain esters of salicylic acid, and as early as 1763, an English clergyman, Edward Stone, reported the antipyretic effects of a preparation from white willow bark (*Salix alba*) which contained salicylic acid. Since those early reports, aspirin has gained wide acceptance as an antipyretic, analgesic, and anti-rheumatic [1].

Acetylsalicylic acid is widely used to relieve low-intensity pain such as headache, myalgia, arthralgia, toothache, and dysmenorrhea. It is also an effective antipyretic. Due to its anti-inflammatory activity, aspirin remains one of the most effective agents for the treatment of rheumatic diseases. Acetylsalicylate in the US are available in 81, 162, 325, and 500 mg dosage. Low dose of 81 mg/day is used for reducing risk of stroke and myocardial infarction, while daily doses as high as 5–6 g have been prescribed for the treatment of rheumatoid arthritis. Aspirin therapy in children is extremely limited because of the risk of Reye's syndrome, a rare syndrome with fulminant liver swelling and encephalopathy. Acetaminophen is the preferred antipyretic for children.

Acetylsalicylate is slightly soluble in water. It is well absorbed after oral administration. Acetylsalicylic acid is metabolized by esterases in GI mucosa, red blood cells, synovial fluid, and blood to salicylates. Pharmacological activities are mostly due to salicylate. The elimination half-life of acetylsalicylate in plasma is approximately 15–20 min. Only about 1% of an oral dose is excreted unchanged in urine. Salicylates are metabolized in the liver and the metabolic products are excreted in the urine; these metabolic pathways are saturable. The metabolites include salicyluric acid, salicyl phenolic glucuronide, salicyl glucuronide, gentisic acid, and gentisuric acid [2, 3].

The clinical pharmacology of aspirin is complex. It inhibits the synthesis of prostaglandins in inflamed tissues preventing sensitization of peripheral pain receptors. The antipyretic effect also is related to an inhibition of prostaglandin synthesis, especially in the hypothalamus. Likewise, the inhibition of prostaglandin synthase prevents the production of thromboxane A2, thereby inhibiting platelet aggregation [4].

GI toxicity is a major problem of prolonged acetylsalicylate therapy. However, in some patients antiplatelet based therapy is needed. In most cases, the antiplatelet regimen is based on low-dose aspirin, a drug that is highly effective in reducing the incidence of cardiovascular events. Dyspeptic symptoms, which may occur with or without associated ulceration and bleeding, may lead patients to discontinue therapy, thus increasing their risk of cardiovascular disease. For patients in whom aspirin is indicated and who are deemed to be at increased risk of upper GI events, concomitant therapy with a proton pump inhibitor (PPI) is currently recommended. These agents are highly effective in reducing the upper GI lesions associated with aspirin therapy and have been associated with increased aspirin adherence [5].

Therapy with acetylsalicylate is measured using serum or plasma salicylate levels for which immunoassays are commercially available. In general, serum salicylate concentrations are measured in patients suspected with salicylate toxicity. It is usually accepted that for analgesic and antipyretic effect serum salicylate levels of below 100 μg/mL are sufficient, but anti-inflammatory effect may require levels of 150–300 μg/mL (15–30 mg/dL). Therapeutic drug monitoring may be initiated for patients taking higher doses of salicylate for pain control for conditions such as rheumatoid arthritis. Prolonged, higher doses are associated with chronic salicylate toxicity characterized by tinnitus, lassitude, and other mental status changes. The most commonly encountered side-effects with therapeutic doses are dizziness, tinnitus, epigastric distress, nausea, and prolonged bleeding time.

Aspirin is contraindicated for patients with severe hepatic damage, vitamin K deficiency, hypoprothrombinemia, hemophilia, or peptic ulcer disease. Therapeutic doses of aspirin have been known to cause hepatic damage. Patients are usually asymptomatic but have elevated AST and ALT. About 5% of patients will also develop anorexia, hepatomegaly, and/or jaundice. Severe salicylate toxicity is initially treated with sodium bicarbonate to render urine alkaline (pH 7.5 or higher) for higher renal excretion and supportive therapy but hemodialysis may also be required for some patients [6].

Severe toxicity resulting in death has been reported following ingestions of 10 g by an adult. According to the 2016 report from the United States National Poison Data System, salicylate exposure occurred 19,401 times with 22 deaths reported [7]. Salicylate toxicity may occur when salicylate level exceeds 300 μg/mL within 4 h of ingestion. However, in one study, the authors observed that a small but significant proportion (3.5%) of patients who developed acetylsalicylate toxicity (salicylate level >30 mg/dL or 300 μg/mL within 4 h of ingestion) had an initially undetectable levels because levels were measured too early after ingestion [8].

References

[1] Desborough MJR, Keeling DM. The aspirin story—from willow to wonder drug. Br J Haematol 2017;177:674–83.

[2] Trinder P. Rapid determination of salicylate in biological materials. Biochem J 1954;57:301.

[3] Dressman JB, Nair A, Abrahamsson B, Barends DM, et al. Biowaiver monograph for immediate-release solid oral dosage forms: acetylsalicylic acid. J Pharm Sci 2012;101:2653–67.

[4] Ghooi RB, Thatte SM, Joshi PS. The mechanism of action of aspirin—is there anything beyond cyclo-oxygenase? Med Hypotheses 1995;44:77–80.

[5] Lavie CJ, Howden CW, Scheiman J, Tursi J. Upper gastrointestinal toxicity associated with long-term aspirin therapy: consequences and prevention. Curr Probl Cardiol 2017;42:146–64.

[6] Pearlman BL, Gambhir R. Salicylate intoxication: a clinical review. Postgrad Med 2009;121:162–8.

[7] Gummin DD, Mowry JB, Spyker DA, et al. 2016 Annual report of the American Association of Poison Control Centers' National Poison Data System (NPDS): 34th annual report. Clin Toxicol 2017;55:1072–252.

[8] Moss MJ, Fisher JA, Kenny TA, Palmer AC, et al. Salicylate toxicity after undetectable serum salicylate concentration: a retrospective cohort study. Clin Toxicol (Phila) 2019;57:137–40.

14.3

Ibuprofen

Chemical properties	
Solubility in H_2O	100 mg/mL, as sodium salt
Molecular weight	206.3
pKa	4.4
Melting point	76 °C
Dosing	
Recommended dose, adult	200–800 mg, 2–4 times/day
Recommended dose, child	5–30 mg/kg/day
Monitoring	
Sample	Serum, plasma
Effective concentrations	15–30 μg/mL
Toxic concentrations	Inconsistent findings
Methods	HPLC, GC
Pharmacokinetic properties	
Oral dose absorbed	85%
Time to peak concentration	1–2 h
Protein bound	95%
Volume of distribution	0.11 L/kg
Half-life, adult	2–4 h
Time to steady state, adult	24–48 h
Half-life, child	1–3 h
Time to steady state, child	24–48 h

Excretion, urine

	% Excreted	Window of excretion	Active	Detected in blood
Parent	<10	24 h	Yes	Yes
2-Hydroxyibuprofen	35	24 h	No	No
2-Carboxyibuprofen	25	24 h	No	No

OH
O
Ibuprofen

Ibuprofen, a chiral compound derived from arylpropionic acid, is a non-narcotic, non-steroidal anti-inflammatory drug (NSAID) used for the treatment of minor aches and pains, reducing fever, and relieving symptoms of dysmenorrhea. It is also used for pain control in inflammatory diseases such as rheumatoid arthritis, osteoarthritis and ankylosing spondylitis. It has also been found effective in inducing closure of a patent ductus arteriosus (PDA) in neonates about the second day of postnatal life when early or therapeutic PDA closure is needed. It has been available as a prescription drug since 1968 and as an over the counter drug since 1984. However, ibuprofen should not be administered in children <12 years old without consultation with a physician [1, 2]. Ibuprofen is believed to exert much of its therapeutic effect by inhibiting cyclooxygenase (COX) enzymes, which in turn inhibits prostacyclin production.

Ibuprofen is available as a single formulation as a racemic mixture or in combination with other drugs such as pseudoephedrine and hydrocodone. The general recommended dose when used as an over the counter product is usually 200 mg. At low doses (up to 800–1200 mg/day) ibuprofen has a good safety profile comparable to acetaminophen. Its analgesic activity is linked to its anti-inflammatory effects and is related to reduction in the production in blood of COX-1 and COX-2 derived prostanoids. Higher prescription doses (1800–2400 mg/day) are employed long-term for the treatment of rheumatic and other more severe musculoskeletal conditions. High-dose ibuprofen has also been found to be effective in slowing the progression of lung disease in children with cystic fibrosis. Typical doses of ibuprofen used in this setting are 20–30 mg/kg, with target serum/plasma concentrations of 50–100 μg/mL [3]. Higher doses ibuprofen use should be medically supervised.

After oral administration, the drug is well absorbed, and peak concentrations are usually achieved within 1–2 h. In the circulation, ibuprofen is strongly protein bound. The half-life of ibuprofen is approximately 1–4 h, but is prolonged in patients with liver cirrhosis. Ibuprofen is extensively metabolized by liver cytochrome P450 (CYP) enzymes. These principally involve cytochrome CYP2C9, CYP2C8 and 2C19, catalyzing the oxidation of the alkyl side chain to hydroxyl and carboxyl derivatives. The major metabolites, 2-hydroxyibuprofen and 2-carboxyibuprofen, are excreted by the kidneys [4].

A number of side effects have been reported following ibuprofen use. Those considered less serious include nausea, epigastric pain, diarrhea, vomiting, dizziness, blurred vision, tinnitus, and edema. More severe reactions include anaphylactic reactions, GI bleeding, seizures, metabolic acidosis, hypotension, bradycardia, tachycardia, atrial fibrillation, coma, hepatic dysfunction, acute renal failure, and cardiac arrest. Despite these possible adverse effects, ibuprofen is considered less toxic compared to salicylates and acetaminophen. Moreover, there are reports of massive overdoses (>8 g) of ibuprofen with little or no clinical effects. The toxic effects of ibuprofen are primarily related to the effects related to accumulation of the two acidic metabolites, 2-hydroxyibuprofen and 2-carboxylibuprofen. However, the great majority of patients suffer no or only mild symptoms. In one series of 1033 enquiries involving ingestion of ibuprofen alone, 705 (65%) patients were asymptomatic; 199 (18%) experienced mild symptoms; and 23 (2%) experienced moderate symptoms. There are only seven case reports of fatal overdose with ibuprofen and in each case there were complicating factors related to other drugs and/or other disease processes. The management of ibuprofen overdose is generally straightforward and can be related to the dose ingested. In addition, toxicity associated with modified release formulations are less compared to the conventional tablets which immediately releases ibuprofen after oral administration. There is at present no reason to be concerned that co-ingestion of ethanol increases the risk of toxicity from ibuprofen overdose [5].

The treatment of ibuprofen toxicity includes gastric decontamination with activated charcoal, and gastric lavage is used in severe cases of severe overdose. Emesis is not recommended, and forced alkaline diuresis is of limited value. Ibuprofen drug levels are more useful in predicting the toxicity than the ingested dose. Unfortunately, unlike acetaminophen and aspirin, few clinical laboratories perform due to its overall safety and unavailability of rapid immunoassays. The nomogram developed by Hall et al. may be helpful in treating overdoses, though the levels may not correlate well with the symptoms [6].

Although ibuprofen is not routinely monitored, therapeutic range is considered as 15–30 μg/mL. Correlation between ibuprofen serum levels with clinical outcome is quite inconsistent. A relatively mild clinical picture despite a very high serum ibuprofen concentration of 1034 μg/mL has been reported, whereas another case with ibuprofen serum concentration of 1050 μg/mL had fatal outcome. Due to high protein binding (>95%), ibuprofen cannot be removed by hemodialysis. However, total plasma exchange is capable of removing highly protein bound drugs from circulation. In one report, the authors described a case of a 48 year-old male with suicidal mono-ingestion of approximately 72 g ibuprofen. Despite an initial rapid spontaneous drop in the total ibuprofen plasma concentration from 550 (4 h after ingestion) to 275 μg/mL within the first 5 h after admission, the patient developed a circulatory failure, refractory to aggressive fluid resuscitation and high doses of vasopressors. The authors used therapeutic plasma exchange (TPE) for extracorporeal elimination of ibuprofen. The ibuprofen serum level was reduced from 275 to 180 μg/mL after TPE, but then unexpectedly increased to 320 μg/mL 10 h after TPE and slowly decreased within the following 20 h to 100 μg/mL. Ibuprofen was completely removed at day 5 after multiple TPE procedures with terminal half-life of ibuprofen estimated to be 17.2 h which is substantially longer than ibuprofen half-life after therapeutic dosage [7].

Methods for analysis of ibuprofen include HPLC and GC. HPLC methods involving liquid phase or solid phase extraction with ultraviolet detection have been described. Reverse phase HPLC, as compared to normal phase, is a common method [8]. Several GC–MS methods involving derivatization have been also described [9]. More recently, liquid chromatography combined with mass spectrometry has also been developed for determination of ibuprofen concentration in human serum [10].

References

[1] Aycock DG. Ibuprofen: a monograph. Am Pharm 1991;31:46–9.
[2] Aranda JV, Thomas R. Systematic review: intravenous Ibuprofen in preterm newborns. Semin Perinatol 2006;30:114–20.
[3] Lands LC, Dauletbaev N. High-dose ibuprofen in cystic fibrosis. Pharmaceuticals (Basel) 2010;3:2213–24.
[4] Rainsford KD. Ibuprofen: pharmacology, efficacy and safety. Inflammopharmacology 2009;17:275–342.
[5] Volans G, Monaghan J, Colbridge M. Ibuprofen overdose. Int J Clin Pract Suppl 2003;135:54–60.
[6] Hall AH, Smolinske SC, Stover B, Conrad FL, Rumack BH. Ibuprofen overdose in adults. J Toxicol Clin Toxicol 1992;30:23–37.
[7] Geith S, Renner B, Rabe C, Stenzel J, Eyer F. Ibuprofen plasma concentration profile in deliberate ibuprofen overdose with circulatory depression treated with therapeutic plasma exchange: a case report. BMC Pharmacol Toxicol 2017;18(1):81.
[8] Canaparo R, Muntoni E, Zara GP, Della Pepa C, Berno E, Costa M, et al. Determination of Ibuprofen in human plasma by high-performance liquid chromatography: validation and application in pharmacokinetic study. Biomed Chromatogr 2000;14:219–26.
[9] Way BA, Wilhite TR, Smith CH, Landt M. Measurement of plasma ibuprofen by gas chromatography-mass spectrometry. J Clin Lab Anal 1997;11:336–9.
[10] Nakov N, Petkovska R, Ugrinova L, Kavrakovski Z, et al. Critical development by design of a rugged HPLC-MS/MS method for direct determination of ibuprofen enantiomers in human plasma. J Chromatogr B Anal Technol Biomed Life Sci 2015;992:67–75.

14.4

Indomethacin

Chemical properties	
Solubility in H_2O	~ 1 mg/L
Molecular weight	357.8
pKa	4.5
Melting point	158 °C
Dosing	
Recommended dose, adult	75–150 mg/day divided into 2–3 doses
Recommended dose, child	1–2 mg/kg/day
Monitoring	
Sample	Serum, plasma
Effective concentrations	0.3–3.0 μg/mL
Toxic concentrations	>5 μg/mL
Methods	HPLC, GC
Pharmacokinetic properties	
Oral dose absorbed	~100%
Time to peak concentration	2 h
Protein bound	99%
Volume of distribution	0.34–1.57 L/kg
Half-life, adult	2–6 h
Time to steady state, adult	24–48 h
Time to steady state, child	120 h

Excretion, urine

	% Excreted	Active	Detected in blood
Parent	<10	Yes	Yes
Glucuronide	60	No	No
Desmethylindomethacin	<5	No	No

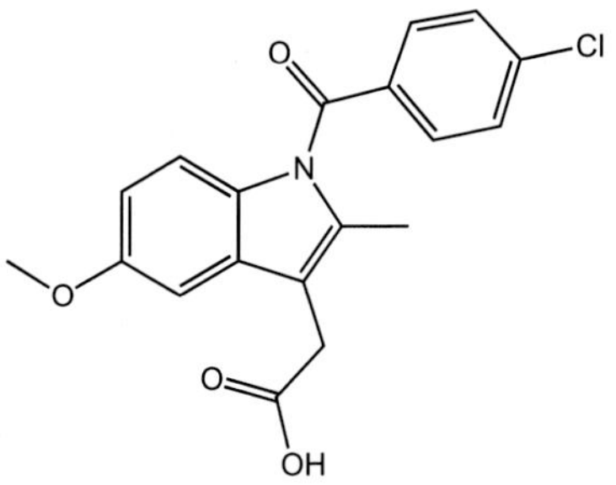

Indomethacin

Indomethacin (Indocin) is a potent nonsteroidal anti-inflammatory drug (NSAID) with antipyretic, analgesic, and anti-inflammatory activity. Its NSAID chemical classification is an indole-acetic acid derivative with the chemical name 1-(*p*-chlorobenzoyl)-5-methoxy-2-methylindole-3-acetic acid. Indomethacin is used for the treatment of rheumatoid arthritis, ankylosing spondylitis, osteoarthritis, and many other inflammatory diseases. It is also used for induction of closure of a patent ductus arteriosus in neonates [1]. The mechanism of action includes inhibition of prostaglandin synthesis mediated through the inhibition of cyclooxygenase (COX) enzymes. Prostaglandins are known to sensitize afferent nerves to induce pain and also serve as mediators of inflammation.

Indomethacin is available in 25 and 50 mg capsules for oral administration and in a 50 mg suppository for rectal administration (Merck). Naproxen is also available in extended release capsules containing 75 mg of pelletized indomethacin, of which 25 mg is immediately released, and the remaining 50 mg is released slowly over several hours. The rate of indomethacin absorption is decreased and delayed when taken with food, but the extent of bioavailability is not affected. Similarly, if taken with an antacid containing aluminum and magnesium hydroxides, peak plasma concentrations are slightly decreased and delayed, but effect is not clinically significant. Indomethacin is rapidly absorbed from the GI tract and following oral administration has virtually 100% bioavailability with peak plasma concentrations following a single dose occurring between 0.9 and 1.5 h in a fasting state. The peak plasma concentrations are dose-proportional and averaged 1.54 ± 0.76 μg/mL, 2.65 ± 1.03 μg/mL, and 4.92 ± 1.88 μg/mL following 25 mg, 50 mg, and 75 mg single doses in fasting subjects, respectively. Although there are no precise data on therapeutic range for the anti-inflammatory effect of indomethacin, a therapeutic range of 0.5–3 μg/mL has been suggested. Indomethacin is 90% serum protein bound. Major indomethacin metabolites are *O*-desmethyl-indomethacin, *N*-deschlorobenzoyl-indomethacin, and *O*-desmethyl-*N*-deschlorobenzoyl-indomethacin and their glucuronides. Metabolites have no antiinflammatory property. Approximately 33% of an indomethacin dose is excreted in feces as demethylated metabolites in unconjugated form; 1.5% is excreted in feces as indomethacin and approximately 60% are excreted in urine as metabolites [1].

Dosage of indomethacin should be reduced by approximately 25% in elderly (69 years of age and older), because half-life is longer (average 3.2 h) in elderly compared to young people (average 2.4 h). Moreover, clearance of indomethacin is much higher in younger people (average: 1.37 mL/min/kg) compared to elderly (average: 0.81 mL/min/kg). As expected area under the curve (AUC) indomethacin is approximately 54% higher in elderly [2].

Contraindications of indomethacin use include hypersensitivity to the drug, co-administration of aspirin and other NSAIDs, perioperative pain in the setting of coronary artery bypass surgery and pregnancy (3rd trimester). In neonates, the contraindications include necrotizing enterocolitis, impaired renal function, active bleeding, and thrombocytopenia. Patients who have overdosed on indomethacin have presented with gastritis, drowsiness, lethargy, nausea, vomiting, convulsions, paresthesia, headache, dizziness, hypertension, cerebral edema, tinnitus, disorientation, hepatitis, and renal failure. The boxed warnings include: indomethacin is associated with an increased risk of adverse cardiovascular events, including myocardial infarction, stroke and new onset or worsening of pre-existing hypertension [3].

Therapeutic drug monitoring of indomethacin is not routinely conducted, and there are no commercially available immunoassays. Indomethacin poisoning is infrequently reported. Sheehan reported two cases of indomethacin poisoning producing non-life threatening symptoms: nausea, vomiting, abdominal pain, anorexia, drowsiness, headache, tinnitus, restlessness and agitation. The terminal elimination half-lives in these two cases were 6.8 h and 2.9 h respectively, which were similar to that found following a therapeutic dose [4]. The commonly used methods for quantification of indomethacin are HPLC and GC–MS [5, 6].

References

[1] Prescott S, Keim-Malpass J. Patent ductus arteriosus in the preterm infant: diagnostic and treatment options. Adv Neonatal Care 2017;17:10–8.

[2] Oberbauer R, Krivanek P, Turnheim K. Pharmacokinetics of indomethacin in the elderly. Clin Pharmacokinet 1993;24:428–34.

[3] Lucas S. The pharmacology of indomethacin. Headache 2016;56:436–46.

[4] Sheehan TM, Boldy DA, Vale JA. Indomethacin poisoning. J Toxicol Clin Toxicol 1986;24:151–8.

[5] Taylor PJ, Jones CE, Dodds HM, Hogan NS, Johnson AG. Plasma indomethacin assay using high-performance liquid chromatography-electrospray-tandem mass spectrometry: application to therapeutic drug monitoring and pharmacokinetic studies. Ther Drug Monit 1998;20:691–6.

[6] Mei C, Li B, Yin Q, Jin J, et al. Liquid chromatography-tandem mass spectrometry for the quantification of flurbiprofen in human plasma and its application in a study of bioequivalence. J Chromatogr B Analyt Technol Biomed Life Sci 2015;993–994:69–74.

14.5

Naproxen

Chemical properties	
Solubility in H_2O	Freely soluble as sodium form
Molecular weight	230.3
pKa	5.0
Melting point	153 °C
Dosing	
Recommended dose, adult	440–660 (as naproxen sodium), OTC use
Recommended dose, child	5–15 mg/kg/day
Monitoring	
Sample	Serum, plasma
Effective concentrations	30–90 μg/mL
Toxic concentrations	>400 μg/mL
Methods	HPLC, GC
Pharmacokinetic properties	
Oral dose absorbed	~100%
Time to peak concentration	1–2 h
Protein bound	99.5%; mostly to albumin
Volume of distribution	0.16 L/kg
Half-life, adult	12–15 h
Time to steady state, adult	96–120 h
Half-life, child	8–10 h
Time to steady state, child	48–60 h

Excretion, urine

	% Excreted	Window of excretion	Active	Detected in blood
Parent	10	96 h	Yes	Yes
Desmethylnaproxen	5	96 h	No	No
Conjugated naproxen	60	96 h	No	No
Conjugated Desmethylnaproxen	25	96 h	No	No

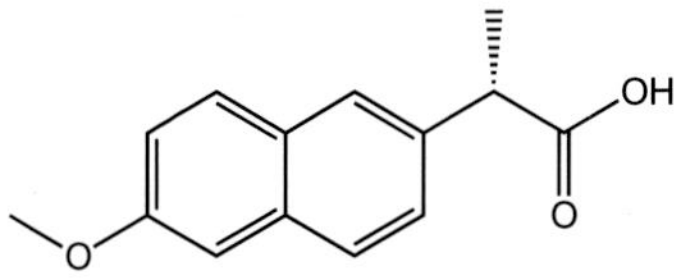

Naproxen

Naproxen [S-(+)-2-(6-methoxynaphth-2-yl)propionic acid] is a stereochemically pure 2-arylpropionic acid class nonsteroidal anti-inflammatory drug (NSAID) with analgesic, anti-inflammatory, and antipyretic properties. It is used in the management of mild to moderate pain, dysmenorrhea, and fever. It is also used in the treatment of rheumatoid arthritis, ankylosing spondylitis, osteoarthritis, and inflammatory diseases. Naproxen has been available as a prescription product in the USA since 1976, and naproxen sodium has been approved for over-the-counter (OTC) use in many countries. Non-prescription dosing is appropriate every 8–12 h, with a maximum total daily OTC dose of 440–660 mg, as approved by local regulatory authorities. Naproxen is commonly available as naproxen sodium 220 mg which is equivalent to 200 mg naproxen. This differs from the prescription dosing regimen, which is usually 500 mg naproxen (550 mg naproxen sodium) two to three times daily with a maximum total daily dose of 1500 mg. One long-acting formula, naproxen sodium CR, is designed so that 30% of the dose is released immediately while the remaining part of the dose, coated as microspheres, is released over a much longer period. Like other NSAIDs, the mechanism of action of the naproxen is believed to be related to reduction of prostacyclin production by inhibition of cyclooxygenase enzymes. Patients treated with naproxen show decreased levels of pro-inflammatory cytokine interleukin-6 and substance P in synovial fluid and plasma. Naproxen analgesic effect may last up to 12 h [1–3].

Naproxen is available in two forms: the free acid or the sodium salt as mentioned earlier. Naproxen sodium dissolves more rapidly in gastric fluid thus producing earlier and higher serum naproxen concentration compared to naproxen. Peak plasma concentrations are achieved in about 1 h with naproxen sodium and 2 h with naproxen. However, the post-absorption phase pharmacokinetics of the sodium salt and the parent drug are identical. Naproxen is readily absorbed after oral administration and the bioavailability is almost 100%. The drug is strongly bound to serum protein (>99.5%). The nonlinear kinetics exhibited by naproxen when daily doses exceed 1000 mg is most likely due to saturation of protein binding and increased renal clearance. The drug readily reaches the synovial fluid and synovial membrane. Naproxen crosses the placental barrier, and minimal transfer occurs to breast milk (about 1% of maternal plasma samples). Approximately 95% of a radiolabeled dose of naproxen is recovered in urine and 3% or less in feces. Approximately 70% of naproxen is excreted unchanged in the urine, and the rest is metabolized to an inactive 6-demethyl metabolite (6-*O*-desmethylnaproxen), probably by hepatic microsomal oxidation mediated by cytochrome P450 (CYP) 1A2 and CYP2C9. The parent compound and metabolite are excreted free or as glucuronide or sulfate conjugates. The elimination half-life is about 12–15 h [4]. Steady state is reached in 4–5 days. Hepatic disease and rheumatoid arthritis can also significantly alter the disposition kinetics of naproxen. Although naproxen is excreted into breast milk the amount of drug transferred comprises only a small fraction of the maternal exposure [5].

Zhou et al. compared pharmacokinetic parameters after administration of 500 mg naproxen (two tablets) as a single dose and administration of 500 mg extended release naproxen. A rapid increase in naproxen plasma concentrations was observed after administration both formulations; 250-mg conventional tablets, with maximum plasma concentration of 87.3 µg/mL observed approximately 2.6 h but after administration of 500 mg sustained released tablet, maximum serum concentration of 47.2 µg/mL was observed at 4.1 h. However, no statistically significant differences between the two treatments were observed for any of the other pharmacokinetic parameters. However, average half-life was slightly longer (19.8 h) after administration of sustained release naproxen compared to regular dosage (17.7 h) [6].

Naproxen is considered safer than aspirin. Toxic reactions most commonly associated with naproxen include nausea, abdominal pain, constipation, headache, dizziness, drowsiness, tinnitus, and skin eruptions. Severe allergic reactions to naproxen have been reported. The use of naproxen is contraindicated for patients who are receiving aspirin or other NSAIDs. Food and Drug Administration warnings include: (1) NSAIDs may cause an increased risk of serious cardiovascular thrombotic events, myocardial infarction, and stroke, which can be fatal. This risk may increase with duration of use. Patients with cardiovascular disease or risk factors for cardiovascular disease may be at greater risk. (2) NSAIDs cause an increased risk of serious GI adverse events including bleeding, ulceration, and perforation of the stomach or intestines, which can be fatal. These events can occur at any time during use and without warning symptoms. Elderly patients are at greater risk for serious GI events. (3) Treatment of perioperative pain in setting of coronary artery bypass graft (CABG) surgery is contradictory.

(4) It is contraindicated in patients who have experienced asthma, urticaria, or allergic-type reactions after taking aspirin or other NSAIDs [1, 7].

Massive naproxen overdose can present with serious toxicity including seizures, altered mental status, and metabolic acidosis. Al-Abri reported a case of a 28-year-old man who ingested 70 g of naproxen along with an unknown amount of alcohol in a suicidal attempt. On examination in the emergency department 90 min later, he was drowsy but had normal vital signs apart from sinus tachycardia. Serum naproxen level 90 min after ingestion was 1580 μg/mL (therapeutic range 25–75 μg/L). He developed metabolic acidosis requiring renal replacement therapy and had recurrent seizure activity requiring intubation within 4 h from ingestion. However, he recovered after 48 h [8].

Fatality from taking naproxen is rarely reported. A 62 year old woman with a confirmed diagnosis of multiple myeloma started on 1000 mg/day on naproxen for treating fever. However, after 10 days of therapy with naproxen she developed acute renal failure with a fatal outcome [9]. Fatal interaction between methotrexate (MTX) and naproxen has also been reported [10]. Combination therapy of NSAIDs including naproxen and MTX sometimes triggers adverse effects such as liver injury, renal failure, GI disorders, and myelosuppression, owing to the reduction of MTX clearance. Inhibition of organic anion transporters (OAT1 and OAT3)-mediated renal uptake of methotrexate by glucuronide metabolites of NSAIDs may be one of the competitive sites underlying complex drug interaction between MTX and NSAIDs [11].

Naproxen levels are not usually monitored and may not correlate well with toxicity; however, the levels may be useful in the patients who do not respond to therapy. Other laboratory tests such as renal function tests, hepatic function tests and coagulation markers are useful. There is no commercially available immunoassay for measuring naproxen concentration in serum or plasma. Common methods for estimation of naproxen include gas chromatography/mass spectrometry after liquid-liquid extraction and derivatization using *N*-methyl-*N*-(trimethylsilyl) trifluoroacetamide [12] or high performance liquid chromatography [13]. Analysis of naproxen in human serum using liquid chromatography combined with tandem mass spectrometry has also been reported [14].

References

[1] Angiolillo DJ, Weisman SM. Clinical pharmacology and cardiovascular safety of naproxen. Am J Cardiovasc Drugs 2017;17:97–107.

[2] James MJ, Cleland LG. Cyclooxygenase-2 inhibitors: what went wrong? Curr Opin Clin Nutr Metab Care 2006;9:89–94.

[3] McGettigan P, Henry D. Cardiovascular risk and inhibition of cyclooxygenase: a systematic review of the observational studies of selective and nonselective inhibitors of cyclooxygenase 2. JAMA 2006;296:1633–44.

[4] Todd PA, Clissold SP. Naproxen. A reappraisal of its pharmacology, and therapeutic use in rheumatic diseases and pain states. Drugs 1990;40:91–137.

[5] Davies NM, Anderson KE. Clinical pharmacokinetics of naproxen. Clin Pharmacokinet 1997;32:268–93.

[6] Zhou D, Zhang Q, Lu W, Xia Q, Wei S. Single- and multiple-dose pharmacokinetic comparison of a sustained-release tablet and conventional tablets of naproxen in healthy volunteers. J Clin Pharmacol 1998;38:625–9.

[7] Solomon DH, Husni ME, Libby PA, Yeomans ND, et al. The risk of major NSAID toxicity with celecoxib, ibuprofen, or naproxen: a secondary analysis of the PRECISION trial. Am J Med 2017;130:1415–22.

[8] Al-Abri SA, Anderson IB, Pedram F, Colby JM, Olson KR. Massive naproxen overdose with serial serum levels. J Med Toxicol 2015;11:102–5.

[9] Shpilberg O, Douer D, Ehrenfeld M, Engelberg S, Ramot B. Naproxen-associated fatal acute renal failure in multiple myeloma. Nephron 1990;55:448–89.

[10] Singh RR, Malaviya AN, Pandey JN, Guleria JS. Fatal interaction between methotrexate and naproxen. Lancet 1986;1(8494):1390.

[11] Iwaki M, Shimada H, Irino Y, Take M, Egashira S. Inhibition of methotrexate uptake via organic anion transporters OAT1 and OAT3 by glucuronides of nonsteroidal anti-inflammatory drugs. Biol Pharm Bull 2017;40:926–31.

[12] Yilmaz B, Sahin H, Erdem AF. Determination of naproxen in human plasma by GC-MS. J Sep Sci 2014;37:997–1003.

[13] Tashtoush BM, Al-Taani BM. HPLC determination of naproxen in plasma. Pharmazie 2003;58:614–5.

[14] Gopinath S, Kumar RS, Shankar MB, Danabal P. Development and validation of a sensitive and high-throughput LC-MS/MS method for the simultaneous determination of esomeprazole and naproxen in human plasma. Biomed Chromatogr 2013;27:894–9.

CHAPTER 15

Antineoplastic drugs

Therapeutic drug monitoring (TDM) is not routinely used for most cytotoxic agents, because few drugs in this group fulfill the prerequisites of TDM—wide intraindividual pharmacokinetic variability, well defined relationship between drug levels and response (both toxicity and efficacy), and narrow therapeutic range. However, for some antineoplastic few drugs, there is a good correlation between blood levels and efficacy as well as toxicity; for such drugs, TDM is practiced routinely for patient management. There are established correlations between methotrexate levels and response to lymphoblastic leukemia, carboplatin levels and response to ovarian cancer, serum concentrations of 5-fluorouracil and clinical response to colorectal, head, and neck cancers [1, 2].

There is a narrow therapeutic window between suboptimal therapy and toxicity in the treatment with certain antineoplastic drugs. Genetic polymorphism in phase I and phase II enzymes are present in the population and may explain in part the variations in the pharmacokinetic parameters of a particular drug between individual patients. The potential for applying pharmacogenetic screening before cancer chemotherapy may have applications with several cytochrome P450 (CYP) enzymes, in particular with CYP2B6 (cyclophosphamide treatment), CYP2C8 (paclitaxel therapy) and CYP3A5 [3]. Moreover, tamoxifen is a prodrug which must be converted into its active metabolite for its pharmacological effects in treating breast cancer. Genetic polymorphisms that confer reduced CYP2D6 activity or concurrent use of CYP2D6-inhibiting drugs may reduce the clinical efficacy of tamoxifen [4].

Immunoassays are commercially available for TDM of methotrexate. For other antineoplastic drugs, high performance liquid chromatographic (HPLC) techniques are usually employed. Schoemaker et al. developed a HPLC protocol for simultaneous measurement of anticancer drug irinotecan (CPT-11) and its active metabolite SN-38 in human plasma after converting both CPT-11 and SN-38 to their carboxylate form by using sodium tetraborate [5]. Determination of docetaxel and paclitaxel concentrations using HPLC with UV detection has also been reported [6]. Titier et al. described

Therapeutic Drug Monitoring Data
https://doi.org/10.1016/B978-0-12-815849-4.00015-3

quantification of imatinib, a selective tyrosine kinase inhibitor, in human plasma using HPLC combined with tandem mass spectrometry (LC/MS/MS) [7]. Recently, Herbrink et al. developed a fast and accurate method for simultaneous quantification of anticancer drugs afatinib, axitinib, ceritinib, crizotinib, dabrafenib, enzalutamide, regorafenib and trametinib in human plasma using LC/MS/MS [8].

Gas chromatography (GC) combined with nitrogen phosphorus detector or mass spectrometry can also be used for determination of specific neoplastic drugs. Kerbusch et al. determined concentration of ifosfamide along with its metabolites 2-and 3-dechloroethylifosfamide in human plasma by using GC coupled with nitrogen phosphorus detector or mass spectrometry without any derivatization. Interestingly, the authors concluded that GC with nitrogen phosphorus detector was more sensitive for analysis of these compounds compared to GC combined with positive ion electron-impact ion trap mass spectrometry [9]. Lekskulchai developed a gas chromatography/mass spectrometry (GC/MS) method for analysis of cyclophosphamide and ifosfamide in urine [10].

References

[1] Lennard L. Therapeutic drug monitoring of cytotoxic drugs. Br J Clin Pharmacol 2001;52(Suppl 1):75S–87S.
[2] Rousseau A, Marquet P. Application of pharmacokinetic modelling to routine therapeutic drug monitoring of anticancer drugs. Fundam Clin Pharmacol 2002;16:253–62.
[3] van Schaik RH. Implications of cytochrome P 450 genetic polymorphisms on the toxicity of antitumor agents. Ther Drug Monit 2004;26:236–40.
[4] Cronin-Fenton DP, Damkier P. Tamoxifen and CYP2D6: a controversy in pharmacogenetics. Adv Pharmacol 2018;83:65–91.
[5] Schoemaker NE, Rosing H, Jansen S, Schellens JH, et al. High performance liquid chromatographic analysis of the anticancer drug irinotecan (CPT-11) and its active metabolite SN-38 in human plasma. Ther Drug Monit 2003;25:120–4.
[6] Zufia Lopez L, Aldaz Pastor A, Armendia Beitia JM, Arrobas Velilla J, et al. Determination of docetaxel and paclitaxel in human plasma by high performance liquid chromatography: validation and application to clinical pharmacokinetic studies. Ther Drug Monit 2006;28:199–205.
[7] Titier K, Picard S, Ducint D, Teihet E, et al. Quantification of imatinib in human plasma by high performance liquid chromatography-tandem mass spectrometry. Ther Drug Monit 2005;27:634–40.
[8] Herbrink M, de Vries N, Rosing H, Huitema ADR, et al. Development and validation of a liquid chromatography-tandem mass spectrometry analytical method for the therapeutic drug monitoring of eight novel anticancer drugs. Biomed Chromatogr 2018;32:e4147.
[9] Kerbusch T, Jeuken MJ, Derraz J, van Putten JW, et al. Determination of ifosfamide, 2 and 3-dechloroethylifosfamide using gas chromatography with nitrogen-phosphorus or mass spectrometry detection. Ther Drug Monit 2000;22:613–20.
[10] Lekskulchai V. Quantitation of anticancer drugs—cyclophosphamide and ifosfamide in urine and water sewage samples by gas chromatography-mass spectrometry. Int J Occup Med Environ Health 2016;29:815–22.

15.1

Busulfan

$CH_3-S(=O)_2-O-CH_2-CH_2-CH_2-CH_2-O-S(=O)_2-CH_3$

Chemical properties

Solubility in H_2O	Very slightly soluble in H_2O or alcohol, soluble in acetone and acetonitrile
Molecular weight	246.304
pKa	n/a
Melting point	287 °C

Dosing

Recommended dose, adult	Conventional dosage (1–12 mg/day) High dose (approximately 16 mg/kg)
Recommended dose, child	0.8–1.2 mg/kg, dependent on child's weight

Monitoring

Sample	Serum, plasma
Effective concentrations	600–900 ng/mL
Toxic concentrations	Not established
Methods	HPLC, LC-MS, GC–MS, immunoassay

Pharmacokinetic properties

Oral dose absorbed	~99%
Time to peak concentration	0.5–5 h (oral)
Protein bound	32%, albumin; 47% RBC
Volume of distribution	0.56–1.4 L/kg
Half-life, adult	2–3 h
Time to steady state, adult	10–17 h
Half-life, child	1.7–2.3 h (oral)
Time to steady state, child	8–12 h

Excretion, urine

	% Excreted	Active	Detected in blood
Parent	<2%	Yes	Yes
~12 metabolites (tetrahydrothiophene, tetrahydrothiophene 12-oxide, sulfolane, 3-hydroxysulfolane, S-methanesulfonic acid, 3-hydroxytetrahydrothiophene-1, 1-dioxide)	~32%	No	Yes

Busulfan (1, 4-butanediol dimethanesulfonate) is a bifunctional alkylating agent that acts in the cell cycle phase. Busulfan contains two easily displaceable methane sulfonate groups on opposite ends of a butane chain. Hydrolysis of these groups produces highly reactive, positively charged carbonium ions capable of alkylating and damaging DNA molecules. Therefore, it can be assumed based on in vitro studies that toxicity of busulfan is correlated with formation of the DNA lesions. Busulfan is also capable of reacting with cysteine molecules on histone proteins, inducing DNA-protein binding. Overall, alkylating activity of busulfan produces alterations of cell replication, DNA damage repair, and gene transcription. Busulfan also increases cellular oxidative stress by reacting with the sulfhydryl groups of the endogenous tripeptide glutathione, either spontaneously or via catalysis by glutathione *S*-transferase enzyme. Increased oxidative stress is one of the major causes of busulfan toxicity [1].

Busulfan exhibits selective cytotoxicity for myeloid cells. It also exhibits an ability to remove sulfur from other compounds, i.e., proteins, polypeptides and amino acids; but it is unclear how much this contributes to its cytotoxic mechanism of action. The bone marrow depressive effects of the drug are dose dependent, allowing for its use in several conditions. At low doses, the drug depresses granulocytopoiesis and thrombocytopoiesis but has little effect on lymphocytes. Low dose busulfan (1–12 mg/day) is used to treat chronic myelogenous leukemia (CML), polycythemia vera, myelofibrosis, and essential thrombocythemia. The treatment of CML using busulfan does not result in a cure but is considered palliative. Patients who lack the Philadelphia chromosome or who are in a blast phase show a poor response to busulfan [2]. Busulfan at high dose (~16 mg/kg) is used in combination with agents such as melphalan or cyclophosphamide as myeloablative conditioning prior to allogeneic or autologous bone marrow transplant for treatment of high-risk lymphoma, relapsed acute leukemia, and multiple myeloma. At high doses of busulfan, full bone marrow suppression is observed [3, 4].

Busulfan is available for oral or intravenous (IV) administration. Although absorption following oral administration is usually complete for most patients, wide interindividual variation of bioavailability in adults and children has been reported. The elimination half-life after IV administration in children (2.46 ± 0.27 h, mean ± SD) did not differ from that obtained for adults (2.61 ± 0.62 h). However, the volume of distribution normalized for body weight was significantly higher in children (0.74 ± 0.10 L/kg-) compared with adults (0.56 ± 0.10 L/kg) [5].

The drug is extensively metabolized by the liver such that <2% of the drug is excreted unchanged. About 32% of busulfan is excreted in the urine as metabolic products. At least 12 inactive metabolites have been identified among them tetrahydrothiophene, tetrahydrothiophene 12-oxide, sulfolane, 3-hydroxysulfolane, *S*-methanesulfonic acid, and 3-hydroxytetrahydrothiophene-1,1-dioxide. Busulfan may induce its own metabolism [4].

Adverse reactions to busulfan include nausea, vomiting, anorexia, weakness, anemia, infertility, hyperpigmentation, amenorrhea, and seizures. The most serious adverse effect reported is bone marrow suppression resulting in severe pancytopenia. The complication is reversible, but recovery may take 1 month to 2 years and requires supportive care. The development of bronchopulmonary dysplasia with pulmonary fibrosis is a rare and usually fatal complication. The histological findings associated with this condition are similar to those following pulmonary irradiation. Onset is varied with symptoms developing as rapidly as within 8 months to as long as 10 years after initiation of therapy. Patients receiving high-dose therapy have also experienced fatal cardiac tamponade and hepatic veno-occlusive disease. Busulfan is considered a teratogen and carcinogen and thus contraindicated in pregnancy [6]. Hepatic sinusoidal-obstruction syndrome and busulfan-induced lung injury in a two and half year old boy post-autologous stem cell transplant has been reported. The outcome was fatal [7].

TDM is not typically used for patients undergoing low-dose busulfan therapies, but the high-dose regimens target a narrow therapeutic window and thus have the combined risks of rejection and toxicity. It is for patients on these protocols that TDM has proven beneficial, with the development of multiple TDM approaches. Chandy and other have shown that the risk of graft rejection is higher when trough concentrations are below 150 ng/mL [8, 9]. Patients whose serum drug concentrations exceeded 900 ng/mL had a greater risk of veno-occlusive disease compared to those with those whose concentrations were lower. TDM is performed using either trough concentrations or pharmacokinetic studies over a 4–6 h period. Trough concentrations between 600 and 900 ng/mL are targeted. Concentrations have also been monitored in oral fluid and cerebrospinal fluid [10, 11]. Plasma and oral fluid busulfan concentration correlate well with one another. The busulfan oral fluid/plasma ratio was 1.09 ± 0.04. The mean concentration of busulfan was 785 ng/mL in plasma and 820 ng/mL in oral fluid. The half-life was 2.31 h in plasma and 2.30 h in oral fluid. Therefore, busulfan TDM in children can be conducted using oral fluid, limiting the need for venipuncture [11]. Recently Choong et al. established that the therapeutic range of busulfan is 600–900 ng/mL. TDM is essential in children receiving busulfan because the authors observed a 4.2-fold interindividual variability in busulfan concentrations after the first dose, with only 28% of children showing plasma busulfan levels within targeted therapeutic range [12].

A number of chromatography-based methods, including both HPLC and GC with or without mass spectrometry detection, have been described for measuring busulfan concentrations [11–14]. An immunoassay method for busulfan exists that can be performed as an ELISA or adapted to some automated clinical chemistry platforms [15]. Specimens include serum and heparinized plasma. EDTA has been reported to interfere with at least one LC/MS method [13]. Samples should be processed immediately after collections as busulfan is reported to undergo in vitro hydrolysis with a degradation half-life of 8.7 h in whole blood at 37 °C, 12 h in plasma, and 16 h in phosphate buffer [1, 16]. However, El-Serafi et al. commented that busulfan was stable up to 50 days after extraction if stored at −20 °C but only 48 h at room temperature [17].

References

[1] Myers AL, Kawedia JD, Champlin RE, Kramer MA, et al. Clarifying busulfan metabolism and drug interactions to support new therapeutic drug monitoring strategies: a comprehensive review. Expert Opin Drug Metab Toxicol 2017;13:901–23.

[2] Buggia I, Locatelli F, Regazzi MB, Zecca M. Busulfan. Ann Pharmacother 1994;28:1055–62.

[3] Willcox A, Wong E, Nath C, Janson B, et al. The pharmacokinetics and pharmacodynamics of busulfan when combined with melphalan as conditioning in adult autologous stem cell transplant recipients. Ann Hematol 2018;97(12):2509–18.

[4] Baselt RC. Clarifying busulfan metabolism and drug interactions to support new therapeutic drug monitoring strategies: a comprehensive review. In: Busulfan in disposition of toxic drugs and chemicals in man. 7th ed. Foster City, CA: Biomedical Publications; 2004. p. 142–3.

[5] Hassan M, Ljungman B, Ringdén O, et al. Busulfan bioavailability. Blood 1994;84:2144–50.

[6] Bishop JB, Wassom JS. Toxicological review of busulfan (Myleran). Mutat Res 1986;168:15–45.

[7] Jain R, Gupta K, Bhatia A, Bansal A, Bansal D. Hepatic sinusoidal-obstruction syndrome and busulfan-induced lung injury in a post-autologous stem cell transplant recipient. Indian Pediatr 2017;54:765–70.

[8] Chandy M, Balasubramanian P, Ramachandran SV, Mathews V, et al. Randomized trial of two different condition regimens for bone marrow transplantation in thalassemia—the role of busulfan pharmacokinetics in determining outcome. Bone Marrow Transplant 2005;36:839–45.

[9] Lindley C, Shea T, McCune J, Shord S, et al. Intraindividual variability in busulfan pharmacokinetics in patients undergoing a bone marrow transplant: assessment of a test dose and first dose strategy. Anti-Cancer Drugs 2004;15:453–9.

[10] Pichini S, Altieri I, Bacosi A, Di Carlo S, Zuccaro P, Iannetti P, Pacifici R. High-performance liquid chromatographic-mass spectrometric assay of busulfan in serum and cerebrospinal fluid. J Chromatogr 1992;581:143–6.

[11] Rauh M, Stachel D, Kuhlen M, Groschl M, et al. Quantification of busulfan in saliva and plasma in haematopoietic stem cell transplantation in children: validation of liquid chromatography tandem mass spectrometry method. Clin Pharmacokinet 2006;45:305–16.
[12] Choong E, Uppugunduri CRS, Marino D, Kuntzinger M, et al. Therapeutic drug monitoring of busulfan for the management of pediatric patients: cross-validation of methods and long-term performance. Ther Drug Monit 2018;40(1):84–92.
[13] Kellogg MD, Law T, Sakamoto M, Rifai N. Tandem mass spectrometry method for the quantification of serum busulfan. Ther Drug Monit 2005;27:625–9.
[14] Abdel-Rehim M, Hassan Z, Blomberg L, Hassan M. On-line derivatization utilizing solid-phase microextraction (SPME) for determination of busulphan in plasma using gas chromatography-mass spectrometry (GC-MS). Ther Drug Monit 2003;25:400–6.
[15] Courtney JB, Harney R, Li Y, Lundell G, et al. Determination of busulfan in human plasma using an ELISA format. Ther Drug Monit 2009;31:489–94.
[16] Balasubramanian P, Srivastava A, Chandy M. Stability of busulfan in frozen plasma and whole blood samples. Clin Chem 2001;47:766–8.
[17] El-Serafi I, Terelius Y, Twelkmeyer B, Hagbjörk AL, et al. Gas chromatographic-mass spectrometry method for the detection of busulphan and its metabolites in plasma and urine. J Chromatogr B Anal Technol Biomed Life Sci 2013;913–914:98–105.

15.2

5-Fluorouracil

Chemical properties

Solubility in H_2O	<0.1 g/100 mL
Molecular weight	130.08
pKa	8.02
Melting point	280–282 °C

Dosing

Recommended dose, adult	300–500[a] mg/kg/day

Monitoring

Sample	Serum
Effective concentrations, adult	2–3 µg/mL (AUC: 20–30 mg h/L most commonly targeted)
Toxic concentrations	>3 µg/mL
Methods	HPLC, LC/MS, GC/MS, CZE

Pharmacokinetic properties

Oral dose absorbed	Oral: 0–80% Topical: 0.5–5.0%
Time to peak concentration	Oral: 0.55 h Topical: 1 h I.V.: 0.45–0.69 h
Protein bound	8–12%
Volume of distribution	0.13–0.37 L/kg

Excretion, urine

	% Excreted	Active	Detected in blood
Parent	<10	Yes	Yes

Safety and efficacy not established in pediatric patients.
[a]Dose is dependent on specific indication.

5-Fluorouracil

5-Fluorouracil (5-FU) is an anti-neoplastic anti-metabolite. Fluoro-pyrimidines, which include 5-FU (Adrucil; Carac; Efudex; Fluoroplex) and capecitabine are widely used chemotherapeutic agents. 5-FU is the third most commonly used chemotherapeutic agent. Since its discovery in the 1950s, it has been used to treat a vast array of neoplasms including carcinomas of the breast, colon, head and neck, pancreas, rectum, and stomach. 5-FU is also used topically for the management of actinic or solar keratoses and superficial basal cell carcinomas. Fluoro-pyrimidines including 5-FU also possesses radiosensitizing properties and are often used in conjunction with external beam radiotherapy [1, 2].

5-FU is a halogenated pyrimidine that interferes with DNA and RNA synthesis and processing. The name 5-fluorouracil derives from the substitution of fluorine for hydrogen at the 5th carbon position of uracil. 5-FU is a prodrug that is enzymatically converted to the active metabolite 5-fluorodeoxyuridine monophosphate. This active metabolite binds to thymidylate synthase, preventing production of thymidine triphosphate. 5-FU treatment causes thymidine triphosphate deficiency resulting in cytotoxicity.

Several mechanisms of drug resistance to 5-FU have been reported. These include the loss of enzymes required to activate 5-FU, amplification of the target enzyme (thymidylate synthase), or production of altered thymidylate synthase that is unaffected by the drug. However, the molecular mechanisms of 5-FU resistance have not been completely clarified. Recently, microarray analyses have shown that noncoding RNAs (i.e. microRNAs and long noncoding RNAs) play a vital role in 5-FU resistance in multiple cancer cell lines. These noncoding RNAs can function as oncogenes or tumor suppressors, contributing to 5-FU drug resistance [3]. To minimize resistance and enhance drug efficacy, 5-FU is often used in combination with other chemotherapeutic agents such as cisplatin and methotrexate [4].

The bioavailability of oral 5-FU can vary significantly (0–80%) depending on saturable first-pass effects. Because of this variability, 5-FU is typically administered intravenously. >80% of the administered 5-FU dose is eliminated by catabolism through dihydropyrimidine dehydrogenase (DPD), the rate-limiting enzyme. DPD activity is found in most tissues but is highest in the liver. Population pharmacokinetic analysis has identified patient co-variables that influence 5-FU clearance. Drug kinetics are significantly reduced by increased age, high serum alkaline phosphatase, length of drug infusion, and low peripheral blood mononuclear cell dihydropyrimidine dehydrogenase (PBMC-DPD) activity. A circadian rhythm in PBMC-DPD activity has been observed, with average peak and trough at 1 a.m.

and 1 p.m., respectively. As a corollary, a circadian rhythm was observed in 5- FU plasma concentration, with a peak at 11 a.m. and a trough at 11 p.m. [5, 6]. The plasma 5-FU concentration in the evening was 1.3-fold higher compared with the morning after the first cycle of treatment with 5-FU. In the second cycle, it was increased by 1.5-fold compared with the first cycle, with relatively small inter-individual variations (23.3% and 16.8%, respectively). The higher plasma concentration of 5-FU in the early phase of treatment may be the key determinant of clinical efficacy, whereas the variations in the plasma concentration of 5-FU owing to the time of day and treatment cycle are small contributors [7].

When administered intravenously, 5-FU is rapidly distributed, yielding peak plasma concentrations ranging from 0.1–1 mM depending on the bolus dose. Continuous intravenous infusion of 5-FU for 1–5 days results in steady state plasma concentrations of 0.5–0.8 μM. However, area under the curve (AUC) of 5-FU concentrations is considered to be the most relevant pharmacokinetic parameter associated to 5-FU-related efficacy and toxicity. An AUC range of 20–30 mg h/L is generally required for successful therapy.

Common toxicities of 5-FU include anemia, leukopenia, thrombocytopenia, alopecia, and nail damage. Additional toxic effects include skin alterations such as hyperpigmentation or skin atrophy [8, 9]. In severe instances, mucosal ulcerations caused by 5-FU treatment can result in shock and death. The predominant toxic effect due to bolus injection of 5-FU is pancytopenia caused by the myelosuppressive effect of the drug. 5-FU is the second most common chemotherapeutic drug associated with cardiotoxicity after anthracyclines, which can manifest as chest pain, acute coronary syndrome/myocardial infarction, or death [10]. Pharmacogenetic variation impacting expression of the DPD enzyme is a factor which can affect the pharmacokinetics and toxicity of 5-FU. Partial and complete deficiency of DPD occurs in 3–5% and 0.1% of the general population, respectively, and may cause severe toxicity in the form of neutropenia, diarrhea, mucositis, and hand-foot syndrome. TDM of 5-FU is strongly recommended to achieve therapeutic success as well as to avoid drug toxicity [9]. Currently LC/MS/MS is used for TDM of 5-FU [9].

References

[1] Grem JL. 5-Fluorouracil: forty-plus and still ticking. A review of its preclinical and clinical development. Invest New Drugs 2000;18:299–313.

[2] Diasio RB, Harris BE. Clinical pharmacology of 5-fluorouracil. Clin Pharmacokinet 1989;16(4):215–37 [Review].

[3] Deng J, Wang Y, Lei J, Lei W, Xiong JP. Insights into the involvement of noncoding RNAs in 5-fluorouracil drug resistance. Tumour Biol 2017;39(4):1010428317697553.

[4] Chang PM, Teng HW, Chen PM, Chang SY, et al. Methotrexate and leucovorin double-modulated 5-fluorouracil combined with cisplatin (MPFL) in metastatic/recurrent head and neck cancer. J Chin Med Assoc 2008;71:336–41.

[5] Milano G, Etienne MC. Individualizing therapy with 5-fluorouracil related to dihydropyrimidine dehydrogenase: theory and limits. Ther Drug Monit 1996;18:335–40.

[6] Milano G, Chamorey AL. Clinical pharmacokinetics of 5-fluorouracil with consideration of chronopharmacokinetics. Chronobiol Int 2002;19:177–89.

[7] Kuwahara A, Kobuchi S, Tamura T. Association between circadian and chemotherapeutic cycle effects on plasma concentration of 5-fluorouracil and the clinical outcome following definitive 5-fluorouracil/cisplatin-based chemoradiotherapy in patients with esophageal squamous cell carcinoma. Oncol Lett 2019;17:668–75.

[8] Joulia JM, Pinguet F, Grosse PY, Astre C, Bressolle F. Determination of 5-fluorouracil and its main metabolites in plasma by high-performance liquid chromatography: application to a pharmacokinetic study. J Chromatogr B Biomed Sci Appl 1997;692(2):427–35.

[9] Macaire P, Morawska K, Vincent J, Quipourt V, et al. Therapeutic drug monitoring as a tool to optimize 5-FU-based chemotherapy in gastrointestinal cancer patients older than 75 years. Eur J Cancer 2019;111:116–25.

[10] Sara JD, Kaur J, Khodadadi R, Rehman M, et al. 5-Fluorouracil and cardiotoxicity: a review. Ther Adv Med Oncol 2018;10:1–18.

15.3

Methotrexate

Chemical properties	
Solubility in H_2O	8.9 mg/mL
Molecular weight	454.46
pKa	4.3–5.5
Melting point	184–204 °C
Dosing	
Recommended dose	Dependent on administration route and clinical indication, with high doses used for treatment of neoplasms. For chronic treatment of immune diseases such as rheumatoid arthritis, typically doses are 15 mg/week orally
Monitoring	
Sample	Serum, plasma (heparin, EDTA), CSF
Effective concentrations	1.0–1000[a] μmol/L
Toxic concentrations	>0.02 μmol/L (1–2 weeks low dose) ≥5 μmol/L (24 h post high dose therapy) ≥0.5 μmol/L (48 h post high dose therapy) ≥0.005 μmol/L (72 h post high dose therapy)
Methods	HPLC, immunoassay
Pharmacokinetic properties	
Oral dose absorbed	50–100%
Time to peak concentration	1–4 h
Protein bound him	41–51%
Volume of distribution	0.75 L/kg
Half-life, adult	8–10 h low dose therapy
Time to steady state, adult	24–48 h
Half-life, child	5–9 h
Time to steady state, child	24–48

Excretion, urine

	% Excreted	Active	Detected in blood
Parent			
(low dose)	40–50	Yes	Yes
(high dose)	60–90		
7-Hydroxy metabolite	1–20	No	Yes
2,4-diamino-*N*-10-methylpteroic	Unknown	No	Yes

[a]Effective dose is dependent on indication.

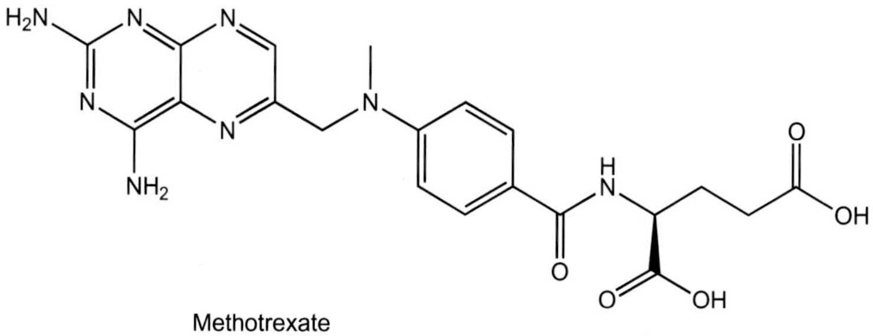

Methotrexate

Methotrexate (4-amino-10-methylpteroylglutamic acid, Rheumatrex, Trexall) is an anti-metabolite used for treating a wide range of neoplastic and non-neoplastic disorders. The mechanism of action of methotrexate is due to its interference with the metabolism of folic acid. After entry into the cell, methotrexate is polyglutamated and then binds with dihydrofolate reductase (DHFR) with an affinity 1000-fold greater than that of folate, thus competitively inhibiting conversion of dihydrofolate to tetrahydrofolate. As a result, biosynthesis of thymidine and purines needed for synthesis of DNA are blocked, thereby also disrupting synthesis of RNA and proteins. This inhibition is useful in chemotherapy, because the drug interferes with the enzymatic reactions occurring in the proliferating tumor cells. Methotrexate is an essential component of therapy for acute lymphoblastic leukemia (ALL) and is active against many types of cancer including non-Hodgkin's lymphoma, osteogenic sarcoma, and medulloblastoma. Additional carcinomas treated by methotrexate include those of the head, neck, lung, breast, cervix, ovaries, testes, and bladder. Methotrexate is also used for medical management of ectopic pregnancy (especially early-stage) and gestational trophoblastic disease including choriocarcinoma. The versatility of methotrexate has warranted a place on the World Health Organization's list of essential medicines.

Occasionally, tumor cells demonstrate resistance to methotrexate. Resistance is attributed to a number of factors including impaired transport of drug into the cell, production of altered forms of DHFR, and decreased ability to synthesize methotrexate polyglutamates. It is also proposed that resistance can occur due to decreased thymidylate synthase activity and increased concentrations of intracellular DHFR. [1].

Methotrexate can be administered as either high-dose or low-dose therapy. Methotrexate doses of least 500 mg/m^2 or higher (up to 33.6 g/m^2) given intravenously are defined as high-dose and are used to treat a variety of adult and pediatric cancers, including ALL, osteosarcoma, and lymphomas. Although methotrexate at a high dose can be safely administered to most patients, it can cause significant toxicity, including acute kidney injury due to crystallization of methotrexate in the renal tubular lumen, leading to tubular toxicity [1]. The unwanted toxic effects of methotrexate are reduced by use of a reduced folate known as leucovorin (folinic acid, citrovorum factor). This "leucovorin rescue" therapy permits methotrexate doses that would otherwise be lethal, and renders the therapy relatively free of toxicity; however, use of high-dose therapy with leucovorin rescue is not without its challenges. Insufficient leucovorin therapy following high-dose methotrexate can lead to significant, irreversible toxicity and a fatal outcome. There are several routes of methotrexate administration including oral and intramuscular, subcutaneous or intravenous injection [2].

For the treatment of non-neoplastic autoimmune conditions such as rheumatoid arthritis, methotrexate is used as a disease modifying agent and is administered chronically at a much lower dose than used for neoplasms. Usual oral dosage is 15 mg once a week. However, if significant improvement is not observed by 3–4 months of treatment, the dosage can be increased up to 20 to 30 mg/week or may be administered subcutaneously for increased bioavailability [3]. Low-dose methotrexate is also used in treating various other autoimmune diseases such as psoriatic arthritis, vasculitis, uveitis, inflammatory bowel disease and lupus erythematosus [4]. Methotrexate is a safe and effective drug for the treatment of psoriasis, an autoimmune disease [5]. Methotrexate is used off-label in treating ectopic pregnancy. In the absence of any contraindication, a single intramuscular injection of methotrexate at a dosage of 1 mg/kg or 50 mg/m^2 is recommended. It can be repeated once at the same dose should the human chorionic gonadotropin concentration not fall sufficiently [6].

When methotrexate enters the circulation, it binds with serum protein (average 41–51%, mostly albumin) [7]. The drug can be displaced from albumin by several drugs, including salicylates, sulfonamides, tetracycline, chloramphenicol, and phenytoin. Metabolism of methotrexate is minimal, although high doses of methotrexate can result in the accumulation of metabolites. These metabolites, such as 7-hydroxymethotrexate, are potentially nephrotoxic. Approximately 40 to 50% of small doses and about 90% of large doses are excreted unchanged in the urine within 24–48 h of ingestion. Excretion involves both glomerular filtration and active tubular secretion.

In cancer chemotherapy, high dose methotrexate is typically administered over a period of 6–24 h. Precautions must be taken to ensure a high urine flow and an alkaline urine pH, so as to prevent precipitation of methotrexate in urine. At the end of methotrexate infusion, and periodically thereafter for 24–48 h, methotrexate concentrations are measured to assure that the disappearance rate of drug from plasma is occurring at a normal rate. At the end of the infusion, the patient is given leucovorin to replenish intracellular stores of reduced folate and attenuate the toxicity secondary to methotrexate [8]. After completion of high dose methotrexate infusion, the target peak value should be equal to >1000 μmol/L. After high-dose therapy, methotrexate shows a biphasic elimination pattern with an average half-life of the first phase of approximately 3 h and a second phase of approximately 23 h [9, 10]. Della-Pria observed a median half-life of 21.7 h in HIV-infected patients receiving high-dose methotrexate therapy (3 g/m^2: intravenous infusion). The authors considered elevated serum levels of methotrexate at 24 h to be 20 μM, while after 48 and 72 h levels, serum levels exceeded 2 and 0.2 μM were considered as elevated [11].

The pharmacokinetics of methotrexate after low-dose therapy has also been investigated. The mean absolute bioavailability after oral administration is about 70–80% after 15 mg/week dosage. Under fasting conditions, maximum plasma concentrations of methotrexate (C_{max}) range between 0.3 and 1.6 μmol/L, and occur 0.75–2 h after administration. Food did not significantly influence the bioavailability of methotrexate but did slightly reduce C_{max} and prolong t_{max} as a result of delayed gastric emptying. The volume of distribution of methotrexate is 0.87–1.43 L/kg. The mean terminal biological half-life of 6–15 h has been observed [12].

Toxicity may occur even after low-dose methotrexate therapy. In one study, the authors observed adverse effects of low dose methotrexate therapy in 27% of rheumatoid arthritis patients treated with this medication. The major side effect of methotrexate was hepatotoxicity and hematological problems. TDM can be used to avoid toxicity after low dose methotrexate therapy. Although it is generally assumed that methotrexate serum levels 0.8–1 μmol/L after 2 h of oral administration of low-dose methotrexate may be associated with adverse effects, the authors utilized receiver-operator curve analysis to conclude that serum methotrexate levels of 0.71 μmol/L may produce adverse effects with a sensitivity of 71% and specificity of 76% in the patients on low-dose maintenance therapy [13, 14]. Immunoassays are commercially available for TDM of methotrexate. FPIA (fluorescence polarization immunoassays) was widely used for a long time for monitoring of methotrexate in serum or plasma. More recently, chemiluminescent microparticle immunoassay (CMIA) for the application on the Abbott Architect analyzer is commercially available for TDM of methotrexate. This assay has good correlation with methotrexate concentration measured by FPIA as well as liquid chromatography combined with tandem mass spectrometry (LC/MS/MS). The authors concluded that CMIA methotrexate assay is suitable for TDM of methotrexate [15]. After absorption, methotrexate is metabolized in the liver to 7-hydroxymethotrexate and in the intestine to 2,4-diamino-*N*-10-methylpteroic acid (DAMPA). DAMPA is known to interfere with methotrexate immunoassays. However, TDM using LC/MS/MS is free from such interference [16].

The primary toxic effects of methotrexate, especially after high-dose therapy, are exerted on bone marrow (decreased blood cell counts) and intestinal epithelium (mucosal ulceration). Patients suffering from these toxic effects are at risk for spontaneous hemorrhage or life-threatening infection; these patients often require platelet transfusions and broad-spectrum antibiotics. The kidneys and liver are also affected by methotrexate. Since the drug is eliminated mainly through the kidneys, patients with impaired renal function may have increased toxicity. High doses of the drug may precipitate out in the renal tubules, causing toxicity. It is also reported that elevated hepatic enzyme levels occur during high dose therapy. Prolonged drug use may lead to liver fibrosis and occasionally to cirrhosis, therefore, renal and hepatic function tests should be routinely monitored in these patients. Other adverse effects associated with methotrexate therapy include alopecia, dermatitis, intestinal pneumonitis, osteoporosis (children), defective oogenesis or spermatogenesis, and teratogenesis. The drug is an abortifacient and should not be used during the first trimester of pregnancy. Toxic concentrations of methotrexate generally depend on the dosing regimen. For low-dose therapy concentrations of 0.02 μmol/L one or 2 weeks after completion of treatment are considered toxic [1].

References

[1] Howard SC, McCormick J, Pui CH, Buddington RK, Harvey RD. Preventing and managing toxicities of high-dose methotrexate. Oncologist 2016;21:1471–82.

[2] Woods R, Fox RM, Tattersall MH. Methotrexate treatment of squamous-cell head and neck cancers: dose-response evaluation. Br Med J (Clin Res Ed) 1981;282(6264):600–2.

[3] Bianchi G, Caporali R, Todoerti M, Mattana P. Methotrexate and rheumatoid arthritis: current evidence regarding subcutaneous versus oral routes of administration. Adv Ther 2016;33:369–78.

[4] Gansauge S, Breitbart A, Rinaldi N, Schwarz-Eywill M. Methotrexate in patients with moderate systemic lupus erythematosus (exclusion of renal and central nervous system disease). Ann Rheum Dis 1997;56:382–5.

[5] Cabello Zurita C, Grau Pérez M, Hernández Fernández CP, González Quesada A, et al. Effectiveness and safety of methotrexate in psoriasis: an eight-year experience with 218 patients. J Dermatolog Treat 2017;28:401–5.

[6] Marret H, Fauconnier A, Dubernard G, Misme H, et al. Overview and guidelines of off-label use of methotrexate in ectopic pregnancy: report by CNGOF. Eur J Obstet Gynecol Reprod Biol 2016;205:105–9.

[7] Paxton JW. Protein binding of methotrexate in sera from normal human beings: effect of drug concentration, pH, temperature, and storage. Pharmacol Methods 1981;5:203–13.

[8] Treon SP, Chabner BA. Concepts in use of high-dose methotrexate therapy. Clin Chem 1996;42:1322–9.

[9] Graf N, Winkler K, Betlemovic M, Fuchs N, Bode U. Methotrexate pharmacokinetics and prognosis in osteosarcoma. J Clin Oncol 1994;12:1443–51.

[10] Arakawa Y, Arakawa A, Vural S, Mahajan R, Prinz JC. Renal clearance and intracellular half-life essentially determine methotrexate toxicity: a case series. JAAD Case Rep 2018;5:98–100.

[11] Dalla Pria A, Bendle M, Ramaswami R, Boffito M, Bower M. The pharmacokinetics of high-dose methotrexate in people living with HIV on antiretroviral therapy. Cancer Chemother Pharmacol 2016;77:653–7.

[12] Grim J, Chládek J, Martínková J. Pharmacokinetics and pharmacodynamics of methotrexate in non-neoplastic diseases. Clin Pharmacokinet 2003;42:139–51.

[13] Gilani ST, Khan DA, Khan FA, Ahmed M. Adverse effects of low dose methotrexate in rheumatoid arthritis patients. J Coll Physicians Surg Pak 2012;22:101–4.

[14] Romão VC, Lima A, Bernardes M, Canhão H, Fonseca JE. Three decades of low-dose methotrexate in rheumatoid arthritis: can we predict toxicity? Immunol Res 2014;60:289–310.

[15] Bouquié R, Grégoire M, Hernando H, Azoulay C, et al. Evaluation of a methotrexate chemiluminescent microparticle immunoassay:comparison to fluorescence polarization immunoassay and liquid chromatography-tandem mass spectrometry. Am J Clin Pathol 2016;146:119–24.

[16] Silva MF, Ribeiro C, Gonçalves VMF, Tiritan ME, Lima Á. Liquid chromatographic methods for the TDM of methotrexate as clinical decision support for personalized medicine: a brief review. Biomed Chromatogr 2018;32:e4159.

15.4

Tamoxifen

Chemical properties	
Solubility in H_2O	Practically insoluble
Molecular weight	371.52
pKa	8.85
Melting point	96–98 °C
Dosing	
Recommended dose	20 mg/day (adult)
Monitoring	
Sample	Serum, plasma
Effective concentration of endoxifen metabolite	≥5.97 ng/mL
Toxic concentrations	Not established
Methods	HPLC, LC/MS
Pharmacokinetic properties	
Oral dose absorbed	97%
Time to peak concentration	5 h
Protein bound	99%
Volume of distribution	34 L/kg (adults)
Half-life, adult	5–7 day
Time to steady state, adult	2 weeks
Half-life, child	Not established
Time to steady state, child	Not established

Excretion, urine

	% Excreted	Active	Detected in blood
Unchanged	<5	Yes	Yes
Endoxifen	<5	Yes	Yes
4-Hydroxytamoxifen	<5	Yes	No

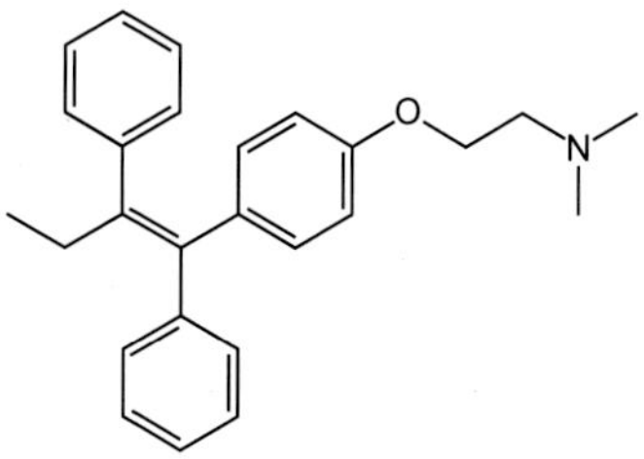

Tamoxifen

Tamoxifen (Nolvadex) is an estrogen receptor (ER) modulator with differing pharmacologic actions in various tissues [1]. Tamoxifen is an antagonist for breast estrogen receptors (ERs), agonist for endometrial ERs, and a partial agonist for bone ERs. The major therapeutic use for tamoxifen is the treatment of ER-positive breast cancers. Tamoxifen is well-absorbed after oral administration, with peak levels around 5 h. Tamoxifen is heavily bound (~99%) to plasma proteins.

The metabolism of tamoxifen is complex, with extensive metabolites generated mainly by cytochrome P450 (CYP) 2D6 and CYP3A4 [2]. Much of the ingested dose of tamoxifen is ultimately eliminated in feces as glucuronide metabolites, with urinary excretion playing only a minor role (<10%). Two of the metabolites of tamoxifen, 4-hydroxytamoxifen and endoxifen (*N*-desmethyl-4-hydroxytamoxifen), show 30- to 100-fold higher affinity for breast ERs than tamoxifen itself. Even though tamoxifen shows some pharmacologic activity on its own, it can be thought of functionally as a pro-drug, with 4-hydroxytamoxifen and endoxifen producing much of the pharmacologic effect.

Therapeutic drug monitoring (TDM) of tamoxifen focuses mainly on the metabolite endoxifen, although some studies have also included measurement of 4-hydroxytamoxifen and other metabolites [3]. One rationale for focusing on endoxifen is that pharmacogenetic variation of CYP2D6, or drug-drug interactions impacting this enzyme, can significantly influence the metabolism of tamoxifen to endoxifen and thereby the therapeutic efficacy of tamoxifen. The relationship between endoxifen plasma/serum concentrations and therapeutic outcome has been explored in large retrospective cohorts of women treated with tamoxifen for ER-positive breast cancer [4]. One threshold proposed from outcome analysis is that endoxifen plasma/serum concentrations should exceed 5.97 ng/mL for effective therapy. This is based on overall sub-optimal outcomes in women with concentrations lower than this. Approximately 80% of women treated with the standard dose of 20 mg tamoxifen per day reach this threshold.

So far, no clear relationship between endoxifen concentration and toxic effects has been elucidated. Active areas of inquiry include use of CYP2D6 genotyping and/or determination of 4-hydroxytamoxifen concentrations and other metabolites to optimize tamoxifen TDM. Endoxifen and other metabolites can be measured in plasma or serum by high-performance liquid chromatography or liquid chromatography coupled to tandem mass spectrometry [5].

References

[1] Martinkovich S, Shah D, Planey SL, Arnott JA. Selective estrogen receptor modulators: tissue specificity and clinical utility. Clin Interv Aging 2014;9:1437–52.
[2] Morello KC, Wurz GT, DeGregorio MW. Pharmacokinetics of selective estrogen receptor modulators. Clin Pharmacokinet 2003;42:361–72.
[3] Groenland SL, van Nuland M, Verheijen RB, Schellens JHM, Beijnen JH, Huitema ADR, Steeghs N. Therapeutic drug monitoring of oral anti-hormonal drugs in oncology. Clin Pharmacokinet 2019;58:299–308.
[4] Madlensky L, Natarajan L, Tchu S, Pu M, Mortimer J, Flatt SW, Nikoloff DM, Hillman G, Fontecha MR, Lawrence HJ, et al. Tamoxifen metabolite concentrations, CYP2D6 genotype, and breast cancer outcomes. Clin Pharmacol Ther 2011;89:718–25.
[5] Heath DD, Flat SW, Wu AH, Pruitt MA, Rock CL. Evaluation of tamoxifen and metabolites by LC-MS/MS and HPLC methods. Br J Biomed Sci 2014;71:33–9.

CHAPTER 16

Caffeine and theophylline

16.1

Caffeine

Chemical properties	
Solubility in H_2O	21.7 mg/mL
Molecular weight	194.19
pKa	0.8
Melting point	236 °C
Dosing	
Recommended dose, infant	20 mg/kg loading and 5 mg/kg of maintenance
Monitoring	
Sample	Serum, plasma (heparin, EDTA)
Effective concentrations, neonate	5–20 μg/mL
Toxic concentrations	>40 μg/mL
Methods	GC, HPLC, Immunoassay, LC/MS
Pharmacokinetic properties	
Oral dose absorbed	100%
Time to peak concentration	0.5–2.0 h
Protein bound	29–43%
Volume of distribution	0.6 L/kg in adults 0.8–0.9 L/kg in infants
Half-life, adult	4–5 h
Half-life, neonate	65 h to 102 h

Excretion, urine

	% Excreted	Active	Detected in blood
Parent	1	Yes	Yes
Paraxanthine	4	Yes	Yes
Methylxanthines	22	Yes	Yes
Methyluric acids	35	n/a	n/a

Therapeutic Drug Monitoring Data
https://doi.org/10.1016/B978-0-12-815849-4.00016-5

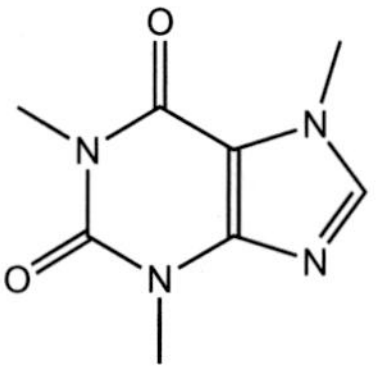

Caffeine

Caffeine (1,3,7-trimethylxanthine) is one of the most widely used drugs. Found in many beverages and food products, caffeine is also widely present in prescription and non-prescription medications in concentrations of up to 250 mg per dose. Caffeine is chemically related to other methylxanthines including theophylline and theobromine. The majority of the world's caffeine is derived from coffee plants. Tea also contains caffeine in amounts approximately ½ the amount found in coffee depending on the brew strength; typical cup of coffee contains ~40 mg of caffeine, whereas espresso or strong drip coffee contain >100 mg/cup. Soft drinks also contain significant amounts of caffeine (~50 mg/serving), whereas energy drinks may contain >140 mg/serving [1].

Caffeine is used therapeutically to treat neonatal apnea, as a stimulant to prevent drowsiness and fatigue, and as an analgesic to relieve headache pain in adults. Caffeine citrate is currently one of the most prescribed medicines in neonatal units for apnea of prematurity. It is the first choice of drug among all methylxanthines because of its efficacy, better tolerability and wider therapeutic index. Caffeine stimulates the respiratory center, sensitizing it to hypercapnia. This leads to increase in mean respiratory rate and tidal volume, improved pulmonary blood flow, better carbon dioxide sensitivity and enhanced diaphragmatic function and breathing pattern. In addition, caffeine also acts as a central stimulant. Caffeine stimulates the myocardium and increases heart rate, cardiac output, stroke volume as well as the mean arterial blood pressure. In kidneys, caffeine increases glomerular filtration rate and produces diuresis. Caffeine also increases basal metabolic rate, enhances catecholamine secretion and alters glucose homeostasis [2].

For medical purposes, caffeine is available as caffeine citrate in both oral and injectable formulations. The most common oral dosing is 20 mg/kg loading and 5 mg/kg of maintenance dose. Caffeine is rapidly and completely absorbed after oral administration and does not undergo first pass metabolism by the liver. Caffeine is hydrophobic and distributes rapidly without tissue accumulation. It is rapidly distributed into the brain, and in preterm infants the levels of caffeine in the cerebrospinal fluid approximate the plasma levels.

In infants the mean volume of distribution (V_d) is 0.8–0.9 L/kg compared to 0.6 L/kg in adults. Metabolism of caffeine occurs in the liver mainly by microsomal cytochrome P450 (CYP) mono-oxygenases (especially CYP1A2) and partially by xanthine oxidase. The major pathway of caffeine metabolism in the preterm infant is N7-demethylation, which matures at about the age of 4 months. In the first weeks of life, caffeine is eliminated mainly by renal excretion. The half-life of caffeine in preterm neonates is very long (65–102 h), and such long half-life is observed in infants up to 38 weeks of age. The rate of metabolism of caffeine is found to be higher in female than male preterm neonates [3].

The peak plasma concentration after oral administration is reached within 30 min to 2 h. The clearance increases non-linearly with increasing post-natal age, reaching a plateau at 120 days, and V_d increases linearly with increasing weight. Renal excretion is the main route of elimination in neonates, where almost 86% of the drug is passed unchanged in urine, whereas in adults only 4% is excreted via renal route. The elimination half-life starts to decrease from birth and reaches the adult values at 60 weeks post-conception age [3].

Metabolism of caffeine is strongly influenced by expression levels of CYP2A1 which vary by age and caffeine use. The half-life of caffeine in healthy adults is approximately 4–5 h. Habitual heavy users of caffeine are faster metabolizers. Caffeine metabolism is faster in cigarette smokers, who typically consume more caffeine than nonsmokers. Children 12 years or younger metabolize caffeine more rapidly than adults, and pregnant women are slower metabolizers, especially in the later stages of pregnancy [4]. Caffeine is extensively metabolized in the liver by demethylation to paraxanthine, theobromine, and theophylline. These metabolites are further metabolized to methylxanthine derivatives.

The current evidence generally supports that consumption of up to 400 mg caffeine/day in healthy adults is not associated with adverse cardiovascular, behavioral, musculoskeletal, reproductive, or developmental effects. In general it is accepted that consumption of up to 300 mg caffeine/day in healthy pregnant women is not associated with adverse reproductive and developmental effects. While more limited data is available for children, the available evidence suggests that up to 2.5 mg caffeine/kg body weight/day is generally safe in young children and adolescents. However, physical dependence may develop with regular intake of caffeine. Acute adverse effects appear following administration of 0.5–1.0 g of caffeine. Following overdosage, central nervous system (CNS), cardiovascular and gastrointestinal effects are commonly observed. The most serious CNS effects are seizures; other prominent effects include restlessness, insomnia, tremor, and nervousness. Cardiac symptoms include atrial tachycardia, premature ventricular contractions, ventricular tachycardia and fibrillation. Potential gastrointestinal symptoms include vomiting and nausea. Co-ingestion of other drugs can result in additive or potentiated pharmacologic effects. Treatment of overdosage includes medical support of respiratory and cardiovascular function. Medical personnel should be prepared to treat life-threatening cardiac arrhythmias and seizures with appropriate medication. For severe poisonings emesis, activated charcoal and saline administration, or sorbitol cathartic are recommended. Hemoperfusion and hemodialysis have been utilized to enhance the elimination of caffeine [5]. Magdalan et al. reported one non-fatal and two fatal caffeine overdoses. Case 1 was a severe intoxication (persistent vomiting, hypotension, tremor), and the concentration of caffeine in the blood was found to be 80.16 μg/mL. The patient was treated using hemodialysis and survived due to significant decline in serum caffeine level. Cases 2 and 3 were fatal poisonings, and recorded levels of caffeine in post mortem blood samples were 140.64 μg/mL and 613.0 μg/mL respectively. In case 2 the patient died 10 min after admission to hospital as a result of sudden cardiac arrest, and in case 3 death occurred in home and was also sudden in nature. The authors warned that ingestion of pure caffeine is associated with a high risk of overdose and the development of serious and even fatal poisoning, and those using pure caffeine may be unaware of these risks. [6].

Immunoassays are commercially available for measuring caffeine concentration in serum or plasma. However routine monitoring of caffeine is not necessary when caffeine is used for the treatment of apnea of prematurity in neonates [7]. In contrast, caffeine monitoring may be helpful in preterm babies with elevated liver enzymes or significantly impaired renal function. Measuring caffeine serum/plasma concentrations is useful in patients overdosed with caffeine [6]. The therapeutic range of caffeine in neonates is 5–20 μg/mL while toxicity is encountered at levels exceeding 40 μg/mL. Yu et al. concluded that therapeutic drug monitoring of caffeine should be considered in critically ill neonates with unexplained adverse effects, such as tachycardia. [8].

References

[1] McCusker RR, Goldberger BA, Cone EJ. Caffeine content of energy drinks, carbonated sodas, and other beverages. J Anal Toxicol 2006;30:112–4.

[2] Shrestha B, Jawa G. Caffeine citrate—is it a silver bullet in neonatology? Pediatr Neonatol 2017;58:391–7.

[3] Abdel-Hady H, Nasef N, Shabaan AE, I N. Caffeine therapy in preterm infants. World J Clin Pediatr 2015;4:81–93.
[4] Doepker C, Lieberman HR, Smith AP, Peck JD, et al. Caffeine: friend or foe? Annu Rev Food Sci Technol 2016;7:117–37.
[5] Wikoff D, Welsh BT, Henderson R, Brorby GP, et al. Systematic review of the potential adverse effects of caffeine consumption in healthy adults, pregnant women, adolescents, and children. Food Chem Toxicol 2017;109(Pt 1):585–648.
[6] Magdalan J, Zawadzki M, Skowronek R, Czuba M, et al. Nonfatal and fatal intoxications with pure caffeine—report of three different cases. Forensic Sci Med Pathol 2017;13:355–8.
[7] Natarajan G, Botica ML, Thomas R, Aranda JV. Therapeutic drug monitoring for caffeine in preterm neonates: an unnecessary exercise? Pediatrics 2007;119:936–40.
[8] Yu T, Balch AH, Ward RM, Korgenski EK, Sherwin CM. Incorporating pharmacodynamic considerations into caffeine therapeutic drug monitoring in preterm neonates. BMC Pharmacol Toxicol 2016;17(1):22.

16.2

Theophylline

Chemical properties	
Solubility in H_2O	8.3 mg/mL
Molecular weight	180.17
pKa	8.6
Melting point	270–274 °C
Dosing	
Recommended dose, adult	Maximum dosage 600 mg/day
Recommended dose, child	16–24 mg/kg/day
Monitoring	
Sample	Serum, plasma (heparin, EDTA)
Effective concentrations, adult	10–20 μg/mL
Toxic concentrations	>20 μg/mL
Methods:	GC, HPLC, immunoassay, LC/MS, CZE
Pharmacokinetic properties	
Oral dose absorbed	88–100%
Time to peak concentration	1–3 h
Protein bound	Approximately 60%
Volume of distribution	0.5 L/kg
Half-life, adult	5–6 h
Time to steady state, adult	15–55 h
Half-life, child	3.5 h (average)
Time to steady state, child	5–40 h

Excretion, urine

	% Excreted	Active	Detected in blood
Parent	15–20	Yes	Yes
1,3 Dimethyluric acid	30–50	No	Yes
3-Methylxanthine	10–20	No	Yes
1-Methyluric acid	10–20	No	Yes

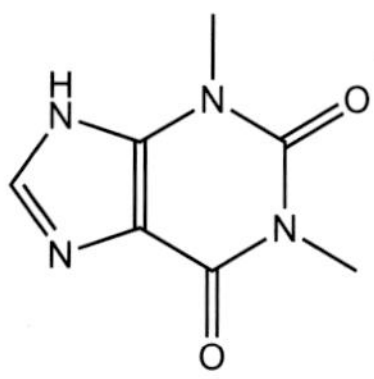

Theophylline

Theophylline (Elixophyllin, Theochron) is a methylxanthine with a long history of use as an inexpensive bronchodilator. Theophylline is still one of the most widely prescribed drugs for the treatment of asthma and chronic obstructive pulmonary disease (COPD) worldwide. Theophylline (dimethylxanthine) occurs naturally in tea, from which it may be extracted. It was first synthesized chemically in 1895 and used initially as a diuretic. Its bronchodilator property was later identified, and it was introduced as a clinical treatment for asthma in 1922. Despite its widespread global use, theophylline has become a third-line treatment for asthma as an add-on therapy in patients with poorly controlled disease. Theophylline has been supplanted for asthma therapy by inhaled β2-adrenergic receptor agonists for bronchodilation and inhaled corticosteroids for antiinflammatory effect. Theophylline is used as an oral therapy (rapid or slow-release tablets) or as more soluble aminophylline, an ethylenediamine salt, which is suitable for oral and intravenous use [1]. Aminophylline, a widely used ethylenediamine salt, contains 85% anhydrous theophylline by weight.

Theophylline is classified as a bronchodilator based on its ability to effectively relax airway smooth muscle. It does this by inhibiting nucleotide phosphodiesterase enzymes, thereby stimulating cAMP and cGMP mediated signal transduction pathways involved in smooth muscle control. In addition, theophylline acts as a competitive antagonist for adenosine receptors. As these receptors can cause bronchoconstriction in asthmatics, it is believed that this antagonist effect also serves to open airways [2].

Theophylline after oral administration is rapidly and completely absorbed. Peak theophylline level is observed approximately 2h after oral administration. Serum protein binding of theophylline is approximately 60% and the V_d is 0·5 L/kg. In adults, 90% of theophylline is metabolized by the CYP enzymes (predominately CYP1A2). In neonates, approximately 50% is excreted unchanged in the urine due to immature hepatic function. Many factors are reported to affect CYP1A2 activity, such as single nucleotide polymorphisms of CYP1A2 and concurrent administration of other drugs Therefore, variability in theophylline clearance may be explained in part by these factors. The elimination half-life of theophylline was 3·5 h in pediatric patients and 5–6 h in adults, and it was prolonged in the elderly. The median (range) estimate values of clearance for pediatric and adult patients were 0·062 (0·0056–0·0949) L/h/kg and 0·053 (0·0493–0·0517) L/h/kg, respectively. The median values of the interindividual variability of clearance were 33·5% in adults and 25·8% in pediatric patients [3]. Increased clearance is seen in children (1–16 year) and in cigarette and marijuana smokers. Concurrent administration of phenytoin, phenobarbital, or rifampicin, which induces activity of liver enzymes, increases metabolism of theophylline. Therefore, in patients taking such medications, higher theophylline doses may be needed. Reduced metabolism is found in liver disease, pneumonia, and heart failure, and doses need to be reduced to one-half with careful monitoring of serum theophylline concentrations. Decreased clearance is also seen with several drugs, including erythromycin, ciprofloxacin, allopurinol, cimetidine, serotonin uptake inhibitors (fluvoxamine), and the 5-lipoxygenase inhibitor zileuton, all of which interfere with CYP1A2 function. Thus, if a patient on maintenance theophylline requires a course of erythromycin, the dose of theophylline should be halved. Although there is a similar interaction with clarithromycin, there is no interaction with azithromycin. Viral infections and vaccinations (influenza immunizations) may also reduce clearance, and this may be particularly important in children [4, 5].

Theophylline adult dose is initially 300 mg per day (given in in divided dosage) which after 3 days may be increased to 400 mg/day, again given in divided dosage. Maximum daily dosage is 600 mg/day. Extended release tablets (400 mg or 600 mg) are also available. Because of close relationship between the acute improvement in airway function and serum theophylline concentrations, therapeutic drug monitoring is very useful. Immunoassays are commercially available for routine monitoring of theophylline. The therapeutic range is 10–20 μg/mL. The dose of theophylline required to achieve therapeutic concentrations varies among patients, largely because of differences in clearance [5].

Minor side effects which may occur early with theophylline include headache, nausea, vomiting, nervousness and insomnia. Chronic or acute overdose may be manifested in the early stages by similar symptoms as well as more serious effects such as tremor, convulsions, tachycardia, cardiac arrhythmias, and cardiorespiratory arrest. Eshleman and Shaw reported a case of fatal theophylline overdose in a 16-year-old asthmatic boy who presented with seizures, respiratory arrest, and a theophylline concentration of 117 μg/mL in serum. Markedly increased catalytic activities of creatine kinase, aspartate aminotransferase, and alanine aminotransferase in serum were also noted [6].

References

[1] Weinberger M, Hendeles L. Theophylline in asthma. N Engl J Med 1996;334:1380–8.

[2] Pesce AJ, Rashkin M, Kotagal U. Standards of laboratory practice: theophylline and caffeine monitoring. National Academy of Clinical Biochemistry. Clin Chem 1998;44:1124–8. [Review].

[3] Ma YJ, Jiang DQ, Meng JX, Li MX, Lee TC, et al. Theophylline: a review of population pharmacokinetic analyses. J Clin Pharm Ther 2016;41:594–601.

[4] Nahata M. Drug interactions with azithromycin and the macrolides: an overview. J Antimicrob Chemother 1996;37:133–42.

[5] Barnes PJ. Theophylline. Am J Respir Crit Care Med 2013;188:901–6.

[6] Eshleman SH, Shaw LM. Massive theophylline overdose with atypical metabolic abnormalities. Clin Chem 1990;36:398–9.

Index

Note: Page numbers followed by *f* indicate figures and *t* indicate tables.

B

C

D

E

F

N

O

P

Printed in the United States
By Bookmasters